FIELD METHODS
IN
MARINE SCIENCE

FROM MEASUREMENTS
TO MODELS

FIELD METHODS
IN
MARINE SCIENCE

FROM MEASUREMENTS TO MODELS

SCOTT P. MILROY

Garland Science
Vice President: Denise Schanck
Associate Editor: Allie Bochicchio
Editorial Assistant: Louise Dawnay
Production Editor: Natasha Wolfe
Illustrator: Oxford Designers and Illustrators
Cover Design: Andrew Magee
Copyeditor: Teresa Wilson
Typesetting: Cenveo, Inc.
Proofreader: Chris Purdon
Indexer: Indexing Specialists (UK) Ltd

Cover image: Retro style marine landscape with underwater view. (©Dmytro Tolokonov/Dollar Photo Club.)

Dr. Scott P. Milroy is the newly appointed Associate Chair for Undergraduate Programs on the Coast within the Department of Marine Science at the University of Southern Mississippi, where he also holds a tenured appointment as Associate Professor of Marine Science. He specializes in marine ecosystems modeling, with particular interests in biogeochemical cycling of nutrients, closed-system microcosms, coral reef ecology, ecological engineering, ecological feedback mechanisms, hydrologic optics, operational models of harmful algal blooms, ecosystem approaches to management, and the use of modeling approaches for evaluating ecosystem services and function. As a teaching and research faculty member in both the undergraduate and graduate schools of marine science, he currently teaches and has provided lectures for such courses as introductory oceanography, field methods in marine science, marine biology of the caribbean, special topics in marine science, marine environmental science, remote sensing of the oceans, ocean dynamics, and ecological modeling.

© 2016 by Garland Science, Taylor & Francis Group, LLC

ISBN 978-0-8153-4476-6

The Library of Congress are should be filled in with:

Milroy, Scott, author.
  Field methods in marine science : from measurements to models / Scott Milroy.
    pages cm
    ISBN 978-0-8153-4476-6
  1. Marine sciences—Research. I. Title.
    GC57.M557 2016
    551.46072'3—dc23
                                    2015024200

Published by Garland Science, Taylor & Francis Group, LLC, an informa business, 270 Madison Avenue, New York NY 10016, USA, and 2 Park Square, Milton Park, Abingdon, OX14 4RN, UK.

Printed in the United States of America

15 14 13 12 11 10 9 8 7 6 5 4 3 2 1

Visit our website at http://www.garlandscience.com

To my family, for all your love and support, and for teaching me that integrity and hard work are the keys to success, no matter what.

To my friends, mentors, and colleagues, your inspiration and good-natured competitiveness help me to stay sharp and to never become complacent.

To my students, for the honor of your trust in my tutelage, it is upon your shoulders that future generations shall stand, and it is from your minds that all that is knowable shall become known.

To my son Ricky, for your patience during all those long days (and nights) that I was not able to run around and play with you, daddy's coming! And boy do I owe your mom big-time.

Which brings me to my amazing wife Helen, for all the sacrifices you have made on my behalf, and for all the joy you bring into my life every single day, I did it all because of you.

Preface

In the marine sciences, the practice of our craft requires an interdisciplinary approach, where we must proficiently integrate the fields of biology, chemistry, geology, and physics, with simultaneous training in mathematics and computer science as well. And although some of us may prefer to pursue our scientific endeavors in the controlled environment of our laboratories, our pursuit of marine science (or, for that matter, any aquatic science) will invariably call us to the field, where the natural world rarely obliges our desire for experimental control. So we are challenged to grow our expertise in these multiple disciplines and conduct our science in nature's laboratory, a rather unforgiving place where only the most difficult and perplexing scientific questions remain.

Field Methods in Marine Science was written as an introductory text to serve as both a training manual and a trusted reference for marine science students and early-career professionals. The book provides the reader with the key conceptual linkages between the theory and practice of science, from the philosophy of the scientific method to practical advice on designing appropriate field experiments for hypothesis-testing and data analysis, and how to transition the reader's research "from measurements to models" and create numerical models as new investigative tools. Unlike many other titles, this text was designed to be broadly applicable to all of the major disciplines (biology, chemistry, geology, mathematics, physics) within the marine sciences, as the fundamentals of field methods and numerical modeling are ubiquitous throughout.

The text is organized into four distinct units: 1) First Principles, 2) Methods of Data Acquisition, 3) Methods of Data Analysis, and 4) Methods of Data Assimilation (Modeling). Each unit is designed as a self-contained plan of study, allowing instructors greater flexibility to select discrete units as they develop curriculum for their particular course. Units 1 and 2 (First Principles and Methods of Data Acquisition) are relevant to traditional field methods and could be taught as a single-semester course in field methods.

Unit 1 (First Principles) begins of course with Chapter 1, The Foundations of Scientific Inquiry, which provides a philosophical justification of the sciences and guides the reader through the stepwise process that forms the basis of all scientific inquiry: the scientific method. Chapter 2, Introduction to Statistical Inference, provides very practical advice and basic training in the clever (and proper) use of statistics to summarize data and to infer meaning from those data. Through the use of several clear examples, the reader is introduced to the methods by which scientific hypotheses can be tested and analyzed in terms of statistical significance.

Unit 2 (Methods of Data Acquisition) focuses on the development of the research prospectus (experimental design) and the proper execution of that experiment in the laboratory or in the field. Chapter 3, Experimental Design, starts with the very basic ingredients that must be included in virtually all lab or field experiments, and builds the reader's confidence by tackling the more subtle nuances of sampling effort and quantitative survey methods. Chapter 4, Oceanographic Variables, provides an overview of how to describe an ever-changing ocean in both time and space, and which essential measures should be most carefully considered, in terms of the biological, chemical, geological, or physical phenomena under investigation. Chapter 5, Common

Hydrologic Census Methods, presents the most commonly used methods for collecting data from the aquatic medium, when the body of water itself is the focus of inquiry. Chapter 6, Census Methods for Benthic Organisms, continues with the general quantitative survey methods first introduced in Chapter 3, but delivers more specific guidance (and more experimental design options) for census methods specifically designed for organisms living in or on the bottom. Not surprisingly, Chapter 7, Census Methods for Pelagic Organisms, shifts the attention to those census methods and data collection strategies best suited for those organisms either drifting or swimming in the water column.

Unit 3 (Methods of Data Analysis) returns to the realm of statistics and numerical analysis, where our understanding of the basics from Unit 1 and our data collection methods from Unit 2 are put to the test. In this unit, the reader is introduced to the most common single- and multivariate analysis methods, including the t test, one-way ANOVA, and various correlation and regression analyses. The contents of this unit could be taught either as a stand-alone data analysis course or as the second semester of a field methods sequence. Chapter 8, Introduction to Univariate Analysis, starts off with the fundamentals of statistical association, using the simplest case where two or more populations can be compared (and tested for significant differences) through the use of only a single variable. Chapter 9, Introduction to Multivariate Analysis, gently guides the reader into the more complex statistical methods used to explore the associative and causative relationships among populations, using several different variables at once.

Unit 4 (Methods of Data Assimilation) represents the culmination of our research, where the data collected from the methods outlined in Unit 2 and the statistical relationships defined with the help of Unit 3 can now provide us with the tools to develop numerical models of the very phenomena we have been investigating. Since the chapters in this unit deal entirely with the basics of numerical modeling, Unit 4 can be used as a resource for an Introduction to Numerical Modeling course, or as a follow-up course (subsequent to the field methods sequence) designed to help students use data acquired through previous field or laboratory work to develop their own numerical models. Chapter 10, Fundamental Concepts in Modeling, leads the reader through an unintimidating prelude to numerical modeling, including a discussion of the fundamental elements to every numerical model, as well as a working example of how to take a conceptual map of complex dynamics (such as the nitrogen cycle) and translate each of its elements into a cohesive numerical model in the subsequent chapters. Chapter 11, Model Structure, includes an intuitive approach of how to define the spatial and temporal constraints of the model, how to populate the variables in our model with measurements we already have on hand (from our previous field research), and how our model can be used to interpolate values (using finite differencing as an example) when we are missing critical data. Chapter 12, Modeling Simple Dynamics, continues with the example of the nitrogen cycle (used in Chapters 10 and 11) and demonstrates to the reader how complex interrelationships can be systematically broken down into several constituent parts, where each can be solved as a much simpler model. Then in Chapter 13, Modeling Complex Dynamics, these simpler models are stitched back together following the conceptual model we first designed back in Chapter 10, thereby re-creating the complex dynamics we originally dissected to make our modeling tasks more manageable. Finally in Chapter 14, Modeling Large System Dynamics, the reader is introduced to the four most prevalent types of models used to model large system dynamics: hydrodynamic, biogeochemical, radiative transfer, and ecological models.

It is hoped that the content and organization of this text shall provide an opportunity for both instructor and student to bridge their gains in knowledge and skills across multiple semesters, and in so doing shall enable students to make use of research and study products acquired from their previous coursework.

The design of this text (as a plan of study) was specifically engineered to have significant implications for science students enrolled in field-based courses and those undertaking honors projects that require theses or directed or independent study. Of course, this text was also designed for the benefit of graduate level and early-career professional scientists, who may not require this text in an academic setting, but may find it to be a valuable tool for their professional development and training.

Acknowledgments

First and foremost, I would like to thank the hardworking folks at Garland Science who have quite literally made this work possible. Gina Almond, you were the first to envision what we could accomplish with this text, and you have my deepest gratitude for the opportunity. Louise Dawnay, little did you know what you were getting yourself into when you agreed to shepherd this book and its author. I am embarrassed to say that I missed just about every deadline I had agreed to meet, but your patience (and subtle impatience) were exactly what I needed to see this through. For a time, even Allie Bochicchio was there to spare you from the incessant cat-herding, and Allie did an incredible job of keeping me on task. Actually, let me extend that compliment to Natasha Wolfe as well, whose diligence and attention to detail during copyediting was uncanny. You ladies run a pretty amazing shop, and I'm honored that this text is a product of your excellence.

The author and publisher would like to thank the many reviewers who provided helpful comments on the original proposal and gave thoughtful feedback and suggestions on the draft chapters:

Mark Baird (University of Technology Sydney); Anne Boettcher (University of South Alabama); Damien Brady (University of Maine); Shannon Bros (San Jose State University); Mark Butler (Old Dominion University); Daniel Hayes (Michigan State University); Kevin Hovel (San Diego State University); Jeffery Hughes (Wellesley College); Daniel Kamykowski (North Carolina State University); Marion McClary (Fairleigh Dickinson University, Metropolitan Campus); Pippa Moore (Aberystwyth University); Peter Mumby (University of Queensland); Heather Schrum (Sea Education Association in Woods Hole); Aswani Volety (Florida Gulf Coast University).

Online Resources

Accessible from www.garlandscience.com, the Student and Instructor Resource Websites provide learning and teaching tools created for *Field Methods in Marine Science*. The Student Resource Site is open to everyone, and users have the option to register in order to use book-marking and note-taking tools. The Instructor Resource Site requires registration and access is available only to qualified instructors. To access the Instructor Site, please contact your local sales representative or email science@garland.com. Below is an overview of the resources available for this book. On the website, the resources may be browsed by individual chapters and there is a search engine. You can also access the resources available for other Garland Science titles.

For students and instructors

Three additional appendices complement the content in the book:

Appendix D. Exercises in Unit Conversion.
Appendix E. Oceanographic Sampling Equipment.
Appendix F. A Compendium of Routine Oceanographic Methodologies.

For instructors

Figures: The images from the book are available in two convenient forms: PowerPoint® and JPEG. They have been optimized for display on a computer.

PowerPoint is a registered trademark of Microsoft Corporation in the United States and/or other countries.

For students

Flashcards: Each chapter contains a set of flashcards that allow students to review key terms from the text.

Glossary: The complete glossary from the book can be searched and browsed as a whole or sorted by chapter.

Contents

Chapter 4: Oceanographic Variables 69

Unit 1
First Principles

Contents

Chapter 1

The Foundations of Scientific Inquiry

"There can be no doubt that all our knowledge begins with experience." – Immanuel Kant

It may seem an overly simplistic question, but very few take the time to really consider: What is science? How does the knowledge gained through scientific inquiry differ from other modes of learning, and is it appropriate to afford greater respect to the sciences? Of course, those are judgments we must each make for ourselves, but it is certainly worth our time to consider such questions honestly.

The foundations of scientific inquiry were laid by ancient mathematicians, logicians, and naturalists stretching back to the dawn of human intelligence and curiosity. Since science is, at its core, a process of learning through trial and error, it is not surprising that a rudimentary understanding of the natural world could be gleaned from evidence acquired by the ancients, using very little technology. Even today, the performance of science is not predicated by the use of expensive, highly technical equipment—good science can be performed using one's own senses and acute intellect. But with each new discovery, scientists were able to draw upon the wisdom and experience of their predecessors to form a deeper understanding of the natural world, and use that information to guide the direction of new inquiries. But what was so critically important to learning was the very process of science itself.

Key Concepts

- Science relies upon the philosophy of empiricism, which states that neither fact nor truth can be known; we must rely instead on evidence.

- All scientific endeavors must be performed according to a strict set of rules, called the scientific method.

- The scientific method does not guarantee accuracy or provide proof; it simply provides an ordered framework for scientific inquiry.

- In order for knowledge to be gained through scientific inquiry, it must be testable through experimentation.

- Experiments are not self-determining—they merely provide evidence that must be analyzed for significance and meaning.

Figure 1.1 René Descartes (1596–1650). One of the most influential philosophers, whose works helped to establish the philosophical underpinnings of the scientific method.

The Difference Between Evidence and Truth

Truth. The concept, at its face, seems simple enough for our minds to grasp, and so we speak of it casually, as if we could demonstrate our mastery of it by being so dismissive of its real significance. But "truth" (as we so flippantly regard it) is a concept that should bring us all to our knees as we consider the awesome enormity of what that word represents. In its grandest sense, truth is represented by the sum total of the universe's facts: integrated, unchanging, and eternal.

Long and laboriously have philosophers struggled with the problem of truth. As it turns out, the problem is humankind's fundamental (in)ability to learn certain truths about the universe we occupy, simply because our method of inquiry—via intellect and experience—is inherently flawed. Interestingly enough, it is not the human intellect that harbors the origin of error—it is instead the knowledge gained through experience that cannot be trusted.

The most persuasive philosophical arguments for this were offered by René Descartes, a seventeenth-century French mathematician and philosopher (**Figure 1.1**). In his *Meditations on First Philosophy*, Descartes offered,

> *"Whatever, up to the present, I have accepted as possessed of the highest truth and certainty I have learned either from the senses or through the senses. Now these senses I have sometimes found to be deceptive; and it is only prudent never to place complete confidence in that by which we have even once been deceived."*

In other words, any evidence gained through the use of our imperfect senses must be considered suspect and never afforded absolute confidence. This philosophical axiom would at first seem to be a fatal blow to all learning. After all, if human knowledge is borne from experience, but these experiences can never be fully trusted, how then are we expected to learn anything? Is it possible to know anything prior to first experiencing it with or through our senses?

Although Descartes would have us questioning everything we can see, hear, touch, taste, and smell, he was also able to demonstrate that there is knowable truth in our universe—truth that is independent of our senses. In an oft-quoted passage from his *Meditations*, Descartes offers what is probably his most famous contribution to philosophy:

> *"If I am persuading myself of something, in so doing I assuredly exist. But what if, unknown to me, there be some deceiver, very powerful and very cunning, who is constantly employing his ingenuity in deceiving me? Again, as before, without doubt, if he is deceiving me, I exist….Ego sum, ego existo."*

Though often mistranslated as "I think therefore I am" (rather than "I am, I exist"), the meanings are essentially the same: regardless of whether we are perceiving our universe rightly or wrongly, the mere fact that we are perceiving the universe at all requires that we exist in the first place.

This is a critically important philosophical point to make, because it establishes the human ability to know truth *a priori*; that is, from the beginning (without experience). Hence, if humankind is capable of knowing this truth, perhaps there are more truths that are knowable and that these truths can be used to make ourselves more perfect instruments of learning.

Immanuel Kant's *Critique of Pure Reason* Picks Up Where Descartes Leaves Off

Although Descartes had firmly established that existence is a necessary prerequisite for all sentient beings in the universe, it wasn't until the late eighteenth century that a Prussian philosopher named Immanuel Kant (**Figure 1.2**) first published his *Critique of Pure Reason* and solidified the philosophical foundations necessary for scientific endeavor. The first of Kant's critical assertions was the *a priori* truth of space:

> *"Space is a necessary a priori representation, which underlies all outer intuitions. We can never represent to ourselves the absence of space, though we can quite well think it as empty of objects. It must therefore be regarded as the condition of the possibility of appearances, and not as a determination dependent upon them."*

Thus, Kant posits that space must exist before any observation can be made, whether that observation is made truly or falsely. Between Descartes and Kant, the necessity of existence begs the *a priori* representation of space (wherein all things that exist must be found).

But the true value of Kant's work, with regard to what is knowable in the context of science, is revealed in his treatment of time and causality. In his *Critique*, Kant asserts,

> *"All appearances are in time; and in it alone…can either coexistence or succession be represented. Now time cannot by itself be perceived. Consequently, there must be found in the objects of perception…the substratum which represents time…and all change or coexistence must be perceived in this substratum, and through relation of the appearances to it."*

So, regardless of whether our observations are false, the mere ability for humankind to perceive of them as being either coexistent or successive would require that time be an additional *a priori* truth within our universe. Beyond the obvious implications of time as an *a priori* truth, this notion of succession, in very logical terms, sets the stage for causality (that is, cause and effect).

But causality implies something else that is absolutely critical to the sciences, something that Kant argues must also be an *a priori* truth of our universe: the truth that order exists. Kant offers the following thought experiment, meant to provide his proof:

> *"Let us suppose that there is nothing antecedent to an event…All succession of perception would then be…merely subjective, and would never enable us to determine objectively which perceptions are those that really precede and which are those that follow."*

In other words, were it not for an ordered universe, our perceptions would never allow us to determine consistent causality. In a universe plagued with chaos, it would be impossible for humankind to learn anything because this notion of "cause and effect" would be utterly meaningless. But in our universe, we do not perceive chaos—we perceive order, even though the "truth" of our perceptions should be vigorously questioned (as Descartes would remind us). And so it is the order in our universe that affords us humans the possibility to describe it through our imperfect perception and intellect. That is where science comes in.

Figure 1.2 Immanuel Kant (1724–1804). A pioneer in philosophical thought, particularly as it relates to the empirical sciences.

Empiricism and the Scientific Method

As we have discussed, the process by which nearly all human knowledge is gained is through experience. But the employment of our intellect requires that we first gather the appropriate evidence through the flawed filter of our senses. This is the practice of **empiricism**, a school of philosophy that attempts to understand universal truths through the acquisition and analysis of evidence. Thus, it seems a straightforward proposition to claim that all true science is empirical. But it should be equally obvious that in the pursuit of truth, science is inherently weakened by the limitations of empiricism. Weakened, yes—powerless, no.

After all, if our ultimate goal is to describe the order within our universe, we need not be perfect—just consistent. Regardless of how badly we bungle our attempts to describe our surroundings, the truth of the universe remains unchanging and eternal. Eventually, if we are consistent in our efforts and refinements, we may eventually converge upon the truth (provided that we ourselves do not create chaos where none previously existed).

To that end, the scientific method was developed to provide a uniform process by which the philosophy of empiricism can be employed in a prescriptive manner, designed to maximize the power of analysis and to constrain error. Again, it is important to remember that error can never be completely eliminated from our scientific pursuits. But if we conduct our imperfect science according to some very strict rules, there is hope that we can reveal the consistency of our error, and therefore remove it (or at least account for its presence). To guarantee that kind of consistency, the scientific method must be followed as a very logical, stepwise procedure (**Figure 1.3**).

Observation

What would at first seem to be the simplest aspect of the scientific method can also be the most significant; after all, observation is the first step on the long road to discovery! And since we can use any combination of our senses (sight, smell, sound, taste, and touch) to make an observation about the world around us, there is no shortage of information streaming into our brains at any given moment. The difficulty actually comes when we try to limit our sensory input to only those perceptions that are pertinent to the phenomenon we wish to investigate.

As a very simple example, let us assume we've gone swimming a number of times, but this is our first time to the beach to swim in the ocean. After the first wave hits us square in the face, we taste for the first time the saltiness of ocean water, which is very different from the taste of water from the swimming pool back home. Regardless of this new salty sensation, your first experience in the ocean is providing you with several other sensations you have never witnessed before in your pool back home: the waves crashing over your head, the sand squishing between your toes, and so on. But if the saltiness of the ocean is truly what has piqued your interest, the scientific method requires that you focus your attention only on those senses, and those bits of sensory information that are relevant to the investigation at hand.

Thanks to Immanuel Kant, we can feel confident that all phenomena within the natural world that are observable must also be ordered—they must fundamentally possess some kind of pattern of "cause and effect." So regardless of whether we have correctly perceived something, causality must exist in all our observations. In other words, there must be something causing our perception of salty ocean water. Explaining that causality—that order—is the very point of science, and it is the goal of the inquisitive scientist. And what do inquisitive scientists do? They ask questions.

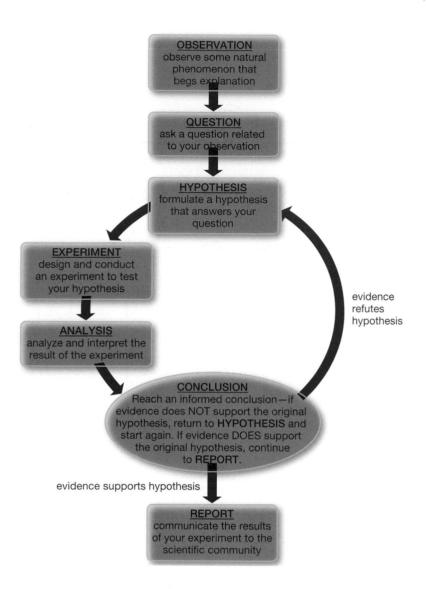

Figure 1.3 The scientific method involves a prescribed technique for investigating natural phenomena, acquiring new knowledge through the use of empirical evidence, and incorporating those discoveries into the body of human knowledge.

Question

It is human nature for us to perceive something unusual or interesting and then immediately ask what, when, which, where, how, and why. In the context of science, we ask questions to explore what is knowable about the world around us. Within the scientific method, it is critical that we ask the right question: it must be a question that is relevant to the observation(s) we just made, and it must be answerable with data. In other words, it must be a question for which we can gather evidence in our attempt to answer it.

Ideally, you should try to construct a question that can be answered in the strictest binary terms: yes or no, positive or negative, black or white. If the phenomenon you wish to explore is far too complex to be explained with a single "yes or no" question, you should try to construct a logical series of binary questions that will allow you to deduce the answer, based on the results of each question answered within the series.

Let us return to our earlier example of experiencing the salty taste of ocean water for the first time. There can be any number of questions we might ask about our observation. Do the waves smacking us in the face (or that sand between our toes) have something to do with the saltiness of the ocean? What about the color of our bathing suit, or the sound of children laughing

Figure 1.4 Within the context of the scientific method, a series of well-designed questions will allow a clever investigator to use one answer as the logical basis to ask further, more in-depth questions. As simple as it may seem, the answer to our first question (A) of whether the ocean tastes salty or not will establish whether the "saltiness" phenomenon even exists and whether it is necessary for us to continue our line of inquiry. If the answer to our first question is "yes," perhaps we can follow up with a question seeking to ascribe that phenomenon to a particular salt like sodium chloride, or NaCl (B). Of course, NaCl is only one kind of salt; other questions may be necessary to explore all of the other salty possibilities (C).

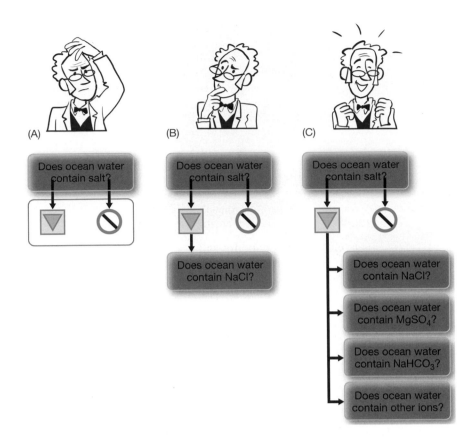

while playing in the sand? Forget about whether any these are truly relevant to the saltiness of the ocean—we have not tested any of them yet, so we can't say for sure. It's more important to consider the usefulness of the questions. The best advice is to "be specific" when asking a scientific question.

It would be more useful (and more logical) to construct a series of very specific, binary questions so that the answer to a preceding question will determine the relevance of subsequent questions (**Figure 1.4**). The simplest (and most relevant) question to ask is simply "Does ocean water contain salt?" If the answer to that question will satisfy our curiosity, regardless of whether it's a "yes" or "no," there would be no reason to continue this line of investigation. If the answer is "yes," we may wish to inquire as to whether the salt in the ocean is sodium chloride (NaCl). If we had a sophisticated understanding of the diversity of salts that occur naturally in seawater, we might wish to expand the question to investigate the presence of many different salts in the ocean, like magnesium sulfate ($MgSO_4$) or sodium bicarbonate ($NaHCO_3$). Of course, if the answer to our original question is "no," we would not need to waste our time exploring the variety of salts in the ocean: that is the value of a wisely ordered series of specific questions.

Hypothesis

Once an observation has been made and a question has been asked, it is our nature to find the answer and attempt an explanation. Even if we know the answer to our question, we still want to know why that is the answer. So, during our application of the scientific method, we as the observers are compelled to consider all the information at our disposal (like our formal education, readings, television shows, personal experiences) in order to make an educated guess, a **hypothesis**, which provides an answer to the original question and therefore explains the observation. Keep in mind that it is not necessary that an initial hypothesis be correct, only that the hypothesis is testable by experimentation. As long as a hypothesis can be tested, any failings in the hypothesis will be revealed in the experimentation and analysis phases of scientific inquiry. It is also important to note that hypotheses that cannot be tested are not invalid or worthless; in fact, they may well be true. But if such hypotheses cannot be tested, they are simply "unscientific."

Within the scientific method, you must have a separate hypothesis for each and every question you have asked. Each hypothesis must also provide an unequivocal answer to the question being asked. For example, your hypothesis for the question "Does ocean water contain salt?" can be either

- Hypothesis 1: Ocean water contains salt.

or

- Hypothesis 2: Ocean water does not contain salt.

Either hypothesis is perfectly valid. What's more important is that they are specific, unequivocal statements: ocean water either contains salt or it doesn't. Of course, we won't know which hypothesis is correct until we perform the experiment and evaluate the data. But if you have crafted an unequivocal hypothesis, the data will provide the evidence to support (or refute) it.

The testability of your hypothesis is of critical importance. Since you will be required to collect data to establish the validity of your hypothesis, you will have to "think ahead" as to what sort of experimental methods and equipment you will have at your disposal, and what sort of data (measurements) will be most useful to you. After all, how could you test the hypothesis "Ocean water contains salt" if you have no means to collect ocean water, or have no equipment to analyze its salt content?

Since most scientific hypotheses will be tested using numerical (that is, quantitative) data, it is often preferable to state your hypothesis as a mathematical statement. This will allow you to use your data as a means to test the mathematical validity of the hypothesis. For instance, the hypothesis "Ocean water contains salt" can be written mathematically as

$$[salt]_{seawater} > 0$$

or as

$$[salt]_{seawater} \neq 0$$

where the concentration of salt in seawater (written as $[salt]_{seawater}$) will always be a positive, nonzero number if any salt is present. Either of these mathematic expressions would be valid; regardless of which hypothesis we chose, the answer to one is simultaneously the answer to the other. If we decided to use the alternate hypothesis instead, that is, "Ocean water does not contain salt,"

$$[salt]_{seawater} = 0$$

this too would be a perfectly valid mathematical expression. Remember that our hypothesis is just an educated guess; we still have no data to support (or refute) any of these hypotheses. But a well-crafted hypothesis should be both testable and answerable, pending the outcome of our experiment (whatever that outcome may be).

Experiment

The proper design and execution of the experiment is perhaps the most difficult aspect of the scientific method. The critical goals of any experiment are twofold: (1) to provide evidence that can be used to affirmatively test the validity of the hypothesis; and (2) to reduce all sources of error introduced by those conducting the experiment (sometimes called experimental **artifacts**). Our challenge as scientists is to design a customized experiment for each and every hypothesis we wish to test, and to conduct such experiments as perfectly as humanly possible.

We know that the natural world is a very complicated place, so it is always in our interest to simplify things as much as we possibly can. In the context of the experiment, we should always take great care to eliminate all conceivable sources of variability in our experiment except for those variables we are specifically testing. As a general rule, this is accomplished by setting up an **experimental group**, where the variable being tested is present and changed (in an orderly fashion) from experiment to experiment. It is also necessary to have a **control group**, where the variable being tested is either absent or is unchanged from experiment to experiment. In this way, comparisons can be made between the experimental group and the control group, based on the measurable differences between the two.

Since we want to limit all sources of variability within our experiment (except for those specifically under investigation), it is often a good idea to create a list of variables that could potentially affect the outcome of the experiment. This step should be taken very carefully and methodically, because variables must first be identified before they can be taken into consideration and controlled within an experiment. Keep in mind that it is not just the natural world that is variable; our chosen experimental method, the equipment being employed to collect and analyze data, and even the inconstant behavior of your research staff can introduce variability if you are not careful.

If we were interested in testing our "Ocean water contains salt" hypothesis, we would need to make a list of every possible variable in our experimental method, which might look something like this:

- Time of collection
- Collection device or method
- Analytical device or method
- Researcher performing collection or analysis
- Water temperature at time of collection
- Water depth at time of collection
- Geographic location of sample

For some of these, we could make a very persuasive argument that they should have no bearing on whether the water being collected and tested actually contains salt, and can therefore be ignored (for example, time of sample collection, water temperature, and/or depth). Others may not directly affect the salt content of the water, but they may affect our ability to confidently perform those measurements (such as the equipment, the methodology, and/or the researcher).

Because human and instrument error are always a factor, they can never be ignored; instead, they must be managed by adhering to a strict collection and analytical methodology. For example, a strict cleaning regimen would be necessary to ensure the collection device (and analytical instrument) do not become contaminated with salt. Likewise, it may be necessary to calibrate the equipment with known concentrations of salt to make sure the instrument is functioning properly, prior to the analysis of field samples. Using the same research staff, or at least having your staff identify which samples they collected and/or analyzed, would also be a useful way to ensure consistency throughout the experiment.

Ultimately, you should control each and every source of variability except for the one variable you wish to test: the salt content of various water samples. In this particular case, you could collect water samples from a variety of different geographic locations to test the hypothesis "Ocean water contains salt." Since our hypothesis presumes that ocean water contains salt, our experimental group would be those samples collected from various locations in the ocean, and our control group would be those samples collected from various locations anywhere except the ocean. In this way, we will have our control group (the water samples not collected from the ocean) to serve as the baseline, against which we can compare the results of our experimental group (that is, our ocean water samples) for salt content.

Analysis

Ultimately, the performance of any experiment will yield data, but what those data mean will still require analysis. Although error can enter the scientific method at any stage, this step in the process is perhaps the most susceptible to **bias** simply because it requires **subjective** (rather than **objective**) analysis of the evidence. During the analysis phase, the investigator is compelled to review all of the experimental data and determine whether the bulk of the evidence either supports or refutes the original hypothesis. Keep in mind that the investigator performing this analysis has also been responsible for the primary observation, the authorship of the initial hypothesis, and both the design and execution of the experimental method—hardly an impartial contributor to the process. This is why it is preferable to phrase the original hypothesis as a mathematical expression. Mathematical equations are inherently objective, as are quantitative data. If the collected data are consistent with the logic of the hypothesis, an objective analysis will naturally follow (and in so doing, we better protect ourselves from unintentional bias).

Regardless of the strength of your hypothesis or the veracity of your data, as human beings we are always prone to making errors in our analyses. In the context of science, this almost always occurs because we have some preconceived notion of how our experiments should turn out, and if our data seem to support our original hypothesis, it is difficult to step back from all of our hard work and look at the results with a truly cynical eye. Results can be easily misinterpreted unless you are very careful to devise an experiment where the results can only be construed in one way.

For example, let us assume we had collected various ocean water samples, meticulously filtered each sample to remove any suspended particles, and then measured the density (ρ) of those samples and compared those results to a sample of ultrapure fresh water ($\rho = 1.00$ g cm^{-3}). In every sample of ocean water, $\rho > 1.00$ g cm^{-3}, providing our evidence that there must be dissolved salt in ocean water, since it is more dense than fresh water: that is our analysis. Although it is perfectly reasonable to assume that the increased density of seawater is caused by dissolved salt, we did not specifically analyze seawater for salt. It is still possible, but implausible, that any other dissolved substance (such as sugars, dissolved free amino acids, pesticides from coastal run-off, etc.) could account for the increased density of seawater.

Unless we analyze specifically for salt ions, we cannot definitively attribute the high density of seawater to dissolved salt.

Conclusion

Although it is often difficult to arrive at a perfectly objective conclusion, this is not to say that scientists cannot be trusted to perform these analyses. In fact, a good scientist should always employ somewhat of an alter ego at this stage, in an attempt to use the data to argue against the initial hypothesis. The mark of a true scientist is to "go where the data take you," without regard to any of your preconceived notions or agendas.

If the data do not support your initial hypothesis, that's perfectly acceptable—you are still practicing good science. Just because your first guess was wrong doesn't mean you should give up; on the contrary, you now have one more piece of the puzzle, which also means you're one step closer to the real solution! To that end, the scientific method will require that you (1) question your initial observations; (2) revise the hypothesis; (3) design and execute new, more revelatory experiments; (4) perform alternative analyses; and (5) repeat as necessary, until the evidence supports your latest hypothesis. That is the beauty of the scientific method: your "failures" are not failures at all—they are used to build the foundation for better hypotheses.

So what happens if we get it right the first time—if our initial hypothesis was carefully crafted and is indeed supported by the evidence? Be careful to remember that there is no such thing as a scientific "fact," nor can a scientific hypothesis ever be "proven." Science (and the scientific method) rely upon the preponderance of evidence in pursuit of the truth, but can a single scientist ever lay claim to discovering the truth? Let's assume for the sake of argument that the truth, through the practice of science, is indeed knowable. It would then stand to reason that our conclusions would stand firm in the face of rugged scrutiny. After all, the truth is the truth, and the truth is never wrong—not one little bit. So let's invite the scrutiny and see how well our scientific conclusions hold up.

Report

Even after all this careful and meticulous work, investigators can never be fully trusted to analyze their own data without bias. Thus, the profession demands that scientists publish their complete research, from observation to conclusion, without omission. This additional scrutiny, performed by the scientific community at large, is the most reliable guarantee of quality assurance and quality control (QA/QC), and will most definitely reveal any mistakes made or biases undisclosed. In this manner, the weakest link in the scientific method (that is, our subjective analyses) is strengthened by the critical review of many, many scholars.

An Example of the Scientific Method in Action

Observation

Sea grasses represent important intertidal and subtidal habitats, providing both food and shelter for a wide variety of marine species, while stabilizing sediments against erosion. Shoal grass (*Halodule wrightii*) is a particularly important species, because it has a global distribution in tropical to subtropical climates, and succeeds in coastal waters of widely variable temperature and salinities (**Figure 1.5**). Shoal grass can even withstand extended periods of exposure during low tide. Although shoal grass can be found at depths of up to 12 m throughout the Gulf of Mexico, its growth within northern Gulf of Mexico waters is conspicuously sparse.

Figure 1.5 Shoal grass (*Halodule wrightii*) is a common species of seagrass, found worldwide in shallow coastal waters with soft (sand or mud) substrates and moderate water clarity. Dense seagrass meadows offer food and protection to a wide variety of marine species, including the microalgae that grow upon the surface of the leaves, giving the mature grass blades a "fuzzy" appearance. ((c) of Hans Hillewaert / CC-BY-SA-4.0.)

So an otherwise widespread and highly competitive species does not seem to grow in northern Gulf of Mexico waters—that is an interesting pattern or phenomenon that begs explanation. Perhaps if we could discover why shoal grass does not grow very well in these waters, there may be some action we could take to improve growth conditions and restore coastal seagrass habitats.

Question

Why does the growth of shoal grass (*Halodule wrightii*) seem to be inhibited in northern Gulf of Mexico waters?

That is the fundamental question, but the scientific method will require us to pose that question in a way that can be tested as a hypothesis (or series of hypotheses). Since shoal grass seems to do quite well in the rest of the Gulf of Mexico, there must be some kind of local stressor—a condition unique to the northern Gulf of Mexico—that adversely affects the growth of shoal grass. Time to make an educated guess as to what that might be.

Hypothesis

H_1—Suspended sediment from river effluent adversely affects the growth of shoal grass.

Based on our research of this particular topic, the Mississippi River (and several other rivers) has a profound impact on several environmental features within coastal waters of the northern Gulf of Mexico. Since shoal grass can survive in wide salinity ranges, the fresh water effluent is not likely to be a factor, but perhaps the sediment load is. Of course, our hypothesis could be completely wrong, but this is a well-reasoned educated guess.

Note that Hypothesis 1 (H_1) is unequivocal—as written, the hypothesis can be easily tested and can have only two possible answers (yes or no). But H_1 only seeks to answer "if" the suspended sediment has an adverse effect on the growth of shoal grass; if our evidence suggests that the sediment does indeed have an effect, H_1 will not be able to answer the "how" or "why." For that, we may need to add a few follow-on hypotheses:

H_2—The deposition of sediment adversely affects the growth of shoal grass.

H_3—The turbidity (that is, the lack of water clarity) caused by suspended sediment adversely affects the growth of shoal grass.

Excellent—we now have three hypotheses to test. If our data support H_1, we will have just cause to explore H_2 and H_3. Note that H_2 and H_3 each will provide further information as to the mode of inhibition: whether it's caused by the actual deposition of sediment upon the shoal grass, whether it's caused by the reduced availability of light, or both!

Experiment

To provide data to test our hypotheses, we might first conduct a survey of several coastal sites throughout the northern Gulf of Mexico (**Figure 1.6**) to measure:

1. The total area of coverage of shoal grass at each site
2. The distance of river outfalls from the Gulf of Mexico
3. The average annual discharge for each of these rivers
4. The average sediment load for each of these rivers
5. A water sample collected from each site
6. The turbidity at each site

Figure 1.6 In order to provide adequate assessment of shoal grass habitat along the northern Gulf of Mexico, we might wish to establish 15 coastal research stations from which we will conduct measurements of (1) live coverage of shoal grass, (2) station distance to nearby river(s), (3) average river discharge, (4) average sediment load, (5) sediment characteristics, and (6) water clarity.

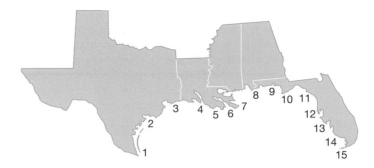

Data collected from items 1–4 of this survey would enable us to establish a quantitative relationship between shoal grass and the potential influence of river effluent; specifically, the amount of suspended sediment in the effluent. If there is an obvious trend between shoal grass coverage and the total sediment load of nearby rivers, our analysis may be able to confirm H_1.

Data provided from item 5 of this survey would enable us to test H_2. Water samples collected from each site could be returned to the lab to measure the particle sizes and settlement rates of the suspended sediment. This would provide additional information as to whether there is a trend between shoal grass coverage and sediment deposition, either as a rate of deposition or as a rate of sediment accumulation.

Data collected from item 6 of this survey would enable us to test H_3. By measuring the turbidity (or water clarity) at each site, we would be able to easily estimate the amount of available light reaching the bottom. This would provide additional information as to whether there is a trend between shoal grass coverage and water clarity.

Note that using previously published data and/or taking measurements at our chosen sites may provide sufficient data to test our hypotheses, without the need for direct experimentation. If such data were available, we would essentially be conducting a survey (or census) of data relevant to our hypotheses and not specifically conducting our own experiment(s).

If we were dissatisfied with the simplicity of one of our hypotheses, we would be required to design our own experiment. This would allow us to perturb the system and monitor the changes in response to that perturbation (using different experimental groups), relative to some baseline condition that serves as the control.

Let's assume our initial survey results strongly supported H_3. If we wanted to know more than just "if" water clarity had an adverse effect on the growth of shoal grass, we could easily design an experiment to actually quantify the growth response of shoal grass to varying light levels (as turbidity).

First, we would need to list all of the experimental parameters that may have unwanted effects on our experimental results. If we had decided that water clarity (turbidity) should be our variable to be tested, everything else has to be controlled:

Test subject:	Sprigs of shoal grass (preferably clones, or at least sections from the same cultivar, to minimize genetic differences in the test subjects)
Static conditions:	Sterile, artificial (or filtered) seawater f/2 growth media (nutrients and minerals)

Sterile quartz sand (as substrate)

5 L opaque black aquaculture vessels

Fixed temperature (25°C)

Fixed photoperiod (14 h light:10 h dark)

Flow-through system (chemo/thermostat)

Variable conditions:

Photosynthetically active radiation (PAR):

25 μmol m^{-2} s^{-1}

50 μmol m^{-2} s^{-1}

100 μmol m^{-2} s^{-1}

250 μmol m^{-2} s^{-1}

500 μmol m^{-2} s^{-1}

1000 μmol m^{-2} s^{-1}

1500 μmol m^{-2} s^{-1}

Control:

Mean solar light intensity (μmol m^{-2} s^{-1} PAR)

at sea level, for mean latitude of the entire study area

Measures:

PAR

Rate of new shoot production

Leaf dimensions (length, width)

Aboveground biomass (ash-free dry weight) per unit area

Belowground biomass (ash-free dry weight) per unit area

Chlorophyll content of leaf tissue (μg chl a/μg C)

An experiment like this would allow us to measure a number of different growth parameters for shoal grass, relative to variable intensities of PAR (Figure 1.7). Then, not only could we use the quantitative data from our experiment to verify the simple hypothesis as stated in H_3, we could also use that same data to determine the complex growth dynamics of shoal grass as a function of PAR intensity (to include optimal light conditions for maximal growth, minimum light thresholds for survival, etc.). Hence, true experimentation (rather than simple surveys of data) will allow a much more powerful application of the scientific method.

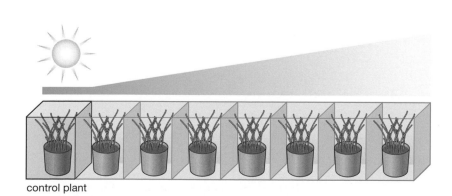

control plant

Figure 1.7 A conceptual design for an experiment to test the effects of increasing light intensity (to represent increasing water clarity) on the growth of shoal grass, *Halodule wrightii*. Although there are many potential factors that could influence the growth of seagrass, the scientific method requires that all other factors be removed, so that our experimental groups are only exposed to changes in one variable: the different intensities of light (25–1500 μmol m^{-2} s^{-1}). Within the control, a fixed light intensity is used to represent the baseline (or "normal") condition, against which all other results will be compared.

Forward in Science, Forward in Truth

If you boil it all down, science is simply defined by the ordered process we use to test hypotheses. And because our hypotheses are little more than educated guesses, we cannot fairly expect our hypotheses to be 100% correct. Thus, true science is about "trial and error." It is about learning—and the learning never ends.

To that end, the most challenging aspect to the scientific method lies in the custom design of an experiment, uniquely fashioned to test our equally unique hypotheses. Obviously, there is no grand experiment that can be applied evenly to all conceivable hypotheses. Instead, we must rely upon the firm conceptual framework of the scientific method and seek inspiration from those scientists who came before us, providing their own well-tested methodologies from which we may borrow. That is the essential goal of this text—to provide you with that guidance, that is, the initial springboard from which you may develop your own ideas and methods.

But remember always the lessons of history's great philosophers, particularly those of Descartes and Kant: truth cannot be found in the sciences. The evidence we glean from the natural world are bread crumbs along the trail of understanding, but these crumbs are mere points of data, not facts (and certainly not truth).

References

Descartes R (1984) Meditations on first philosophy. In The Philosophical Writings of Descartes, Vol. I–II (Cottingham J, Stoothoff R, & Murdoch D, trans), pp 12–23. Cambridge University Press.

Kant I (1965) Critique of Pure Reason. (Smith NK, trans). St. Martin's Press.

Further Reading

Bacon F (2011) Novum Organum Scientiarum (New Instrument of Science). (Spedding J, Ellis RL, & Heath DD, eds). Cambridge University Press.

Cohen MR & Nagel E (2002) An Introduction to Logic and the Scientific Method. Simon Publications.

Gauch Jr. HG (2002) Scientific Method in Practice. Cambridge University Press.

Hume D (1978) A Treatise of Human Nature. (Nidditch PH & Selby-Bigge LA, eds). Oxford University Press.

Locke J (1979) An Essay Concerning Human Understanding. (Nidditch PH, ed). Oxford University Press.

Chapter 2

Introduction to Statistical Inference

"Facts are stubborn, but statistics are more pliable" – Mark Twain

Empiricism requires that we scientists learn about our surroundings by conducting experiments, gathering evidence, and analyzing the results. Although this is by no means a perfect method, it is made more perfect when we conduct ourselves in a critical, unbiased manner and devote ourselves to the pursuit of absolute objectivity. And within the realm of human endeavor, mathematics is about as close as we can come to that lofty goal.

So it should come as no surprise that the sciences have always relied upon the virtue of mathematics to make their own reputation more sterling. Thanks to logic and probability theory, mathematicians have been able to demonstrate that under certain controlled conditions, a small subset of measurements can be used to represent the larger universe from which those measurements were taken. That is exactly what makes science both possible and practical—the fact that we are able to describe order and causation in a natural system without having to gather every conceivable bit of information about that system. Surely, the whole is greater than the sum of its parts, but we can still use some of those parts to discover something new. But to do that, we'll have to learn a bit about statistics first.

Key Concepts

- Any scientific study that relies on numerical data can be tested for using a combination of descriptive and inferential statistical methods.

- Descriptive statistics include those mathematical analyses that can only describe the data that have been collected.

- Inferential statistics rely on the precision and accuracy of the measured data to make predictions about the system from which they were taken.

- Any research hypothesis can be translated into the language of statistics in an effort to either accept or reject the hypothesis being tested.

- Statistical tests are very robust in their capacity to test hypotheses, but they can never provide absolute proof.

Different Types of Data

As we have seen in our review of empiricism and the scientific method, data are the currency of science. Without data of sufficient quantity and quality, we swiftly discover that our ability to test the suitability of our chosen experimental method, the appropriateness of our original hypothesis, and the fundamental conclusions of our study have been severely compromised. But that's the problem—the determination of what constitutes "sufficient" data. Fortunately, an elementary review of mathematics, and the associated rules of simple arithmetic, will serve us quite well in this regard.

Essentially all of the data you may gather in the laboratory or in the field will fall into one of three general classifications: meristic, metric, or categorical data. **Meristic** (or **discrete**) data are those measures that are represented by integers. Although these measures are **quantitative** in nature, they are incapable of conveying fractional information and can be somewhat limiting insofar as numerical analyses are concerned. Such data are best utilized for simple counts, frequencies, or assessments of a binary state (such as species presence = 1 or absence = 0 within a particular habitat).

Metric (or **continuous**) data are those measures that are represented by real numbers; as such, these data include fractional information. Also quantitative in nature, such data are best utilized for those attributes that naturally fall along some continuum of measurement (for example, the body length of a sardine in centimeters or the density of seawater in grams per milliliter). Metric data are inherently more precise than meristic data and are broadly considered to be the most versatile data for numerical analysis. However, as these data are much more dependent on precise measurement, they are also more susceptible to measurement and rounding errors.

Categorical data are those measures that are represented not by numbers, but by some subjective quality as defined by the researcher (such as hurricane intensities categorized as weak, moderate, or intense). As a **qualitative** measure, such data are far more difficult to analyze in an objective manner and are typically disfavored in scientific studies and statistical analyses. However, it is possible for the clever researcher to convert qualitative data into quantitative data, so long as meristic or metric data are used to objectively define the limits of the categories being used. For example, the Saffir–Simpson scale of hurricane intensity (**Figure 2.1**) is used not only to convey very easily understandable information about hurricane intensity (as a series of meristic categories ranked from 1 to 5), but statistical analyses may also be used for the metric data (by using the sustained wind speed) that are inherent in all modern hurricane intensity classifications that use this scale.

Although these three data classifications are applicable in the general sense, when we consider "attribute-specific" data, it is often helpful to define our measurements as being either scale, ordinal, or nominal. To illustrate the differences between these types of measurements, let's assume we were investigating the sedimentary environment at multiple locations, and we decided to study the different sizes and shapes of the sediment grains to accomplish our goals (**Figure 2.2**). These are certainly features that we can measure, but it is important to understand what type of measurements we are taking.

Scale measurements are those that can be ordered according to a continuous scale, typically with a natural, meaningful metric. In the context of science, scale measurements are the ideal type of measurement, because they

category	winds	damage
1	74–95 mph	minimal
2	96–110 mph	moderate
3	111–130 mph	major
4	131–155 mph	extensive
5	>155 mph	catastrophic

Figure 2.1 The Saffir–Simpson scale uses metric data of sustained wind speed to create a more easily understandable ranking system to define categories of hurricane intensity. This is an example of how metric data can be used to create categorical data, thus allowing the researcher to present and analyze the same fundamental data, but in different ways.

are completely objective in nature. For example, if we were measuring sediment grain sizes, we would expect the diameter of each sediment grain to fall along a continuous length scale (from <0.002 mm for clay particles, all the way to >2 mm for sand grains, and beyond). If we wanted to sort our grain size measurements, it would be a very simple task to order them according to increasing particle diameter. Different researchers might choose to categorize the sediment type using a subjectively defined size class (like "clay" or "sand"), but the fundamental measurement of sediment grain diameter is a natural, meaningful metric of length that is, in itself, immune to subjective definition.

Ordinal measurements are those that represent categories within a logically defined (or inherent) ranking. With regard to sediment grains, we might be able to define their overall shape by some logically defined ranking based on how oblong (or round) the grain shape is. That would allow researchers to define the degree of "sphericity," from low to high. If we assign an artificial number to what we've defined as low sphericity (1) versus high sphericity (3), we can still order the sediment grains according to our ranking system. The same could be done for the angularity of the sediment grain, but neither represents a "natural" metric. Even if we analyze sediment type by using the diameter of the grain to define our ranked size classes, ordinal measures are inherently subjective (but at least we're making a logical attempt to define them as objectively as we possibly can).

Nominal measurements are those that represent categories that cannot possibly be ranked in an ordered fashion. As a general rule, these types of measurements are almost useless in a scientific context and should be avoided whenever possible. However, just as we discussed with regard to categorical data, it may be possible to create a system by which nominal measurements can indeed be ranked (thereby "transforming" them into ordinal measurements). For instance, if we had initially graded our sand grains according to their predominant color, we might end up with a jumbled list of brownish-red, yellowish-gold, and olive-black color categories. Although these categories may not at first allow any kind of ranking, if we digitally scanned the images of these sediment grains, our computer would render these colors according to a fixed RGB scale (with each color channel ranging from 0 to 255). Now we have transformed our measurements of color from nominal to ordinal, thereby allowing us to order our sediment grains according to a three-dimensional scale of color (RGB) and along a ranking scale of 0 to 255 in each color channel. This also brings us a bit closer to defining color as an objective, rather than subjective, measure.

Every Measurement is Limited by Imperfections in Precision and Accuracy

Despite our best efforts as scientists, the natural world has a very frustrating habit of introducing error in just about everything we try to do. Even if we wanted to do something that is conceptually simple (like measure the diameter of a grain of sand), we are fundamentally limited in our capacity to do so. That limitation is borne from two confounding facets of reality that we must confront every time we seek to take a measurement: the notion that neither the **precision** nor the **accuracy** of our measurements can ever be perfect.

No matter what we choose to measure, we must do so by utilizing an instrument specifically designed for that purpose. From a philosophical perspective, instruments of human design and fabrication are inherently flawed. Unfortunately, the imperfections of our measuring devices will ultimately

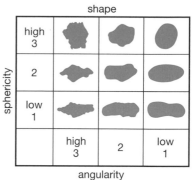

grain diameter	type	
256 mm and up	boulders	
64–256 mm	cobbles	gravel
2–64 mm	pebbles	
0.0625–2 mm	sand	
0.002–0.0625 mm	silt	
0.002 mm and smaller	clay	

Figure 2.2 Sediment grain diameter, type, and shape (angularity and sphericity) are common measures taken by sedimentary geologists. These measures also serve as an excellent example of how scale and ordinal measurements can be represented within the same dataset.

ⓘ MAKE YOUR MEASUREMENTS COUNT

In the context of science, the usefulness of your data will greatly depend upon whether you are taking scale, ordinal, or nominal measurements. As a general rule, the most objective measures are also the best, meaning scale measurements provide the best data while nominal measurements are the least useful.

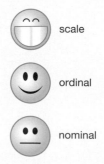

translate to an imperfect measure. When we consider the precision of an instrument, we are limited to the smallest division on the instrument's scale of measurement. For a typical handheld ruler, the smallest division might be 1 mm. Although it is certainly possible that the sand grain we are measuring might be fractionally smaller than 1 mm, we are limited by the **resolution** of our instrument and can only measure to the nearest millimeter. Thus, the precision of our measurement will be likewise limited to ± 1 mm (provided that we did not **interpolate** between the divisions on our instrument's scale).

Although it is impossible to escape the fact that every instrument is inherently imprecise, we can employ our technology to design and construct new instruments of ever-increasing (but never perfect) resolution. A simple way to analyze your instrument's ability to provide sufficiently precise measurements is to determine its **absolute uncertainty** (***AU***), which is simply defined as half the instrument's resolution. This value can then be used in Equation 2.1 to determine the **relative uncertainty** (***RU***) of any measurement taken with that instrument:

$$RU = \left(\frac{AU}{measured\ value} \right) \times 100\% \qquad (2.1)$$

AU directly reflects the overall resolution of the instrument, but *RU* offers a more general way to evaluate the information content of a measurement. Generally speaking, the lower the *RU* of a measured value, the more **significant figures** it contains (and that's a good thing, because a lower *RU* is indicative of increased precision, which will be necessary to achieve greater accuracy in our measurements).

Few scientists would be comfortable admitting such a thing, but the quest for accuracy, though noble, is sadly unachievable in science. This notion of accuracy is more fitting in a philosophical context, as accuracy is essentially truth, in the grandest sense of the word. Whenever we seek to measure some aspect of the natural world, we seek a perfectly accurate measure—the truth. Forever just out of reach, we design instruments of greater and greater precision as we court the truth, but with imprecision comes error, and through error accuracy and truth escape.

Overly dramatic? Perhaps. But remember: as a scientist, it is critical that you remain a vigilant skeptic. To understand the limitations of the scientific method, both in theory and in practice, is to be better prepared to describe those phenomena that have escaped the abilities and intellect of scientists whom you now follow.

The 30–300 Rule is Used to Determine the Proper Instrument of Measure

At some point in your research, you will be forced to choose which instrument of measure you wish to use. Instruments with very high resolution will of course provide very precise measurements, but then the question becomes: what level of precision is "good enough" for the task at hand? As an example, let's say we chose to use a metric ruler (±1 mm resolution) to measure the length and width of seagrass leaf blades (**Figure 2.3**). A length measurement of 97 mm would allow us to use Equation 2.1 to calculate an RU of only (0.5 mm / 97 mm), or 0.5%. If our width measurement were only 3 mm, our calculated RU for the width measurement would be substantially larger: (0.5 mm / 3 mm), or 16.7%. The dramatic difference in uncertainty between our length and width measurements certainly calls into question

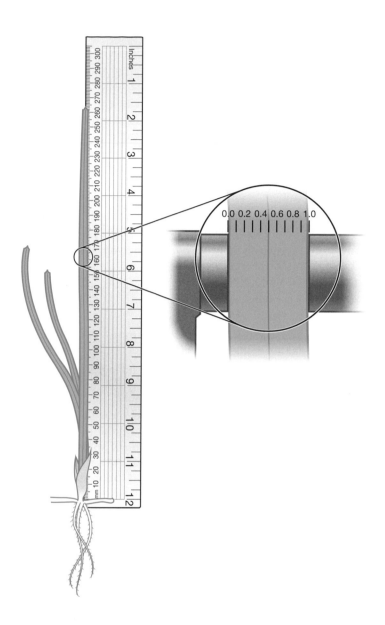

Figure 2.3 The resolution of your instrument will define both the precision and relative uncertainty of every measurement. Take care not to use the same instrument to quantify very different scales of measurement (for example, the length and width of a single blade of seagrass).

whether we should be using an instrument with greater resolution to measure the widths of the seagrass blades and therefore improve the precision of those measures.

As a common rule of thumb, field researchers often use the "30–300 rule" to determine the most appropriate level of precision necessary when taking measurements. The 30–300 rule simply states that your chosen instrument should provide between 30 and 300 unit steps between the minimum and maximum measured values. For example, if our shortest measured blade length was 48 mm and the longest was 104 mm, we would have (104 − 48 =) 56 unit steps at 1 mm resolution. Since that falls within the 30–300 range of acceptability, we would be well advised to use a ruler with ±1 mm resolution to measure leaf-blade lengths. However, if our minimum and maximum measured blade widths were, respectively, 1 mm and 5 mm (that is, only 4 unit steps), it would be clear that our chosen ruler does not possess sufficient resolution. If we instead used a stereoscope with a micrometer plate with ±0.1 mm resolution, our measurements would range between 1.0 mm and 5.0 mm: that's 50 unit steps at 0.1 mm resolution, satisfying the 30–300 rule.

The Rules of Significant Figures Define the Precision of Any Measurement

One of the most difficult aspects of using metric data stems from the fact that any real number can vary to an infinite number of decimal places, but our measurements will be limited to only a few significant figures within that infinite string of decimal places. In other words, significant figures are those digits that contribute to (and essentially define) the precision of a particular measurement. In practical terms, the rules for identifying the appropriate number of significant digits require that

- All exact numbers (such as counts) have infinite significant digits.
- All nonzero digits are significant.
  ($\pi = 3.14159$ has 6 significant digits, as written)
- Any zero digits located anywhere between nonzero digits are significant.
  (1.002205 has 7 significant digits)
- Any leading zero digits are not significant.
  (0.0067 has 2 significant digits)
- All trailing zero digits in numbers containing a decimal are significant.
  (0.006700 has 4 significant digits)
- Any trailing zero digits in numbers without a decimal are ambiguous.
  (12,500 has either 3 or 5 significant digits)

In that last example, it would be impossible to know the significant digits in the value 12,500 unless we also knew the resolution of the instrument that was used to determine that value. If the precision of measurement is ±100, the value 12,500 would have 3 significant digits. However, if the precision of measurement is ±1, it would have 5 significant digits. In order to avoid confusion, it is always best to cite numbers in **scientific notation**. If the value 12,500 were written instead as 1.25×10^4, it would clearly possess only 3 significant digits. If we wrote 12,500 as 1.2500×10^4, the "trailing zero" rule for decimal numbers would clearly indicate the value possesses 5 significant digits.

Because significant digits are linked so intimately with precision, strict adherence to these rules is one of the most important steps we can take to minimize error, especially when performing arithmetic. This is especially true if we are using measurements obtained from several different instruments, each with its own limit of precision (**Technical Box 2.1**).

For Every Measure there is Error

As we have discussed, the process of science is not borne from data, but from the original observations made (our perception) and the question we ultimately seek to answer (our hypothesis). Notwithstanding the clarion warnings of Descartes regarding the fidelity of our own senses, the strength of science lies in the fundamental fact that if something is capable of being perceived, it is also capable of being measured. Fallible as the scientific method may be, we are not completely powerless in our ability to suppress error and uncertainty in its practice.

If we were capable of perfect knowledge, our measurements would define the true properties of the universe, with unfailing accuracy. As we discussed earlier, what is truly unfailing is the inescapable introduction of error as a consequence of our own fumbling, the imperfection of our instruments of measure, and the variability that exists in the natural world. Therefore, the cloud that obscures our view of the truth, by corroding the accuracy of our measurements, is a simply conceived but infinitely complex error function.

HOW TO APPLY THE ARITHMETIC RULES OF SIGNIFICANT FIGURES

To prevent rounding errors, it's always a good idea to keep as many digits as possible while performing intermediate calculations. But when the calculations are done, always cite the final answer according to these rules:

For multiplication or division:
The final result should have the same number of significant digits as the measurement with the least number of significant digits.

For addition or subtraction:
The final result should have the same number of decimal digits as the measurement with the least number of decimal digits.

If both rules are applicable, the multiplication/division rule takes precedence.

TECHNICAL BOX 2.1

An Example of the Proper Use of Significant Figures in a Calculation

Ol' Dusty is interested in measuring the slope of the beach at his study site, and decides on an antiquated but tried-and-true surveying technique whereby two poles of equal length are used in the vertical (**Figure 1**). Since the poles are exactly the same length, any change in beach slope between positions 1 and 2 can be easily measured (as b), so long as Dusty consistently sights the top of pole 2 on the distant horizon. If Dusty has a trusty sidekick to measure the distance between positions 1 and 2 (as a), he can use the Pythagorean theorem and some basic trigonometry to compute the slope-angle (θ).

While Dusty marked his survey poles with 1 cm resolution, his trusty sidekick brought a nonmetric tape measure with 1/16-inch resolution. To perform the calculations necessary to determine θ, Dusty must first convert inches to centimeters to make sure his units are consistent throughout the calculations:

$$a = 15'11\frac{7''}{16}$$

$$b = 9 \text{ cm}$$

Ignoring significant digits:

$$a = \left(\frac{15 \text{ ft}}{1}\right)\left(\frac{12 \text{ in}}{\text{ft}}\right)\left(\frac{2.54 \text{ cm}}{\text{in}}\right) + \left(\frac{11 \text{ in}}{1}\right)\left(\frac{2.54 \text{ cm}}{\text{in}}\right) + \left(\frac{7 \text{ in}}{16}\right)\left(\frac{2.54 \text{ cm}}{\text{in}}\right)$$

$$a = 457.2 \text{ cm} + 27.94 \text{ cm} + 1.11125 \text{ cm}$$

$$a = 486.25125 \text{ cm}$$

$$\theta = tan^{-1}\left(\frac{9 \text{ cm}}{486.25125 \text{ cm}}\right) = tan^{-1}(0.018508949) = 1.060363634°$$

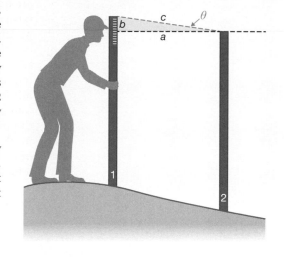

Figure 1 An old surveying technique used to calculate beach slope (θ) involves the use of two vertical rods of standard height, set apart from each other by some distance (a). The angle (θ) is then determined by the height difference from some level reference (b) and a little trigonometry.

So when (and where) should Dusty round off? Or do we really believe Dusty could measure θ to 10 significant digits with such a crude surveying device?

Since the measurement with the least number of significant digits (and the least number of decimal digits) is 9 cm, θ should be rounded to a single significant digit; therefore, $\theta = 1°$.

Keep in mind that Dusty's instrument of measure was a crude one, with only ±1 cm resolution (that is, 1 significant figure). Since our final answer must possess no more than a single significant figure, it stands to reason that any calculated result $0.5° < \theta < 1.5°$ would be rounded to yield $\theta = 1°$ according to the arithmetic rules of significant figures. That leaves a lot of room for error! Perhaps Dusty should have paid more attention to the 30–300 rule and chosen a more precise instrument of measure.

Though we may not be capable of eliminating human frailty from the practice of science, at least we have some control over it (and can therefore define the limits of that uncertainty). Of course, our ability to control variability in the natural world is not so straightforward. In our study of the natural world, we seek not only to measure its basic properties, but also to discover the source of any changes to those properties. And since change implies cause and effect, we are compelled to engage our intellect to describe order in our universe.

That is perhaps the defining purpose of the sciences, and the best argument for a mathematical analysis of order in our universe. It is no mistake that the world's greatest philosophers were originally mathematicians by trade, and it is likewise no mistake that mathematics should play such a central role in the sciences. Mathematics grants us the essential ability to understand order in our time.

More often than not, it is impossible for us to see the whole picture while we are trying to investigate order in the world around us. That means we usually have to take a more pointed look a specific element—just one piece of the puzzle—and try to formulate an explanation of the larger whole. But that, in and of itself, presents a problem. Aristotle said it first, best, and most famously: "The whole is greater than the sum of its parts." And although that may be true, it is an untenable position in the practical world of science. Since it is impossible for us to possess perfect awareness of all the natural factors that could affect our measurements, we are hopeless to know the "truth of the whole" (what we might call Aristotelian holism). So we are reluctantly forced to accept the inherent limitations of reductionism; that is, that we are only capable of measuring some small part of reality and must assume that it faithfully reflects the larger reality we seek to define.

In the world of science and mathematics, we call that "subset analysis." In a few very special cases, mathematicians can use a subset of numbers to prove a mathematical truism. In the natural world, which is plagued by far more uncertainty, we use subset analyses to determine the applicability (or appropriateness) of our conclusions in a much larger context, reaching far beyond our original subset. Here you enter the world of statistics and probability; a world where nothing can be proven, but where our intellect can find order in nature, as supported by evidence of acceptable significance.

In the world of statistics, we can choose one of two general paths. If we wish to describe only the subset of data that was actually measured (the **sample**), we would use **descriptive statistics**. Descriptive statistics cannot be used to infer any information from other subsets of data for which no measurements were taken, nor can descriptive statistics be used to form conclusions about the larger **population** of data from which the subset was taken. To do that, we must employ **inferential statistics**—a host of mathematical techniques that are used to make predictions about other subsets (or the larger population) of data based upon the subset for which we have data. Let's begin by focusing on descriptive statistics, and what those numbers can tell us about our data.

Descriptive Assessments of Central Tendency are Used for Accurate Representations of the Variability in a Sample

With so much talk of uncertainty and error, statistics do offer some good news. As an interesting consequence of number theory, those properties that have been measured are almost completely unimportant; the numbers that represent those properties are what's important in statistics. What makes statistics such a powerful tool in the sciences is its ability to provide numerical justification for the reliability and completeness of our data (which also helps us to design stronger laboratory and field methods).

Let us imagine that all the data we can gather—the full complement of measurements that can been taken to define some property in nature—is best represented by an archery target (**Figure 2.4**). The exact center of that target is the one, true value that exists in reality. As we align our bow and fire our arrow at the target, we are besieged by many sources of error, such as an errant breeze, an imperfection in the shaft of our arrow, or the poorness of our aim: all of these will conspire to deflect our arrow from its true path, causing it to strike the target off-center.

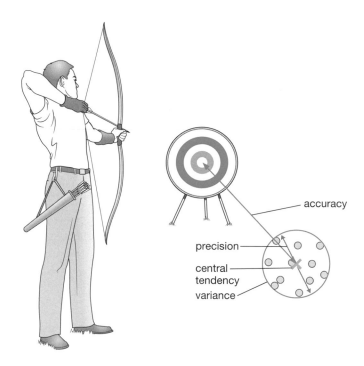

accuracy

precision

central
tendency

variance

Although it is certainly possible, it is highly unlikely that a single shot will strike the bull's-eye. Of course, our chances of hitting the bull's-eye at least once are greatly improved as we take more and more shots—as we take more and more measurements. The better our method (and the better our equipment), the more consistently we will strike the target, and the tighter the grouping of our arrows. By analyzing the **central tendency** of our arrows and how tightly our arrows are grouped (that is, the **variance**), we can say something intelligent about our accuracy and how much error is revealed by the spread of our arrows.

In descriptive statistics, we can determine the central tendency of our measurements in three general ways: by calculating the mean, median, and mode. The most common of these is the **mean** ($\overline{X}$), defined as

$$\overline{X} = \frac{\Sigma X}{N} \qquad (2.2)$$

where ΣX is the sum of all measurements (X) taken to describe a particular property and N is the total number of measurements taken in the effort to describe that property. In most cases, the mean is the most appropriate measure of central tendency for metric data as demonstrated in **Example Box 2.1**.

The **median** is defined as the midpoint of a range of measures that have been organized in **rank order**. Any odd number of observations will have a natural median, but an even number of observations will require the interpolation of an artificial median. This is typically done by taking the simple average of the two values surrounding the theoretical midpoint in the distribution. In instances where the same value is repeated within the dataset, each value must be included in the rank order when determining the median (as demonstrated in **Example Box 2.2**). Note that the determination of the median depends solely upon whether the number of observations (N) is odd or even; beyond that, the median is completely insensitive to the magnitude of N.

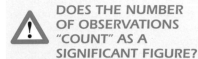

DOES THE NUMBER OF OBSERVATIONS "COUNT" AS A SIGNIFICANT FIGURE?

In the example given, our mean of 35.4 has three significant digits. But what about N? Shouldn't our answer only have one significant digit, like N? After all, it was used in Equation 2.1 to calculate the mean—or are we breaking the rules?

Recall that the rules of significant digits apply specifically to our measurements. N is not really a measurement of the property we are investigating in this example; in fact, N is an exact number because it represents the exact number of measurements taken. Thus, N has an infinite number of significant figures (so we can essentially ignore it).

EXAMPLE BOX 2.1

Determination of the Mean

| 12.2 | 23.8 | 54.7 | 17.8 | 35.2 | 68.9 | $N = 6$ |

$$\text{Mean} = \left(\frac{212.6}{6} \right)$$

$$\text{Mean} = 35.4$$

Note that when calculating the mean, it is not necessary to rank the data in any logical order. In this case, the mean value of the six measurements above is in fact 35.43333333; however, our rules for significant digits indicate that all of our measures possess three significant digits, so our calculated mean should possess only three significant digits as well ($\bar{X} = 35.4$).

EXAMPLE BOX 2.2

Determination of the Median

| 87.4 | 88.2 | 89.6 | | $N = 3$ |

$$\text{Median} = 88.2$$

| 87.4 | 88.2 | 89.6 | 90.2 | $N = 4$ |

$$\text{Median} = \left(\frac{88.2 + 89.6}{2} \right) = 88.9$$

When calculating the median, repetitive measures must be included in the rank order of all values.

| 87.4 | 88.2 | 89.6 | 89.6 | 90.2 | $N = 5$ |

$$\text{Median} = 89.6$$

The **mode** is simply defined as the most frequently appearing value within the range of measurements. Although the mode is quite useful in defining frequencies of occurrence, it is also the least stable measure of central tendency, as changes in the mode are highly likely during sample-to-sample comparisons. Modal analyses may also lead to the determination of multiple modes within the same sample, as demonstrated in **Example Box 2.3**.

EXAMPLE BOX 2.3

Determination of the Mode

| 15 | <u>20</u> <u>20</u> <u>20</u> | 25 | 35 | 35 | | $N = 8$ |

$$\text{Mode} = 20$$

| 15 | <u>20</u> <u>20</u> <u>20</u> | 25 | <u>35</u> <u>35</u> <u>35</u> | $N = 9$ |

$$\text{Modes} = 20 \text{ and } 35$$

Although it is certainly possible that all three assessments of central tendency will yield similar results, it is far more likely that they will differ from each other. In these cases, it is worthwhile to examine not just the central tendency of our data, but its **variance** as well. Luckily, there are several varieties of statistical software available that can be used to import our data and render it graphically so we can literally see what the totality of our data look like, and analyze how they are distributed around the central tendency.

As we saw in the analogy of the archer in Figure 2.4, when we take a look at the distribution of our data, there should be a higher probability that a particular measurement will fall close to the central tendency. That also means that any measurements in our dataset that are quite distant from the central tendency should be few and far between. By simply looking at the patterns of how our data are distributed around the central tendency, we can decide how best to proceed in our statistical analyses of that data.

"Normal" Data Can be Plotted As a Gaussian Curve

Within the context of statistics, if the mean, median, and mode all share the same value, we say that the data are **normal**. When plotted, normally distributed data will possess the classic bell shape—this is what's known as a **Gaussian curve** (Figure 2.5). What defines the Gaussian (normal) curve is that it is symmetrical about the mean ($\bar{X}$) and that its first **standard deviation** (±1s) is the distance between $\bar{X}$ and the inflection point on the curve. In a perfectly normal distribution, 68.2% of all the data will fall within one standard deviation (±1s) of $\bar{X}$, 95.4% will fall within ±2s, and 99.6% will fall within ±3s.

The standard deviation is the most frequently used estimate of variability in the sample data, as it represents the basic tendency of the sample data to depart from the sample mean. In mathematical terms, the standard deviation s is calculated using Equation 2.3:

$$s = \sqrt{\frac{\Sigma d^2}{N-1}} \qquad (2.3)$$

where d represents the departure of a measured value X from the mean ($d = X - \bar{X}$) and Σd^2 is the sum of all calculated values of d^2 in the distribution. As in Equation 2.2, N is simply the total number of measurements in the dataset.

As a practical matter, the magnitude of the standard deviation s is an excellent indicator of just how well each measurement within the distribution comports with the mean. From an experimental point of view, a small standard deviation is indicative of low variability, which is exactly what we want. Let's assume for the sake of argument that our central tendency is in fact

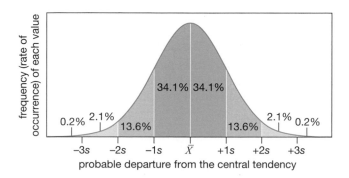

Figure 2.5 Among normally distributed data, all but 0.4% of the data are described by the classic Gaussian curve, ±3s. This is particularly powerful as a descriptor of data, as this mathematical relationship can be used to make statistical inferences about any particular measure within the data (in relation to where it falls along the Gaussian curve). (Courtesy of Jeremy Kemp / CC-BY-2.5.)

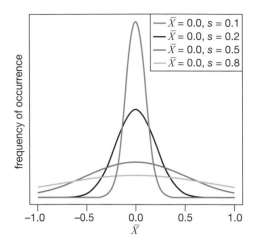

Figure 2.6 Although all four of these Gaussian curves exhibit the same mean, their distributions are hardly equal. When standard deviations are very small (the preferred case in the sciences), the data are distributed very near to the sample mean (a leptokurtic condition). As the magnitude of the standard deviation increases, so too does the dispersion of the data (a platykurtic condition). (Courtesy of Matthieu / CC-BY-SA-3.0.)

 THE PROPER WAY TO CITE DATA

To better convey the true variability of data, it is always advisable to cite both the sample mean and the standard deviation. For the data in Figure 2.6, we would cite these findings as

$\bar{X} = 0.0 \pm 0.1$

$\bar{X} = 0.0 \pm 0.2$

$\bar{X} = 0.0 \pm 0.5$

$\bar{X} = 0.0 \pm 0.8$

the "true" value of what we're trying to measure in nature. If we took multiple measurements, and every single measurement hit the bull's-eye, every measurement would be exactly equal to the central tendency, there would be no variability in our measurements, and our standard deviation would be zero. Alternately, if we fired an arrow exactly 10 cm to the right of the bull's-eye, and the next arrow was exactly 10 cm to the left, the central tendency would still be right smack in the middle of the bull's-eye, but there would be a lot more variability in our measurements, as evidenced by a large standard deviation.

This is exactly why it is common practice in the sciences to cite both the mean and the standard deviation of each sample, as it is the combination of these two measures of dispersion that will reveal the most information about the reliability of the data. The smaller the standard deviation, the more closely our measurements converge on the central tendency, which in turn provides us with greater confidence that our measures are indeed accurate. Sometimes it is more helpful to visualize the relationship between the central tendency and the standard deviation by analyzing the **kurtosis** of the Gaussian curve (**Figure 2.6**).

Another useful measure is the **standard error of the mean**, which represents the standard deviation of the error in the sample mean, relative to the true mean of the population. In other words, the standard error can be viewed as an indicator of how well (or how poorly) the mean of your sample represents the larger population from which the sample was taken. Mathematically, the standard error (*SE*) is calculated using Equation 2.4:

$$SE = \frac{s}{\sqrt{N}} \tag{2.4}$$

where *s* is the standard deviation of the sample and *N* is the number of observations within the sample. Just as lower values of the standard deviation *s* are indicative of greater precision in our measurements, lower values of *SE* provide increasing confidence that our sample mean is an accurate estimate of the true population mean.

Instead of looking at the dispersion of the data as a whole, sometimes it is helpful to determine how many standard deviations a particular value deviates from the mean. This can easily be accomplished by using Equation 2.5 to convert any measurement of *X* into a **z score**, where

$$z = \left(\frac{X - \bar{X}}{s} \right) \tag{2.5}$$

You can use *z* scores to determine the position (or performance) of a particular value in relation to the rest of the data in terms of the standard deviation (**Example Box 2.4**).

The Coefficient of Variation Can be Used to Measure Variability

Another estimate of variation that is particularly useful in the sciences is the **coefficient of variation** *cv*, which is calculated using Equation 2.6 and represents the relative magnitude of the standard deviation to the mean as the ratio:

$$cv = \left(\frac{s}{\bar{X}} \right) \cdot 100\% \tag{2.6}$$

EXAMPLE BOX 2.4

Using z Scores to Determine Single-Value Deviations from the Mean

Amy and Joe are thoroughly enjoying their Marine Field Methods class, despite their somewhat lackluster performance on the exams:

Test	$\overline{X}$	s	Amy $\overline{X}$	Joe $\overline{X}$	Amy z	Joe z
1	75	14	97	82	1.6	0.5
2	40	15	31	14	−0.6	−1.7
3	22	4	28	20	1.5	−0.5
4	144	38	148	194	0.1	1.3
$\overline{X} =$	70.25		76.0	77.5		
$\Sigma z =$					+2.6	−0.4
$\overline{z} =$					+0.65	−0.1

If the instructor merely considered each student's mean score on the exams, it would appear as though Joe ($\overline{X} = 77.5$) did slightly better than Amy ($\overline{X} = 76.0$). However, upon further analysis of their z scores, it is obvious that Amy's performance is much more consistently above the mean (as indicated by the positive z score). Amy's overall performance in the class was +0.65 standard deviations above the mean, whereas Joe consistently underperformed compared with the rest of the class (averaging −0.1 standard deviations below the mean).

Since the area beneath a normal curve contains 100% of the data, the number of standard deviations from the mean also defines the percent of data that fall above, or below, that standard deviation. That is why it is more important to consider the percentile ranking of a student's performance on a standardized test, rather than the raw score itself: z scores are what make that analysis possible.

Since a small standard deviation is the preferred condition, a small cv is likewise indicative of low variability and is therefore preferred above large cv values.

Perhaps the simplest measure of data dispersion is the sample **range** R, simply calculated by Equation 2.7 as

$$R = largest\ value - smallest\ value \qquad (2.7)$$

Although it can be a useful measure in certain circumstances, the range is generally considered to be an inferior measure of dispersion because it lacks information about the distribution of data that falls within that prescribed range.

Another simple way to examine the dispersion of data is to calculate the sample **variance**, which is related to the standard deviation by the simple arithmetic function described by Equation 2.8, where

$$variance = s^2 \qquad (2.8)$$

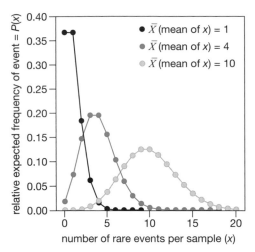

Figure 2.7 As a probability distribution, Poisson curves can become normal as $\bar{X} \to x$ as the number of true counts (x) increases. (Courtesy of Skbkekas / CC-BY-SA-3.0.)

TRANSFORMING YOUR DATA ISN'T CHEATING— IT'S NECESSARY!

At first, it may seem as though data transformations are a way to "cook the books." Before you get too troubled by this notion, remember that statistics utilize the variance and central tendency of the data, regardless of what those data are.

So long as mathematical transformations are applied uniformly to the dataset, they will not change the information content of your data in the least. Consider this: if you converted a temperature reading from degrees Celsius to Kelvins, you would have to apply the mathematical function

$$K = {}^{\circ}C + 273.15$$

The data have been mathematically transformed, but because the transformation has been applied uniformly to the entire dataset, the information contained within those data remains unchanged.

From a mathematical perspective, it is easy to see why a small standard deviation is preferred, since the variance is an exponential function of s. A low standard deviation (and therefore, a low variance) indicates that all the values in the sample are tightly centered about the mean.

What makes the variance so special is that it has been proven by mathematical **axiom** that the distribution of any sample can be described if it is normal and both the mean and variance are known. This also means that if any two normally distributed samples possess the same mean and the same variance, they are considered to be statistically identical to each other. That critical relationship is the foundation upon which inferential statistics has been built—it allows researchers to infer relationships between a subset of data and the larger population from which the subset was gathered.

Data Distributions That Aren't Normal Often Follow a Poisson Distribution

While normally distributed data are the bread and butter of statistics, real data are almost never normal. Very often, data instead follow a **Poisson** distribution (**Figure 2.7**), a geometric curve that represents the true count of independent or rare events (x) that occur in a specified interval of time against the expected mean $\bar{X}$ over that same interval of time. Poisson distributions are generally expected when your data are meristic (that is, discrete measurements represented by integers).

Since the Poisson curve describes how often an event will occur relative to its expected rate of occurrence, it is as much a probability distribution as it is a frequency distribution (rather than a classic data distribution). However, if the data being gathered are time sensitive or possess some probability of being included (or excluded) from the sample, the results might resemble a Poisson curve rather than a Gaussian curve. If the variance of your sample is equal (or nearly equal) to the mean, it is a Poisson distribution.

Another common type of frequency distribution is the **chi square**, which most often resembles the extreme case of $\bar{X} = 1$ in the spectrum of Poisson curves depicted in Figure 2.7. The chi-square distribution is indeed a special case and is most often used to test statistical hypotheses, or the "goodness of fit" between a measured distribution and a theoretical distribution.

Nonnormal Datasets Can be Normalized Using Data Transformation

Although a more accomplished statistician may find Poisson and chi-square distributions to be useful in their own right, most scientific applications of statistics are best served if the data are both continuous (metric) and normal. Frequently, nonnormal datasets can be normalized to better resemble a Gaussian curve through the use of **data transformation**.

Data transformation is a simple process by which all the values within a nonnormal dataset are recalculated according to a uniformly applied mathematical function and then analyzed for their new resemblance to the Gaussian curve. Although the number of potential transformation methods is only as limited as the imagination of the researcher, Equations 2.9–2.12 are just a few common transformation methods that seem to work best (and should be attempted first).

$$X' = log_{10}(X + n) \tag{2.9}$$

The log transformation in Equation 2.9 helps to make **skewed** data more symmetrical and is most appropriate for data that are either related to some measure of biological growth or for counts, especially when the variance of the counts is larger than the mean. If zeroes in the dataset ($X = 0$) should be counted, use the condition $n = 1$; otherwise, $n = 0$.

$$X' = X^k + n \qquad (2.10)$$

The square root transformation described by Equation 2.10 is also useful for biological counts, particularly when the researcher wants to give less weight to numerically abundant species. To count zeroes ($X = 0$), use $n = 0.5$ rather than 1; otherwise, $n = 0$. The default value $k = 1/2$ represents the square root of X and should always apply, unless the researcher wishes to transform counts into presence/absence data (in which case, use $k = 1/3$). To convert from a Poisson distribution to a Gaussian distribution, use $n = 0$ and $k = 2/3$.

$$X' = sinh^{-1}X \qquad (2.11)$$

The arcsinh transformation described by Equation 2.11 is most useful when the dataset is dominated by zeroes.

$$X' = sin^{-1}\left(\frac{\sqrt{X}}{100}\right) \qquad (2.12)$$

The arcsin transformation defined in Equation 2.12 should only be used when X is proportional data, given as percentages.

The Central Limit Theorem Can be Employed to Increase the Predictability of Your Data Distributions

One last little bit of good news before we dive into the realm of inferential statistics and put all this information to good use: just about any data you could conceivably gather from the natural world can be considered normal, so long as you gather enough data. Although that may not sound like a terribly exciting revelation, it is a fundamental axiom of mathematics and probability that as we increase the number of observations (N) in each computed mean ($\bar{X}$), the distribution of those averages becomes more and more Gaussian. Though not a rigorous mathematical definition, this is what's known as the **central limit theorem** and its significance in the sciences cannot be overstated (**Example Box 2.5**).

This may seem like quite a fuss, just to get our data to behave more normally. But remember that much of what we will show by using inferential statistics ultimately depends not on our actual measurements but on the distribution of our data. So the normality of our data factors quite heavily on what statistical analyses we can (and can't) perform.

Attendant to the central limit theorem is the tendency of our data to become more normal as the number of observations (N) increases (**Figure 2.8**). This is a very important consideration because the task of deciding how much data is "enough" is not an easy one. Since the very practice of science requires us to perform subset analysis, we can't measure everything—we are forced to limit ourselves according to some decision as to just how big our subset should be. Not an easy thing to do, especially when you're trying to investigate new (or underexplored) phenomena.

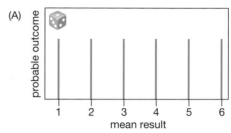

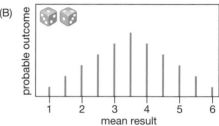

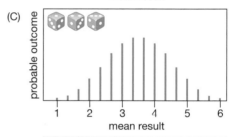

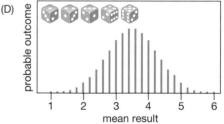

Figure 2.8 Assuming we are not using loaded dice, the result of each die roll should be random. In fact, each specific result has a one-in-six chance of occurring. If we were using a single die (A), the central tendency of our results would not be normally distributed at all. If we instead rolled two dice (B) and calculated the mean result, the probability distribution of our results would look more normal. As we increase the number of dice (C–D), the distribution of data becomes more and more Gaussian—an elegantly simple demonstration of the central limit theorem.

EXAMPLE BOX 2.5

Using the Central Limit Theorem to "Create" Normal Data

As an example, let's say we randomly selected 30 seagrass blades from a seagrass meadow and wanted to investigate the central tendency and standard deviation of the measured lengths (see Figure 2.3). As we saw in our earlier example of the 30–300 rule, the range of seagrass blade lengths was 56 mm (48 mm to 104 mm) (Table 1).

Table 1 Blade Length of Randomly Selected Seagrasses ($N = 30$) within a Subtidal Shoal Grass (*Halodule wrightii*) Meadow

Sample	length (mm)	Sample	length (mm)	Sample	length (mm)
1	104	11	95	21	78
2	59	12	98	22	100
3	68	13	48	23	101
4	49	14	96	24	104
5	94	15	73	25	103
6	94	16	104	26	66
7	75	17	99	27	63
8	102	18	62	28	62
9	75	19	101	29	94
10	80	20	69	30	48
$\bar{X}_{\text{length}} = 82$ $s = 19$					

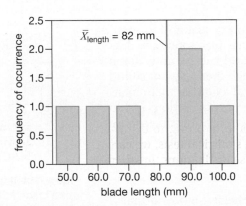

Figure 1 Distribution of seagrass blade-length data pooled into one bin of $N = 30$ observations.

Note that the measures of blade length as depicted in **Figure 1** do not appear to be normally distributed about the central tendency (in this example, $\bar{X}_{\text{length}} = 82$ mm). However, if we binned our measurements into random groups (six groups of $N = 5$) and calculated the means for each of those groups, our results would be

$$\bar{X}_{1-5} = 75 \qquad \bar{X}_{11-15} = 82 \qquad \bar{X}_{21-25} = 97$$

$$\bar{X}_{6-10} = 85 \qquad \bar{X}_{16-20} = 87 \qquad \bar{X}_{26-30} = 67$$

If we then analyzed the descriptive statistics of these means, we would see in **Figure 2** that the data appear much more normally distributed, with a greater number of observations grouped more tightly about the calculated mean.

$$\bar{X}_{Means} = 82 \qquad s = 10$$

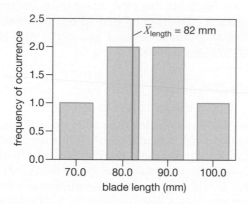

Figure 2 Distribution of the same seagrass blade-length data randomly pooled into six bins of $N = 5$ observations.

This demonstrates how the central limit theorem can be used to transform an otherwise nonnormal dataset into one that exhibits a more Gaussian distribution. This is only possible if

1. The result of one measurement is not dependent upon the result of another;
2. The data are randomly selected when binned into groups; and
3. The sample size in each of the new bins is equal for all bins (for example, six bins of $N = 5$).

It is also important to note that the normality of our transformed data will improve as we increase the number of observations (N) in our bins. So 6 bins of $N = 5$ would yield a more normal distribution than 10 bins of $N = 3$.

Inferential Statistics: A Brief Primer

Now that we have laid the necessary groundwork, it is time to leave the theoretical realm behind and actually try to apply some of what we have learned. As we proceed, keep in mind that the variety of statistical methods available to the scientific community is vast. Since a full treatise on statistics is far beyond the scope of this text, readers are strongly encouraged to familiarize themselves with additional resources on this subject.

That being said, what follow are a number of helpful examples of the simplest, most common statistical methods used in the sciences. As we work through these examples, we shall utilize the same dataset to illustrate how the same data can be analyzed in a variety of different ways. Ultimately, the fundamental goal of science is to distinguish chaos from order; to distinguish pure chance from "cause and effect." By using statistical tests, we can demonstrate just how confidently we can make those distinctions, in strict mathematical terms.

The Probability Level Will Define Which Results Are Statistically Significant, and Which Are Not

So what's your chosen probability level? Although that line would definitely earn some confused looks at your next cocktail party, it is nonetheless a very important question to ponder. The **probability level** (α) is a threshold value chosen by the investigator used to define when a statistical result can be considered "significant" (in other words, when a result is unlikely to occur by chance). If you were testing your data for a particular cause-and-effect relationship, $\alpha = 0.05$ (5%) simply means that you would expect that relationship to occur only 5 times out of 100 (5%) due to pure chance.

For most scientific applications, $\alpha = 0.05$ is the preferred probability level. However, there are instances where it might be advisable to have a stricter

 WHEN IT COMES TO SAMPLE SIZES, THE MAGIC NUMBER IS 30

More is better. When it comes to collecting scientific data, it's hard to argue with that statement. But it's also a statement of no real help to us. What we really want to know is "How much data is enough?"

Using the central limit theorem as inspiration, we can demonstrate that a minimal sample size of 30 ($N \geq 30$) is a good general rule of thumb. What we're essentially looking for here is a subset of data that are sufficiently representative of the larger population. True in all cases? Of course not. But when in doubt, shoot for $N \geq 30$.

probability level. For instance, if you were testing a new design in automobile brakes, a failure rate of 0.05 (5 out of every 100 brake pads) could be absolutely tragic. In such cases, you might wish to use a probability level of 0.01 or 0.001 (or even lower).

The Use of Confidence Intervals and Outliers Helps in the Analysis of Statistical Significance

Once we have established the preferred probability level of our investigation, we can use that same value to define the number of standard deviations within which our normally distributed data should fall. If $\alpha = 0.05$ is chosen, we are using that value to define the threshold of acceptable variance within our data. This also has tremendous implications for the predictability of our data, because we can use Equation 2.13 to establish that

$$CI = \overline{X} \pm \frac{d_\alpha \cdot s}{\sqrt{N}} \tag{2.13}$$

where our **confidence interval** (*CI*) describes the accepted probability that our sample mean $\overline{X}$ accurately represents the true mean of the population from which it was sampled. For Equation 2.13, use $d_\alpha = 1.96$ when $\alpha = 0.05$. Any single measurement that exceeds the confidence interval also falls outside the range of values we would expect for 95% of our data (**Figure 2.9**). In the context of statistics, we may consider those values to be "significantly different" from the dataset from which they were taken.

Sometimes, there may be values that dramatically exceed the accepted variance in the data. As a matter of convenience, $\pm 3s$ is the accepted threshold to determine the presence of what are called **outliers** in the data. Data that are at least three standard deviations from the mean (or possess $z \geq 3$) exceed the range of values in 99.6% of the data (see Figure 2.5). Because outliers represent extreme values in the dataset, they also represent very significant departures from the "background" distribution of data.

As a practical matter, outliers in the dataset must be handled very carefully. Although outliers can certainly be due to extremely rare or atypical natural events captured in the dataset, they are more commonly attributed to experimental or data-recording errors. Therefore, it is usually a good idea to analyze your data for the presence of any single measurement that exceeds the $\overline{X} \pm 3s$ threshold.

If outliers are identified and can be shown undeniably to have arisen from experimental or data-recording errors, those values should be purged from the dataset and new descriptive statistics should be calculated from the cleansed dataset (**Example Box 2.6**). If it is impossible to attribute human error to the presence of outliers, it may be useful to perform the same statistical analyses twice: once including all outliers, and again with all outliers removed. The effects of these outliers on the final statistical results should then be reported, allowing the reader to draw the conclusions.

The Strength of Our Statistical Analyses Will Depend Heavily on Our Assumptions of the Data

As we have already discussed in this chapter, the use of statistics requires that we make all sorts of assumptions about our data—the most fundamental assumption is that the subset of data we have collected is representative of the larger population. There are a few more assumptions that we must address, particularly if we wish to place any confidence in the results of our statistical tests. With rare exception, the statistical tests most commonly

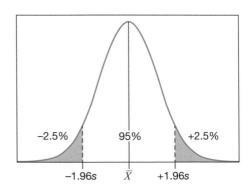

Figure 2.9 A graphical representation of normally distributed data, indicating the 95% confidence interval of the data distribution ($\overline{X} \pm 1.96s$), assuming $\alpha = 0.05$. The shaded areas of the curve indicate those data that fall outside the 95% confidence interval (*CI*).

EXAMPLE BOX 2.6

Using z Scores to Identify (and Purge) Erroneous Outliers in a Dataset

Richard has been measuring the lengths of emergent seagrass blades in a dense patch of shoal grass (*Halodule wrightii*) in a subtidal seagrass meadow. After three hours of monotony in the hot sun, Richard is bound to make a few mistakes. Sure enough—a few of his observations were recorded in centimeters when they should have been entered in millimeters.

128	145	132	121	122	128	131	126
124	12.1*	142	137	129	127	121	127
143	139	126	11.8*	123	134	141	132
12.4*	131	144	126	124	128	129	128

Janet is handling the postprocessing and is clever enough to analyze the data for outliers by calculating z scores (see Equation 2.5). Her initial results indicate three possible outliers:

$N = 32$ $\qquad$ (12.1) z score = –3.01

$\bar{X} = 120$ mm $\qquad$ (11.8) z score = –3.02

$s = 36$ mm $\qquad$ (12.4) z score = –3.00

Although Janet might be tempted to "correct" Richard's mistakes, it's better to simply eliminate the three outliers and reanalyze:

$N = 29$

$\bar{X} = 131$ mm

$s = 7$ mm

Note that both the central tendency and standard deviation of the data have changed quite a bit, now that the outliers have been purged from the dataset. Perhaps most importantly, we can see how much more improved our standard deviation has become.

employed in the sciences require that we first certify that our data agree with the following:

Assumption of independence

Assumption of single population representation

Assumption of normality

Assumption of homoscedasticity

Independence

Under most conditions, it is important that the values that comprise our dataset have been randomly selected from the larger population and that those values are independent of each other. An independent sample is one that, when its value is measured, will not influence (or be influenced by) the value of any other sample.

For example, the measured length of one seagrass blade is not likely to influence the length of any other seagrass blade (particularly if we took care to always measure blades from separate plants). This would be a perfect example of our length measurements being truly independent of each other and therefore satisfying the assumption of independence. Likewise, we would expect that the leaf-width measurements would also be independent of each other.

We may not be so confident in the independence of our data if we measured multiple leaves from the same seagrass plant. Perhaps the length of one leaf would be influenced by the length of another leaf on the same plant, because of an uneven distribution of nutrients, self-shading, or any other scenario where one leaf might impose an influence on the growth (length) of a partner leaf. Why take the chance? If you can design a sampling strategy that maximizes sample independence, the strength of your statistical analyses will be better for the effort.

Single Population

Since our data are supposed to represent the larger population of values from which they were collected, it is critical that we are sure the sampled population is the same for every observation. That means that we must take great care to avoid sampling across multiple populations. In our seagrass example, that means we would want to collect measurements from random individuals within the same general location. If we were to collect our data from three different sites and try to analyze them all together to yield some grandiose perspective on seagrass growth, we would be violating our "single population" assumption. Of course, we could still use statistics to compare seagrass growth between these three sites; we'd just have to be careful to organize our data so that our measurements were grouped into three single populations.

Normality

As we have discussed previously in this chapter, normality merely refers to the dispersion of data following a Gaussian curve. Thanks to the central limit theorem, we can often assume our data can be rendered normal if the number of observations in our dataset is robust ($N > 30$). Recall also that several different data transformation methods may be employed to ensure normality. Keep in mind that there are several statistical tests that do not assume (and therefore do not require) that your data are normally distributed, so even nonnormal data can be used for statistical analyses.

Homoscedasticity

This assumption applies only when comparing two (or more) populations with each other, and refers to the analysis of their variances. If the variances of the populations are equal (or very nearly equal), the assumption of **homoscedasticity** is upheld. Recall that the variance is simply the standard deviation multiplied by itself (see Equation 2.8), so populations whose standard deviations are equal (or very nearly equal) are also covered by this assumption. Keep in mind that the central tendencies of the populations need not be equal; in fact, they can be very different from each other. What is critical to uphold this assumption is that the dispersion of data about the central tendency must be similar.

If none of these assumptions are violated, we can use **parametric** tests to analyze our data. Parametric tests are a special family of statistical methods that generally offer greater power, accuracy, and precision because of the inherent "quality" of the data—quality that has been authenticated by our confirmation that the data are independent, from a single population, normal, and homoscedastic.

If any of these assumptions have been violated, we must instead use **nonparametric** tests to analyze our data. Since nonparametric tests are free from the specific assumptions about data distribution, they are still quite useful but generally considered to be a less powerful alternative to parametric tests. In the natural sciences, it is best to perform parametric tests whenever possible, but nonparametric tests offer a suitable alternative if your data cannot be made to conform with the assumptions outlined above.

Formulating a Testable Hypothesis is Critical to Both the Scientific Method and the Use of Inferential Statistics

As we learned in Chapter 1, the ability to formulate a testable hypothesis is the very backbone of the scientific method. A research hypothesis is tested in the laboratory or in the field by a variety of well-designed (and well-executed) experiments that will produce data—these data are then used to test a **statistical hypothesis**. It is critical that these two hypotheses are linked in such a way that the acceptance (or rejection) of the statistical hypothesis will likewise allow the investigator to either accept (or reject) the research hypothesis. Because there are only two logical conditions of the hypothesis (either true or false), the statistical hypothesis must be structured in such a way as to yield a binary result.

This is done by translating the research hypothesis into a statistical hypothesis, which can be phrased in two ways: the **null hypothesis** (H_o) and the **alternative hypothesis** (H_a). The null and alternative hypotheses have to be mutually exclusive of each other, so that if one of them is statistically demonstrated to be true, the other hypothesis is automatically negated. Because of this relationship between the two, we can use statistics to test either the null or the alternative hypothesis; regardless of which one we choose to analyze first, the result of the statistical test will simultaneously answer both hypotheses.

A proper null hypothesis (H_o) should be defined in such a way that it: (1) negates H_a; (2) denies any relationship between the dependent and independent variable(s) in the study; (3) is assumed to be true unless the results of a statistical test of significance allow it to be rejected; and (4) is phrased so that if H_o is true, the research hypothesis (H_a) cannot possibly be true. That last point is key, and it is why it is called the null hypothesis—if H_o cannot be rejected by a statistical test of significance, we must assume it is true and that our research hypothesis was incorrect (or incapable of verification). As a practical matter, the null hypothesis generally states that there are no significant differences when making some kind of comparison.

By contrast, the alternative hypothesis (H_a) is the research hypothesis we're trying to "prove" with statistics. H_a should always be defined so that it (1) negates H_o; (2) confirms some relationship between the dependent and independent variable(s) in the study; and (3) is assumed to be false unless a statistical test of significance can confirm otherwise. In most cases, the alternative hypothesis is structured so that it implies there are significant differences between two (or more) compared populations, variables, means, variances, or what have you.

The mutually exclusive definitions of H_o and H_a will ultimately lead us to a logician's showdown, which is exactly the point. Since H_o is the condition that is assumed to be true by default, we cannot accept H_a (and therefore reject H_o) unless we have significant mathematical support for doing so—that is the whole purpose of inferential statistics in scientific research.

A PROPERLY DEFINED HYPOTHESIS IS CRITICAL TO SUCCESS

Translating the research hypothesis into a valid statistical hypothesis is a critical step, and is perhaps the most confusing aspect of inferential statistics. If there are errors in how you define the null (H_o) versus alternative (H_a) hypothesis, the results of your statistical tests will be useless.

Take great care that H_o and H_a can only be confirmed with a simple "yes/no" or "true/false" answer, and that H_o and H_a are mutually exclusive—that is, it is impossible for both of them to be true (or for both of them to be false, for that matter).

Table 2.1 Confirmation Matrix of Statistical Conclusions for Comparisons Between the *p*-Value and α

p-value > α	*p*-value $\leq$ α
Accept H_o	Reject H_o
Reject H_a	Accept H_a
Conclusion: No significant differences exist, or variability in the data can be explained as random "noise"	Conclusion: Significant differences do exist, or variability in the data cannot be dismissed as chance occurrences

WHAT A *P*-VALUE REALLY MEANS

Remember that α represents the absolute limit for a result to be significant. Although the line has to be drawn somewhere, you would not want to make a critical decision while teetering on the edge of that decision boundary. For $\alpha = 0.05$,

If $p \leq 0.05$, there is sufficient statistical evidence to reject H_o and accept H_a.

If $p \leq 0.01$, there is strong statistical evidence to reject H_o and accept H_a.

If $p \leq 0.001$, there is overwhelming statistical evidence to reject H_o and accept H_a.

Since statistics is all about probability, we can use our accepted probability level (α) to define the probability value (or **p-value**) in our statistical tests that will indicate when we can define any differences as being statistically significant. In other words, whenever a statistical test returns a *p*-value $\leq \alpha$, we can claim that the results are significant enough to accept H_a (**Table 2.1**).

In order for a *p*-value to be deemed significant, it must be no greater than the chosen α. Since the *p*-value returned by all statistical tests will naturally range from 0.00 to 1.00, the realm of significant *p*-values (if $\alpha = 0.05$) will be 0.00–0.05; if we had chosen an α of 0.01, significant *p*-values would range from 0.00 to 0.01. Remember that we're living in the land of probabilities here, so there's always a remote chance of erroneously assigning significance when none truly exists. If we chose an α of 0.01, we would have a 1 in 100 (1%) chance of assigning statistical significance to our result and mistakenly accepting H_a. However, as $p \rightarrow$ zero, the stronger our statistical evidence of significance.

Because our choice of α is subjective (and because no statistical test is 100% infallible), there is always a possibility that we may come to an erroneous conclusion regarding H_o and H_a. When a null hypothesis is mistakenly rejected, that is called a **Type I error** and essentially means that the statistician has erroneously found differences where none truly exist. Fortunately, the researcher can reduce the likelihood of committing a Type I error by simply choosing a smaller α.

By contrast, a **Type II error** occurs when the null hypothesis is upheld when it should have been rejected. In practice, Type II errors are much more difficult to manage, because the likelihood of this error depends on a number of factors that may be beyond the researcher's control, such as the sample size or the variability of the data (evidenced by large standard deviations). Increasing the sample size is a straightforward way to reduce the chance of committing a Type II error, but researchers rarely have the ability to reduce the overall variability of their data. Regardless, it is impossible to completely eliminate the possibility of committing a Type I or Type II error. We must instead try our very best to minimize the likelihood of these errors as much as possible, and keep in mind that our statistical conclusions are a prediction of "correctness," not a guarantee.

The Most Basic Statistical Tests are Used to Test for Equality (or Inequality) Between Two Populations

Although there are a multitude of different statistical tests available, they are all essentially used with the same goal in mind: to make comparisons and to determine the relative significance of any differences found. Conceptually, it really is that simple.

The simplest comparison that can be made is to test whether one population is indistinguishable from another. Said another way, we are interested in testing whether the difference between two populations can be dismissed as natural variability (or chance). For example, we would expect that the length of seagrass blades would naturally vary, some leaves being shorter and others longer. If those differences can be explained by natural variation, we would expect that same amount of natural variation to exist in other seagrass populations (in other words, the variation between the two populations should be equal, making them indistinguishable from each other). However, if there is some other factor influencing seagrass blade length, those effects should manifest themselves differently in one population when compared to the other.

To investigate those possible differences, we would have to represent those populations by some consistent measure using descriptive statistics, such as the central tendency, standard deviation, or variance for each population within our comparison. If we chose to compare the variances, the simple mathematical expression of this comparison could be written as

$$H_o: s_n^2 = s_o^2$$

$$H_a: s_n^2 \neq s_o^2$$

In this example, our null hypothesis (H_o) states that the variance within one population (s_n^2) is the same as the variance in the population to which it is being compared (s_o^2); in other words, our two populations possess equal variances and are therefore indistinguishable from each other (when using the variance as our measure of comparison). Since we have to define H_a in a way that is contrary to H_o, we simply have to state the alternate hypothesis: that the variances are not equal. This basic test of equality is known as a **two-tail test**, where the default condition (H_o) always assumes that the compared measures are equal.

Keep in mind that we can use any type of descriptive statistics in our comparisons. Instead of comparing the variances, we could just as easily compare the standard deviations or the central tendencies. If we chose to compare the population means, it would require that we slightly modify our stated hypotheses, so that

$$H_o: \overline{X_n} = \overline{X_o}$$

$$H_a: \overline{X_n} \neq \overline{X_o}$$

Of course, comparisons of any measure of central tendency are valid, so the medians or modes could be compared in exactly the same fashion. But in this example, we are now testing whether the mean of some population ($\overline{X_n}$) is equal to (or indistinguishable from) the mean of another population to which it is being compared ($\overline{X_o}$).

If our statistical analysis yields a result that indicates $p \leq \alpha$, we could use Table 2.1 to reject H_o and accept H_a: a result that indicates that the two populations are not equal. Because of the logical consequences of these results, the two-tail test should always be the first statistical test employed. If the results of your statistical analysis indicate that you cannot accept H_a, then the measures are mathematically indistinguishable from each other, H_o is upheld, and there is nothing left for you to do—your statistical analyses are complete.

However, if the results of the two-tail test indicate that the measures are not equal, there exists just one of two other possibilities: either $\overline{X_n} > \overline{X_o}$ or $\overline{X_n} < \overline{X_o}$. Each of these represent a special case of the **one-tail tests**, which are strictly defined as follows:

Right-tail test:

$$H_o: \overline{X_n} \leq \overline{X_o}$$

$$H_a: \overline{X_n} > \overline{X_o}$$

Left-tail test:

$$H_o: \overline{X_n} \geq \overline{X_o}$$

$$H_a: \overline{X_n} < \overline{X_o}$$

Mathematically, it is impossible for both the right- and left-tail tests to yield $p \leq \alpha$ simultaneously. So, because of the logical structure of these one-tail tests, only one of them shall be necessary (chosen at the investigator's discretion) if a two-tail test has already been performed. Essentially what that means is that the result of one will automatically determine the result of the other.

Choosing the Right Statistical Method

Now that we've covered the basics, it's time to put it all together and formulate an ordered checklist of steps to be followed as a standard recipe for any statistical analysis, whether those analyses shall require parametric (**Figure 2.10**) or nonparametric (**Figure 2.11**) tests. How you construct your research hypothesis and how you choose to collect measurements in the lab or in the field will have a tremendous influence on your ability to use inferential statistics, so it is always wise to refer to the following checklist when first designing your lab/field method:

- ☑ Formulate a testable research hypothesis that requires a binary answer (yes/no; true/false).
- ☑ Determine whether you require quantitative (metric or meristic) or qualitative (categorical) data.
- ☑ Research the most appropriate statistical tests you wish to use and how to use them.
- ☑ Acquire and learn how to use a statistical software package appropriate for your goals.
- ☑ Calculate the descriptive statistics of each population you wish to analyze:
  - – Central tendency (see Equation 2.2)
  - – Standard deviation (see Equation 2.3)
  - – Standard error of the mean (see Equation 2.4)
  - – Coefficient of variation (see Equation 2.6, if needed)
  - – Range (see Equation 2.7)
  - – Variance (see Equation 2.8)
  - – Confidence interval (see Equation 2.13, if needed)
- ☑ Test for outliers (z score; see Equation 2.5). Recalculate descriptive statistics if outliers were removed.

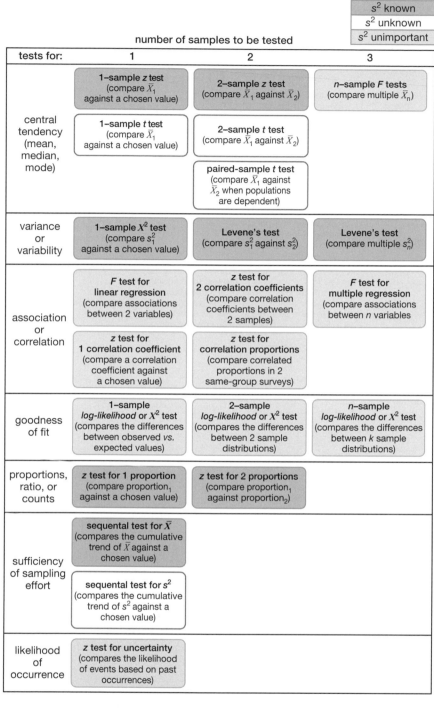

number of samples to be tested

	s^2 known
	s^2 unknown
	s^2 unimportant

tests for:	1	2	3
central tendency (mean, median, mode)	**1–sample z test** (compare $\bar{X}_1$ against a chosen value)	**2–sample z test** (compare $\bar{X}_1$ against $\bar{X}_2$)	**n–sample F tests** (compare multiple $\bar{X}_n$)
	1–sample t test (compare $\bar{X}_1$ against a chosen value)	**2–sample t test** (compare $\bar{X}_1$ against $\bar{X}_2$)	
		paired-sample t test (compare $\bar{X}_1$ against $\bar{X}_2$ when populations are dependent)	
variance or variability	**1–sample X^2 test** (compare s_1^2 against a chosen value)	**Levene's test** (compare s_1^2 against s_2^2)	**Levene's test** (compare multiple s_n^2)
association or correlation	**F test for linear regression** (compare associations between 2 variables)	**z test for 2 correlation coefficients** (compare correlation coefficients between 2 samples)	**F test for multiple regression** (compare associations between n variables
	z test for 1 correlation coefficient (compare a correlation coefficient against a chosen value)	**z test for correlation proportions** (compare correlated proportions in 2 same-group surveys)	
goodness of fit	**1–sample log-likelihood or X^2 test** (compares the differences between observed vs. expected values)	**2–sample log-likelihood or X^2 test** (compares the differences between 2 sample distributions)	**n–sample log-likelihood or X^2 test** (compares the differences between k sample distributions)
proportions, ratio, or counts	**z test for 1 proportion** (compare proportion$_1$ against a chosen value)	**z test for 2 proportions** (compare proportion$_1$ against proportion$_2$)	
sufficiency of sampling effort	**sequential test for $\bar{X}$** (compares the cumulative trend of $\bar{X}$ against a chosen value)		
	sequential test for s^2 (compares the cumulative trend of s^2 against a chosen value)		
likelihood of occurrence	**z test for uncertainty** (compares the likelihood of events based on past occurrences)		

Figure 2.10 A simplified classification of the parametric statistical tests most commonly used for data analysis in the natural sciences. Note that the use of these parametric tests assumes that the data are independent, from a single population, normally distributed, and of equal variances. If any of these assumptions have been violated, the use of nonparametric tests (see Figure 2.11) is indicated. (From Kanji GK [1999] 100 Statistical Tests. With permission from SAGE Publications Ltd.)

☑ Verify data in each population are

   – Independent

   – From a single population

   – Normal

   – Homoscedastic

☑ Determine whether parametric or nonparametric tests are warranted.

☑ Choose the desired probability level (α).

☑ Explicitly state the null (H_o) and alternative (H_a) hypotheses for the two-tail test.

☑ Explicitly state the null (H_o) and alternative (H_a) hypotheses for both the right-tail and left-tail tests.

Figure 2.11 A simplified classification of the nonparametric (distribution-free) statistical tests most commonly used for data analysis in the natural sciences. (From Kanji GK [1999] 100 Statistical Tests. With permission from SAGE Publications Ltd.)

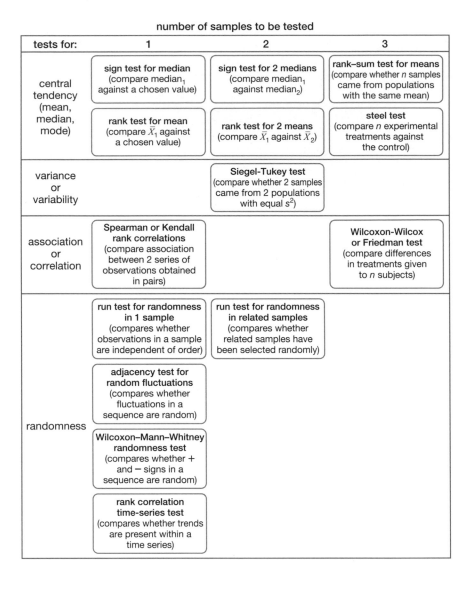

References

Bakus GJ (2007) Quantitative Analysis of Marine Biological Communities: Field Biology and Environment. John Wiley and Sons.

Jones ER (1996) Statistical Methods in Research. Edward R. Jones.

Kanji GK (1999) 100 Statistical Tests. SAGE Publications.

Keeping ES (1995) Introduction to Statistical Inference. Dover Publications.

Linton M, Gallo Jr PS and Logan CA (1975) The Practical Statistician: Simplified Handbook of Statistics. Wadsworth Publishing Company.

Further Reading

Cassel C-M, Särndal C-E, & Wretman JH (1997) Foundations of Inference in Survey Sampling. John Wiley and Sons.

Cochran WG (1977) Sampling Techniques, 3rd ed. John Wiley and Sons.

Dorofeev S & Grant P (2006) Statistics for Real Life Sample Surveys: Non-Simple-Random Samples and Weighted Data. Cambridge University Press.

Dytham C (2011) Choosing and Using Statistics: A Biologist's Guide, 3rd ed. Wiley-Blackwell.

Jaisingh LR (2006) Statistics for the Utterly Confused, 2nd ed. McGraw-Hill.

Keller DK (2006) The Tao of Statistics. SAGE Publications.

Kenny DA (1979) Correlation and Causality. John Wiley and Sons.

Mandel J (1964) The Statistical Analysis of Experimental Data. Dover Publications.

Newman I & Newman C (1977) Conceptual Statistics for Beginners. University Press of America.

Salkind NJ (2007) Statistics for People Who (Think They) Hate Statistics: The Excel Edition. SAGE Publications.

Steiner F (ed) (1997) Optimum Methods in Statistics. Akadémiai Kiadó.

Thompson SK (1992) Sampling. John Wiley and Sons.

Unit 2
Methods of Data Acquisition

Contents

Chapter 3

Experimental Design

"An experiment is a question which science poses to Nature, and a measurement is the recording of Nature's answer." – Max Planck

In many ways, there is no more difficult aspect to science than the design and execution of a well-conceived experiment. Experimentation lies at the very heart of scientific inquiry; so much so, it would almost seem a rather mundane task. But nothing could be farther from the truth—the design of our experiment, as it turns out, requires us to be both philosopher and mathematician; stalwart logician and inspired visionary. Not an easy role to pull off.

As we have already discussed, the sole purpose of any scientific experiment is to provide us with the necessary data, so that we may test our original contemplations in order to establish some pattern of correlation or causation. Although the scientific method provides us with a stepwise guide to do just that, it is up to us to structure our experiments so that they provide us with objective data of sufficient precision and accuracy that we require to perform our analyses.

The question is, how? Our universe is filled with an infinite number of complex curiosities, any number of which we might seek to describe. But if that be the case, how can we possibly hope to arrive on the proper course of action, perfectly designed to reveal those details? Well, let's begin with the basics first.

Key Concepts

- All scientific experiments can be loosely categorized as either a method comparison, single-element assessment, multi-element correlation, or synthetic analysis.

- The single most helpful guide to experimental design is a thorough review of relevant scientific literature.

- The usefulness of inferential statistics relies heavily on the establishment of sufficient sample sizes and appropriate sampling methods.

- Several standard field methods exist that can be duplicated (or modified) to fit the needs of the researcher, either within predefined boundaries (as a plot method) or without any spatial limits (as a plotless method).

Starting with the Basics of Experimental Design

Of course, the first order of business for any field research program is to define the overall goals of the project; to use our hypothesis to envision what the outcomes of our research should be. It is of great value to consider, in very concrete terms, what the research products shall be, as a way to "reverse engineer" the most appropriate field program specifically designed to deliver the goods.

Although the natural world, in all its complexity, presents us with a limitless smorgasbord of research opportunities, we need not be paralyzed with the enormity of the infinite possibilities before us. To move forward in the planning phase of our experiment(s), it is necessary that we focus our attention on defining just a few, very specific objectives that our research is intended to accomplish. After all, what is it that you intend to demonstrate with your research? This seems like a simple question, but it requires that you think very seriously about what you are most interested to explore in your research, and what you can (and cannot) realistically hope to complete in the course of that research. In essence, the research **objective** is an explicit definition of the critical goal(s) you intend to accomplish using a scientific approach.

When defining our specific objectives, it is critical that we structure them in such a way that we can then create hypotheses to test different aspects of the larger research goal. For each hypothesis we wish to test, we will also have to determine which measures will be necessary (that is, what data we must collect) in order to conduct those tests. Those data, the equipment used to collect them, and the method by which they are collected are all dependent on your chosen experimental method. Thus, it is impossible to design an appropriate experiment without first defining the specific research objective (and the attendant hypotheses) that will ultimately allow the researcher to make sense of the data.

In concept, the preliminary steps in experimental design are very straightforward:

1. Define the larger goal(s) of your research as an objective (or list of objectives).
2. Devise a hypothesis to test a specific assertion, relevant to your research objective(s).
3. Design an experiment that will allow you to test that specific hypothesis.

In reality, this is more difficult to do because you must wade through all the interesting possibilities to explore and settle on just a few, very specific research objectives you intend to accomplish. Then you must logically construct a number of testable hypotheses that are required to provide you with sufficient evidence in support of (or in conflict with) your assertions. And for each hypothesis, you must create an experiment designed for the sole purpose of testing that specific hypothesis.

As an example of how we might first approach the daunting task of experimental design, let us consider shoal grass (*Halodule wrightii*; see Figure 1.5), a commonly occurring species of seagrass that we used as earlier examples in Chapters 1 and 2. Let's say we were interested in studying shoal grass, but weren't quite sure what kind of research project we'd like to perform. Our first step would be to define the objective of our research; that is, to define what we are specifically interested in.

In a very general sense, research objectives usually fall into one of these themes:

- Method comparison
- Single-element monitoring and assessment
- Multiple-element correlation or causation
- Complex synthetic analysis

While a field research program may contain more than one of these themes within the research objective, it is really only necessary (and more common) to have one. Although there are always exceptions, method comparisons and single-element assessments are generally considered to be less challenging (and are therefore more appropriate for early-career researchers). Since we had earlier decided to use shoal grass as the general subject of our field research, let us continue that example to see what sort of research objectives we might be able to define, using these themes as guidance.

Method Comparison

These are projects designed to test the accuracy (or veracity) of existing research methods. The simplest form of method comparison is to merely repeat a preexisting experimental protocol to confirm earlier results as a type of method replication. And although method replication may be lacking in the "sizzle" department, it is the foundational method of science—it provides the confirmation of consistent results and is the best vehicle for reducing any bias and/or error in earlier works.

Of course, the history of science is rife with examples of how new and improved methodologies, either borne from inspired researchers or from ingenious inventors of new scientific equipment and technology, led to a complete revision of what was once held as scientific dogma. Hence, method comparisons that focus on the augmentation and improvement of historic methodologies are always of great value to the scientific community. Occasionally, method comparisons that expand the number of pertinent variables, or the temporal/spatial scope of those methodologies, employed by earlier researchers can also provide valuable insight.

Let's say we were interested in performing a method comparison as the general theme of our objective, and we wanted to use shoal grass. In Chapter 2, we used shoal grass to demonstrate the 30–300 rule, as it applied to the measurement of leaf-blade length and width using a metric ruler with 1 mm resolution (see Figure 2.3). In that example, we were able to show that 1 mm resolution was sufficient for measuring leaf-blade lengths, but was insufficient for measuring leaf-blade widths (according to the 30–300 rule). However, if you wanted to avoid using microscopic analysis of leaf-blade widths, you might be motivated to demonstrate that the use of a metric ruler is a perfectly sound method for measuring leaf dimensions of shoal grass, regardless of the 30–300 rule.

If that is your objective, you would essentially be advocating a comparison between the use of a metric ruler (1 mm resolution) and the use of microscopy (0.1 mm resolution), to see if there were significant differences in the leaf width (*lw*) measurements collected using these two different methodologies. So, now that you have a clear objective, you could formulate at least one testable hypothesis designed to meet that objective:

$$H_o : \overline{X}_{lw-ruler} = \overline{X}_{lw-scope}$$
$$H_a : \overline{X}_{lw-ruler} \neq \overline{X}_{lw-scope}$$

In this example, the mean leaf widths measured with a ruler ($\overline{X}_{lw-ruler}$) could be tested against the mean leaf widths measured microscopically ($\overline{X}_{lw-scope}$) to see if any significant differences exist. Of course, we won't know the answer to that question until we perform the experiment and gather the data. But regardless of the outcome of the statistical test, the answer will fulfill our research objective.

Now that we have our hypothesis clearly stated, our next step would be to design an experimental method that will provide us with the data we need to test that hypothesis. In this example, we would need access to shoal grass specimens, the equipment necessary to perform these measurements, and a clear understanding of what variables we were measuring (in this case, the leaf-blade widths). As you can see from this example, it is absolutely essential to first define the research objective, and then the hypotheses, before we can think about the design of our actual experiments. The same thought process (objective → hypothesis → method) is required for the following themes, but more sophisticated research objectives will naturally require more hypotheses to test, which in turn require more sophisticated methods for data acquisition and analysis.

Single-Element Monitoring and Assessment

These are projects that seek only to quantify (or categorize) the measurement of a single variable over the spatial and/or temporal scale of the investigation, without really engaging in any correlation or causation analyses of those measurements (a scientific endeavor sometimes referred to as "monitoring"). Of course, all monitoring projects require that a sufficient methodology exists for measuring the variable of interest, with appropriate accuracy and precision.

Although most monitoring projects typically measure more than one variable at a given place or time, if these variables are not analyzed with any relational context, the data are simply collected and cataloged as single variables, disconnected from each other and constrained to their place and time of collection. That is not to say that monitoring and assessment projects are not a critical part of field research. On the contrary, they provide the bulk of the raw field data that other investigators may use to provide analyses of complex interactions within the natural system from which the data were initially collected. But the mere collection of data, without much thought to the analysis of those data, serves a very shallow research objective.

Multiple-Element Correlation or Causation

These projects build on the foundations laid by proven methodologies and the acquisition of pertinent field data, all in an effort to connect at least two measured variables (and oftentimes more) and provide the context of their connectivity. The "connectivity" of field data is usually viewed through the prism of **correlation** or **causation**.

Correlation does not require that the researcher fully understand how the variables are connected, merely that the quantity (or quality) of one variable has a measurable effect on another (that is, one variable is dependent on another, by some unknown mechanism). If we had measured two different variables, A and B, and discovered that B was dependent on A, we could state the relationship logically as

$$A \rightarrow B, \quad \text{or} \quad \text{if } A \text{ then } B$$

If we were interested only in establishing the correlation as our research objective, we would not engage ourselves in determining why B is affected by *A*—it would be enough to simply demonstrate that the two variables are related to each other in a very predictable manner. By contrast, determining causation is a much more challenging task, because our research objective would then require us to determine the pertinent mechanism(s) at work in the natural system, affecting *B* as a function of *A*. Naturally, the hypotheses needed to test correlation or causation are quite different from each other, so our methodologies will be quite different as well (but more on that later).

Complex Synthetic Analysis

A **synthetic analysis** requires that we consider each of the described, complex relationships (preferably as causative) and combine them into a grand unifying concept of precisely how complex systems function, and how each of the variables within that system is connected. Obviously, this is a tremendously challenging task and often requires years (if not decades) of foundational research—comparing and perfecting several different methodologies of measurement, acquiring huge volumes of field data over multiple spatial and temporal scales, and describing the most critical correlations and causative relationships within the overall system.

So daunting is this task that it typically cannot be accomplished using traditional field studies. Instead, we must rely on the construction of numerical models (Chapter 10) to mathematically describe and test these complex interrelationships using high-performance computing resources to "simulate" natural systems in their entirety.

Proper Experimental Design Requires Significant Preparation

At this point in the process of designing our field program, we should have a general idea of the phenomenon we wish to investigate and a preliminary hypothesis we will ultimately test, using the results of our experiment(s). To that end, we should also have made some basic decisions regarding just how involved our research program will be. Do we wish to compare different methodologies, or merely repeat them? Do we want to engage in a monitoring project to gather raw data for separate variables, or will we investigate the connectivity of those variables, using a variety of correlation and causation analyses?

Before we get too far down the road of designing the perfect field program, it is important for us to take the time for an honest assessment of our operational constraints. This will require that we exercise due diligence in our planning and design of the best plan for the limitations we are forced to accommodate. This due diligence will require us to make some early decisions, informed by our thoughtful consideration of

- Literature review
- Data requirements
- Research strategy
- Scope of investigation
- Equipment and facilities needs
- Preliminary site survey(s)
- Sampling effort

Literature Review

Because the process of science relies so heavily on earlier works, it is absolutely critical that any new scientific endeavor include a thorough review of the published literature, particularly as it relates to the specific research objectives you shall pursue in your proposed field program. Just as there is no sense in re-inventing the wheel, there is no sense in performing research on a particular topic that has already been thoroughly investigated (unless you wish to challenge those earlier findings and/or conclusions).

A competent literature review has the additional benefit of providing deeper context to the topics you wish to explore, and establishes to the scientific community that you have "done your homework" and are fully informed as to the most current research in your particular field of interest. In the planning process, this can be a tremendous boon, as there may be other investigators in your field who can provide advice and inspiration with regard to your choices of the most appropriate objectives, methodologies, and scope of your proposed research.

Data Requirements

These are critical decisions that must be settled very early in the process of designing your experiment(s), as the data collected will determine the fundamental value of your research. As we discussed earlier in Chapter 2, the type of data you collect will determine what sort of numerical analyses shall be possible, especially in the context of statistics. You must also be prepared to make the critical decisions of precisely what variables you wish to measure, taking care to limit your investigation to only those variables that are germane to the objectives of your research. That is a task much more easily said than done, which is yet another reason to review the current body of knowledge on such matters in the scientific literature.

Research Strategy

On a very basic level, the research strategy employed in field studies can be reduced to a simple dichotomy: Shall the proposed research rely on the acquisition of data through passive observation of the natural system, or does the investigator intend to become an active driver of the observed phenomenon by perturbing the natural system, through true experimentation? Typically, observational studies are far easier to perform, simply because it is so challenging to establish the appropriate experimental controls in a perturbed natural system. There may also be some serious legal impediments to manipulating natural systems, as most regional and national regulatory agencies are reluctant to allow "unnatural" changes in an otherwise natural system. However, there are often very specific questions that can only be answered through active perturbation of the system, and monitoring its response thereafter.

Scope of Investigation

It is also critical to recognize that all field research programs are limited by geographic and temporal constraints that shall define the overall scope of the investigation. Every research program must have its beginning and its end, but drawing those boundaries can be a challenging task in itself. Again, you must focus on the geographic and temporal scales that are most relevant to your research objectives and limit your investigation according to those boundaries; otherwise, your research workload will become unmanageable.

It is also important to note that these geographic and temporal constraints not only define the boundaries of the overall project, but also define the boundaries of the actual method(s) employed in the field for data acquisition. In the field, very rarely are we afforded the luxury of **synoptic** data; that is, data gathered instantaneously across a broad geographic range. With the exception of remotely sensed data from aircraft or satellite, field data are more typically gathered from several field stations, located quite distant from each other geographically. Add to this the expiration of time as we transit from one research station to another, and we soon recognize that the scope of our investigation can introduce significant spatiotemporal bias—the more grandiose the scope, the more significant the bias.

Equipment and Facilities Needs

Accomplishing any objective requires the expenditure of some combination of time, money, and effort; scientific research is no different. Resources that are critical to the research effort must be assessed honestly, and if the budget cannot support the proposed research, there are only three possible solutions: either reduce the cost/scope of the investigation, seek additional funding from alternate sources, or abandon the proposed research. And although the investments required in human capital, equipment, facilities, travel costs, and consumable supplies to conduct the field research are the most straightforward expenses to consider, do not underestimate the costs associated with the analysis phase of the research (which should include any publishing costs as well).

Preliminary Site Survey(s)

If at all feasible, a preliminary site survey is always indicated once the fundamental experimental design has been sketched out. Not only does this allow the investigator to perform a "dry run" of the proposed research in the field as practice, but there are invariably unforeseen complications in the field that will allow for some last-minute revisions to the method, station location(s), or sampling frequency. Preliminary surveys are also critical in defining the sampling effort that shall be required to meet the research objectives.

Sampling Effort

One of the most difficult aspects of any scientific endeavor, but particularly in field research, is determining the minimum required effort for the maximum benefit—deciding when enough data have been collected for the task at hand. In the context of statistics, sample sizes of 30 or more will afford a sufficient number of observations to allow statistical analyses, as a general rule of thumb. However, there is no guarantee that a sample size of 30 is an appropriate subset of the larger population from which it was taken, so any predictions made from small sample sizes are always dubious.

It is far more important to consider whether our subset of data is truly representative of the larger population we are trying to describe in our research. If our data are normally distributed, we can estimate how many samples are "enough" by using the Cochran method, which uses Equation 3.1 to calculate the variance-to-mean ratio

$$n_o = \frac{d_\alpha^2 \cdot s^2}{E^2 \cdot \overline{X}^2} \qquad (3.1)$$

as an estimate of the minimum number of required samples (n_o). As we can see from the example given in **Example Box 3.1**, the sample size (n_o) is determined as a function of the mean $(\overline{X})$ and standard deviation (s) of the population being investigated, based on the chosen significance level $(d_\alpha = 1.96$ when $\alpha = 0.05$, from Equation 2.13) and the relative predicted error (E).

Keep in mind that the Cochran method assumes that our data are normally distributed and that the preliminary survey was conducted using random sampling. If our data more closely resemble a Poisson (or any other non-Gaussian) distribution pattern, we can instead use the Krebs method (see Equation 3.2), which defines the minimum number of required samples as

$$n_o = \left(\frac{d_\alpha}{E}\right)^2 \cdot \left(\frac{1}{\overline{X}}\right) \qquad (3.2)$$

EXAMPLE BOX 3.1

Determining Sample Size Using the Cochran Method

A preliminary site survey involved 30 measurements of soluble lead (Pb^{2+}) in the estuarine waters near a textiles manufacturing plant. Based on this small random sampling effort, our estimate of the mean concentration of Pb^{2+} was 1.30 nM, with a standard deviation s of 0.89 nM. If our chosen significance level α is 0.05 ($d_\alpha = 1.96$) and we want to make sure our estimate of the Pb^{2+} mean concentration is within 10% ($E = 0.10$) of the true mean,

$$n_o = \frac{(1.96)^2 (0.89)^2}{(0.10)^2 (1.30)^2} \approx 180$$

the Cochran method would indicate that at least 180 random samples are needed for an adequate study of lead contamination in this particular estuary.

If you wish to reduce the number of samples indicated by the Cochran method, it may be necessary to loosen your restrictions on the relative predicted error by increasing the value of E. Reducing the variance s^2, perhaps through the use of better-performing analytical equipment or by improving the collection method, will also reduce the number of samples needed. Of course, additional observations (beyond the initial 30) may also reduce the variance (and therefore reduce the number of samples indicated).

If the Krebs method is used to determine the minimum sample size, we need not worry about whether our data are normally distributed. Although that may at first seem like somewhat of an advantage, it is that same "lack of normality" that makes it more difficult to predict how much data is sufficient with regard to our sampling effort. As a result, the Krebs method usually indicates that a greater sampling effort is needed to offset the inherent uncertainty in data that are not normally distributed (as demonstrated in **Example Box 3.2**).

A much less technical approach is to simply plot the accumulation of new information against the total effort, such as the number of samples taken,

EXAMPLE BOX 3.2

Determining Sample Size Using the Krebs Method

Recall from our earlier example that the mean concentration of Pb^{2+} at our study site was 1.3 nM. If we determined the minimum number of required samples using Equation 3.2,

$$n_o = \left(\frac{1.96}{0.10}\right)^2 \cdot \left(\frac{1}{1.3}\right) \approx 296$$

The Krebs method indicates that 296 measurements of Pb^{2+} are required, instead of the 180 suggested by the Cochran method. This is because the Krebs method assumes that the data are not normally distributed, so it is harder to predict the distribution of nonnormal data (hence the higher indicated sample size).

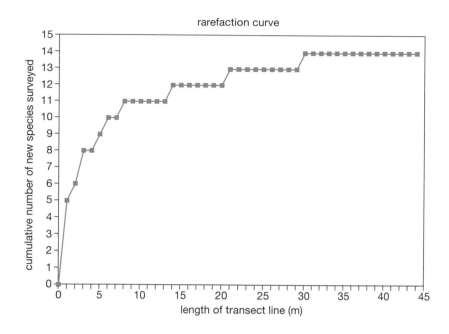

Figure 3.1 A typical rarefaction curve, indicating the accumulation of new information on the y-axis, relative to the expenditure of effort on the x-axis. In this example, the new information gained is the cumulative number of new species surveyed, and sampling effort is represented by the length of the transect used in the survey. If these data were collected from a preliminary site survey to assess species richness, the investigator could reasonably conclude from the rarefaction curve that there is no benefit to extending the transect lines beyond 30 m. (Adapted from the US Department of the Interior, National Park Service [2001] Virgin Islands National Park Coral Reef Monitoring Manual.)

number of replicates performed, and so on. (**Figure 3.1**). This is occasionally referred to as a **rarefaction curve**, especially in the context of ecological studies. During initial sampling, the slope of the resultant line will be positive and quite steep. In subsequent samples, the gain of new information decreases and the slope of the resultant line will trend asymptotically toward zero. When the slope of the line reaches zero, there is no further gain of information regardless of how many more samples are taken, thus establishing the number of samples needed to sufficiently describe the population under investigation.

Sample Size Determination is the Key to Success in Every Research Endeavor

Determining what is the "appropriate" sample size (n_o) for laboratory and field studies is not a task to be taken lightly. Regardless of whether n_o is determined using the Cochran or Krebs (or any other rarefaction) method, there are four critical factors that must be known to provide the best estimate for n_o:

1. Difference threshold (the magnitude of the difference you wish to test between groups)
2. Standard deviation of the population (for continuous variables only)
3. Relative error ($1 - E$, usually 0.9 by convention)
4. Significance level (α, usually 0.05 by convention)

Dichotomous Variables Can be Used to Determine n_o

Dichotomous variables are those expressed as rates of occurrence or the proportion of a specific outcome. Generally speaking, these are variables that possess only one of two possibilities as a binary condition that are mutually exclusive of each other (like present vs. absent, dead vs. alive, male vs. female). In practice, these can most easily be illustrated by comparing a specific occurrence in an experimental group compared to the control. For example, if we wished to test for the occurrence of death in the control versus a treatment (experimental) group, we could determine n_o by using the proportion of dead specimens in the control (p_c) relative to the proportion of dead specimens in the experimental group (p_e). Thus, 10% death in the control versus 50% death in the experiment would yield $p_c = 0.10$ and $p_e = 0.50$. According to the Fleiss function (see Equation 3.3), we can determine n_o as

$$n_o = C\frac{p_c q_c + p_e q_e}{d^2} + \frac{2}{d} + 2 \tag{3.3}$$

where $q_c = 1 - p_c$, $q_e = 1 - p_e$, and $d = |p_c - p_e|$. The value C depends on the relative error E and significance level α chosen by the investigator, according to the following table.

$1-E$	$\alpha = 0.05$	$\alpha = 0.01$
0.8	7.85	11.68
0.9	10.51	14.88

For a relative error of 10% ($1 - E = 0.9$) and a significance level α of 0.05,

$$n_o = 10.51 \times \frac{(0.10)(0.90) + (0.50)(0.50)}{(0.40)^2} + \frac{2}{0.40} + 2 \approx 29.33$$

Thus, a minimum sample size of 30 is indicated for each of the control and experimental groups.

Continuous Variables Can Also Be Used to Determine n_o

Continuous variables are those expressed as values that vary along some continuous gradient or range (like the concentration of Pb^{2+}, or the leaf width of a blade of shoal grass). The most common statistical tests utilizing continuous variables are those that compare group means (although any measure of central tendency would suffice). If comparisons are made between independent groups, sample size can be calculated by the Snedecor and Cochran function (see Equation 3.4) as

$$n_o = 1 + 2C\left(\frac{s}{d}\right)^2 \tag{3.4}$$

where s represents the population's estimated standard deviation and d represents the magnitude of the difference the investigator wishes to test between the two groups (and C is determined in the same manner described by Equation 3.3).

As an example of the Snedecor and Cochran function, let's say we wanted to test the differences in leaf-blade lengths of shoal grass (*Halodule wrightii*) growing in two different regions: one considered to be a pristine coastal region, while the other is influenced by runoff from a nearby river. The sediment load from the nearby river does seem to affect water clarity, so perhaps we'd like to test whether the turbid waters affect seagrass growth when compared to the seagrass at the pristine site (using leaf-blade lengths as our test variable). So how many seagrass blades should we measure?

Recall from our earlier example in Chapter 2 that we have already performed a preliminary survey of shoal grass leaf lengths ($N = 29$) and found $\overline{X} = 131$ mm and $s = 7$ mm (see Example Box 2.6). To detect a 5% difference in the mean leaf lengths between the pristine and affected sites, we employ

Equation 3.4 and use $d = (131 \text{ mm})(0.05) = 6.55$ mm, and $C = 10.51$ to limit our relative error E to 10%, so

$$n_o = 1 + 2(10.51)\left(\frac{7\,\text{mm}}{6.55\,\text{mm}}\right)^2 \approx 25$$

Hence, we would need to conduct a minimum of 25 observations at each site. Since we've already collected 29 observations from the pristine site, our work there is done; all we need to do now is collect a minimum of 25 observations from the affected site. Since it is always best to use the same number of observations when making statistical comparisons, you would be wise to go ahead and collect 29 observations at the affected site.

Note that if the groups being tested are paired (that is, dependent on each other, such as pre- and post-treatment comparisons), sample size is instead computed as

$$n_o = 2 + C\left(\frac{s}{d}\right)^2 \tag{3.5}$$

An Introduction to Quantitative Sampling Methods

Once we have reached some early decisions regarding our project goals, we must turn our attention to the actual method(s) we shall employ in the field to acquire our data. Although there are many different quantitative sampling methodologies in the literature that we can utilize, a vast majority of them are designed for application in terrestrial systems. Aquatic systems, quite literally, add a new dimension of complexity to any quantitative sampling method, as we must account for **heterogeneity** in both the horizontal and the vertical dimensions (in Super-3D, so to speak).

In the aquatic environment, studies limited to the **benthos** will require very few changes to the traditional terrestrial methodologies that exist, as they are both defined predominantly by horizontal, rather than vertical, variability. However, if a fully three-dimensional sampling regime is necessary to resolve three-dimensional variability, there are relatively simple ways to expand traditional two-dimensional sampling methodologies to include the vertical dimension (which we shall discuss shortly). For now, let us review the classic terrestrial methods to provide the context and conceptual foundation for any modifications we may wish to make to our chosen method.

In addition to the basic question of whether we must consider a two- or three-dimensional sampling regime, it is also important to consider whether we shall employ a plot or a plotless method. **Plot** methods are those whereby the data are acquired from areas (or volumes) bounded by strict geographic boundaries. The main benefit to plot methods is that the variables being measured are inherently relatable in terms of their spatial distribution (such as coverage per acre, or counts per m³). Although plot methods typically yield data that are most representative of the population from which they were sampled, they can also be quite labor and time intensive to collect.

In contrast, **plotless** methods are those that do not constrain data to arbitrary areas or volumes; rather, data are gathered from discrete points in space or along **transect** lines and are most commonly related in terms of frequency, proximity, or point-to-point distances. Because plotless methods are not constrained by boundaries, they are often easier to conduct in the field, especially in dynamic regions where establishing semipermanent boundaries may be impractical. However, plotless methods are generally considered to be inferior to plot methods since plotless methods make it impossible to return to the exact same location to conduct repeated measures over time.

Basic Plot Methods

Since the scope of any investigation is limited by geographical constraints to some extent, it shall be necessary for us to define the geographic boundaries of the project as a whole. Unless we seek to perform a complete census of all pertinent measures within the confines of that boundary, we will rely instead on our ability to take multiple subsamples that are representative of the larger population from which they came. By choosing to employ a plot method, we are simply choosing to place spatial boundaries on those subsamples.

For us to be able to relate our data to the true dynamics of the system under investigation, it is absolutely critical that our subsamples are representative of the larger population from which they were gathered. When choosing an appropriate plot method, we must take care to consider all of the following:

- Sampling effort
- Plot shape
- Plot size
- Random sampling

Sampling Effort

Whether or not we have expended sufficient effort in our sampling is largely determined by the number of samples taken (n) for a given measure, for each group or at each site. As we have already discussed, there are a multitude of ways to determine the most appropriate number of samples that are required for a given study (see Equations 3.1–3.5).

Plot Shape

Since data collected from all plot methods are inherently related to the unit area (or unit volume) within the boundary of the plot, it is usually easiest to construct the plot as a regular quadrilateral (that is, a square). Traditionally, most plots are designed as **quadrats**, either square or rectangular in shape, primarily for their ease of use in the field. The regular geometry of the quadrilateral makes it a very simple device to scale by subdividing the quadrat into a number of smaller, constituent quadrats (**Figure 3.2**). Similarly, any quadrat can be used to define a fixed-area grid, containing any number of equidistant points depending on the desired scale of measurement.

However, as an interesting consequence of geometry, it is best to maximize the "area-to-perimeter" ratio of the quadrat, so that irregularly shaped elements are more likely to be contained within the area of the quadrat (rather than falling along the edge of the quadrat). Deciding whether an element should be counted as "in" or "out" of the quadrat can occasionally lead to a positive **bias** in the data, sometimes called the **edge effect**. We can generally minimize this bias by choosing quadrat dimensions that maximize the quadrat area and minimize the quadrat perimeter (edge). This is done simply by using Equation 3.6 to analyze the area-to-perimeter (A:P) ratio:

$$\text{Quadrat } A{:}P = \frac{L \cdot W}{2L + 2W} \tag{3.6}$$

Figure 3.2 A nested quadrat, when placed over the substrate, serves as a versatile instrument for conducting a plot census. This device is especially versatile in the field, as various plot methods can be employed using either the entire quadrat area or a specified number of randomly selected sub-quadrats. Note that the same device can also be used to gather gridded, point-intercept data at each of the "cross-hairs." (Courtesy of the Griffith School of Environment & Australian Rivers Institute, —Coast & Estuaries. Brisbane Australia.)

Using this simple calculation, it is easy to demonstrate that not all quadrats are created equal. As an example, let us assume we wish to employ a quadrat of fixed area, such as 4 m². Let us now compare the A:P ratios of a 2×2 quadrat compared to a 1×4 quadrat:

$$2 \times 2 \text{ Quadrat } A{:}P = \frac{2 \cdot 2}{2(2) + 2(2)} = \frac{4}{8} = 0.5$$

$$1 \times 4 \text{ Quadrat } A{:}P = \frac{1 \cdot 4}{2(1) + 2(4)} = \frac{4}{10} = 0.4$$

Thus, it is always preferable to use quadrat dimensions that yield the highest possible value of A:P (to minimize the edge effect).

For any circle, Equation 3.6 can be rewritten as

$$\text{Circle } A{:}P = \frac{\pi r^2}{2\pi r} = \frac{r}{2} \tag{3.7}$$

where r is the unit radius of the circle. If we continue with our analysis of the best possible quadrat, we can show that a circle of 4 m² area must possess a radius of approximately 1.128 m:

$$A = \pi r^2 \quad \therefore \quad r = \sqrt{\frac{A}{\pi}} = \sqrt{\frac{4\,\text{m}^2}{\pi}} \approx 1.128\,\text{m}$$

If we use Equation 3.7 to analyze our circular "quadrat," we will find

$$\text{Circle } A{:}P = \frac{r}{2} = \frac{1.128}{2} \approx 0.56$$

Because of this interesting geometric relationship, we can demonstrate that it is impossible to improve on the circular quadrat in terms of minimizing its edge effect (thus making the circle the best shape for any size quadrat).

Plot Size

Although it is necessary to independently determine the appropriate number of samples (n_o) as an indicator of sufficient sampling effort (as previously discussed), it is easy for us to recognize that plot size will have a complimentary effect on that effort. For example, we might expect our field data to differ if we surveyed a total area of 100 m² by using 100 quadrats 1 m² in size, 50 quadrats 2 m² in size, or 25 quadrats 4 m² in size. Here too we can seek guidance from the 30–300 rule (see Chapter 2), which would advise against the use of 25 quadrats 4 m² in size.

Although readers are encouraged to perform a thorough literature review for suggestions regarding the appropriate plot size for their specific study, quadrat sizes of 0.1 m² to 1 m² are typical for most ecological applications (although it may be necessary to scale the quadrat up or down, depending on the size of the elements being surveyed). If it is possible for the investigator to perform a preliminary site survey, an appropriate plot size can be quickly and easily estimated by using the "3 & 3 rule" (not to be confused with the 30–300 rule).

The 3 & 3 rule simply requires each of three different quadrat sizes to be chosen (as the investigator's "best guess") and used in triplicate to randomly survey the same location. The survey data from each quadrat size are then analyzed by calculating the simple **variance-to-mean ratio** (**VMR**) using Equation 3.8, and the quadrat size that yields the lowest *VMR* is the size that should be selected for all further surveys:

$$VMR = \frac{s^2}{\overline{X}} \tag{3.8}$$

Although the 3 & 3 rule is usually sufficient for most general survey methods, it may not provide sufficient resolution for those studies that seek to establish density-dependent, spatial relationships (such as geographical distribution patterns as a result of competitive exclusion).

To detect these sorts of associative patterns, Sanjerehei presents the simplest general method to determine plot size. According to his work in ecological modelling, Sanjerehei was able to show that the most appropriate circular quadrat diameter $(\overline{d})$ for detecting an association between members of two different populations $(\phi$ and $\theta)$ is dependent on the spatial distribution of ϕ and θ, and can be determined simply as

$$\overline{d} = \sqrt{1/m_{\phi\theta}}, \quad \text{and} \tag{3.9}$$

$$m_{\phi\theta} = (m_\phi + m_\theta) \tag{3.10}$$

where m_ϕ and m_θ represent the population densities of ϕ and θ, respectively, per unit area.

As written, the Sanjerehei method is applicable regardless of whether the members of ϕ and θ are uniformly or randomly distributed (**Figure 3.3A, B**). If members of ϕ and θ instead follow an aggregated pattern (**Figure 3.3C**), certain adjustments must be made when calculating m_ϕ and m_θ. For example, if duplicate instances of ϕ are in contact with or overlap each other, they must be treated as a singlet when determining m'_ϕ:

$$m'_\phi = m_\phi - \left(\frac{\% \, \phi \text{ as duplicates}}{2} \times m_\phi \right) \tag{3.11}$$

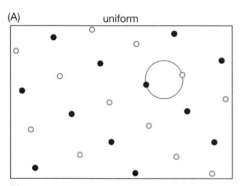

(A) uniform

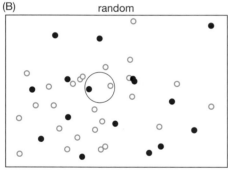

(B) random

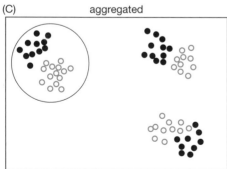

(C) aggregated

Figure 3.3 Idealized spatial distribution patterns of two different populations, typical of (A) uniform, (B) random, or (C) aggregated distributions. Note that both the uniform and aggregated distribution patterns indicate a spatial relationship between members of population ϕ (open blue circles) and those of population θ (closed black circles), which is somehow associative.

If ϕ is present as a significant aggregation, each aggregation must be treated as a singlet when determining m_ϕ'' :

$$m_\phi'' = \left(\frac{m_\phi}{\text{mean } \phi \text{ per aggregation}} \right) \qquad (3.12)$$

Of course, the same analyses (see Equations 3.11–3.12) must also be done to determine m_θ' and m_θ'' for the members of population θ as well.

To demonstrate how plot size $(\bar{d})$ is determined using the Sanjerehei method, let us assume that the areas depicted in Figure 3.3 each represent 24 km^2 (4 km $\times$ 6 km). In the example of a uniform distribution (see Figure 3.3A), we have 12 members of population ϕ (blue) and 12 members of population θ (black); thus, $m_\phi = 0.5$ per km^2 and $m_\theta = 0.5$ per km^2, respectively. Using Equation 3.10, $m_{\phi\theta}$ is simply 1.0 per km^2. Using Equation 3.9,

$$\bar{d} = \sqrt{1/1.0 \text{ km}^{-2}} = 1.0 \text{ km}.$$

In the example of a random distribution (see Figure 3.3B), the distribution of ϕ and θ is a little more complicated. Although there don't seem to be any significant aggregations, there are at least two instances where individuals among population ϕ overlap each other, and one instance of overlapping among population θ. Let us first calculate m_ϕ (28 per 24 km$^2 \approx 1.167$ per km^2) and m_θ (14 per 24 km$^2 \approx 0.583$ per km^2). Since we had only two instances of duplication (overlap) among the 28 members of ϕ (7.14%) and only one instance of duplication (overlap) among the 14 members of θ (7.14%), we must utilize Equation 3.11 to calculate m_ϕ' and m_θ' as

$$m_\phi' = \frac{1.167}{\text{km}^2} - \left(\frac{0.0714}{2} \cdot \frac{1.167}{\text{km}^2} \right) = \frac{1.125}{\text{km}^2}$$

$$m_\theta' = \frac{0.583}{\text{km}^2} - \left(\frac{0.0714}{2} \cdot \frac{0.583}{\text{km}^2} \right) = \frac{0.562}{\text{km}^2}$$

Using Equation 3.10, $m_{\phi\theta}$ is $m_\phi' + m_\theta'$, or 1.687 per km^2. Using Equation 3.9,

$$\bar{d} = \sqrt{1/1.687 \text{ km}^{-2}} = 0.77 \text{ km}.$$

For the aggregated distribution (see Figure 3.3C), we first calculate m_ϕ (35 per 24 km$^2 \approx 1.458$ per km^2) and m_θ (27 per 24 km$^2 \approx 1.125$ per km^2). Since we have significant aggregation, we must then calculate the mean number of ϕ and θ per aggregation (11.67 and 10.33, respectively) and use Equation 3.12 to calculate m_ϕ'' and m_θ'' as

$$m_\phi'' = \frac{1.458/\text{km}^2}{11.67} = \frac{0.125}{\text{km}^2}$$

$$m_\theta'' = \frac{1.125/\text{km}^2}{10.33} = \frac{0.109}{\text{km}^2}$$

Using Equation 3.10, $m_{\phi\theta}$ is $m_\phi'' + m_\theta''$, or 0.234 per km^2. Using Equation 3.9,

$$\bar{d} = \sqrt{1/0.234 \text{ km}^{-2}} = 2.07 \text{ km}.$$

Random Sampling

Unless the field survey is specifically designed to locate rare or concealed objects (or events), most field methods will be heavily dependent on the

Figure 3.4 Gaming dice are an excellent, low-tech solution for generating random numbers in the field, so long as you use the results from single rolls. Two 10-sided dice are a perfect way to generate random percentages or even probabilities from 1 to 100.

premise that the subpopulation being surveyed is representative of the larger population. In order to best assure ourselves that the data captured in our quadrats are truly reflective of the larger system from which they were taken, our quadrats must be placed randomly throughout the system.

As silly as it sounds, "purposeful randomness" is one of the most effective ways of minimizing bias from any field survey. Unfortunately, it is also one of the more difficult aspects to put into practice while in the field. For any investigation, we are compelled to select the most appropriate method and the most appropriate site to test our hypotheses; it is only natural that we would seek to maintain that sense of control, whether it be on the placement of our quadrats or determining where our "starting points" will be. However, once the spatial limits of our investigation are defined, we must relinquish control over some of the decision-making and rely instead on some sort of system that can aid us in determining the random placement of our plots within the boundaries we have defined.

In most applications, it is easiest to define the larger study area according to a gridded map, where each subregion is assigned a unique numerical value. Random number generators (or random number tables) are then used to generate a series of random numbers that indicate which subregions should be surveyed. Depending on the remoteness of the study site, a low-tech solution may be a better option: just ask an old pen-and-paper gamer for their gaming dice (**Figure 3.4**). Now there's a simple random number generator for you to use.

Simple Random Sampling is the Foundation of Most Field Methods

The easiest plot method to employ in the field is simple random sampling, where the entire study area is divided according to a regularly spaced grid pattern, without taking into consideration any of the ecological gradients that may be present in the system. That is, the placement of each plot is determined purely by chance, and it is assumed that any inherent differences between plots shall be captured in the data so long as enough random quadrats are analyzed.

A common practice in most simple random sampling methods is to use a series of nested plots, where the overall study area is first divided into a regular, large-scale grid so that the number (and location) of each quadrat can be randomly determined and surveyed. Then each of the randomly selected quadrats may be divided into even smaller quadrats, to determine the number (and location) of each minor quadrat to be surveyed (**Figure 3.5**). This method can be further adapted to determine "point contact" data by measuring only those objects located at the vertices of four contiguous minor quadrats. This variant method can also be used to yield simple estimates of percent cover. For example, the large quadrat depicted in Figure 3.5 contains a total of nine vertices. If four of nine vertices were in contact with the same object or species, it would be a simple task to estimate the percent cover as 4/9, or ~44.4%.

Although this method is quite simple to execute in the field and is usually sufficient for comparisons between different study areas, its very design makes it difficult to perform comparisons within a single study area. For that, we would need to design a stratified plot method.

Stratified Random Sampling Can Be Used to Organize the Study Area into Horizontal or Vertical Strata

The stratified random sampling method simply seeks to divide the study area according to inherent ecological gradients, or **strata**, so that a simple random

Figure 3.5 A simple random sampling method might require that the entire area first be gridded out on a map, and the location of where the larger quadrats should be placed can then be determined randomly. Once the quadrat has been placed, the investigator may choose to survey either the entire quadrat, any number of randomly determined minor quadrats, or simply at the "cross-hairs" formed by four contiguous minor quadrats (the point-contact method).

sampling method can be employed in each of the defined strata, separate from each other. In essence, this method simply divides the study area into two or more regions that are exclusive of each other according to some discriminating feature (**Figure 3.6**). In most aquatic systems, the ecological gradients present in the horizontal dimension can be quite different from those in the vertical. In practice, it is usually easier to treat the horizontal and vertical strata separately from each other (although it is certainly possible to define complex three-dimensional strata, if that is the investigator's desire).

Because survey data are organized according to the strata from which they were gathered, it is possible to perform comparisons between strata. With the use of statistics, this sampling method is particularly powerful and versatile, as comparisons can be made between different strata in the same geographic area, or between similar strata in different geographic areas.

Transect Interval Sampling Is a Variation of the Belt/Line Transect Method

Unlike the simple random sampling method, where the overall study area is divided into equally sized subdivisions, investigators utilizing transect

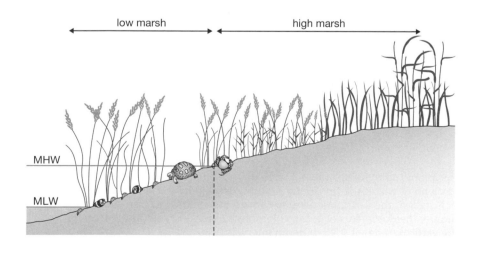

Figure 3.6 A coastal marsh can be defined by those regions found above mean high water (MHW) and mean low water (MLW), according to the local tidal cycle. As a result of differential tidal inundation, a natural ecological gradient is established. This relationship can be used to define different marsh strata, where sampling conducted in the "high marsh" area (above MHW) can be conducted separately from the "low marsh" area (below MHW). (From Laresen C et al. The Blackwater NWR Inundation Model. US Geological Survey, Open File Report 04-1302.)

Figure 3.7 Transect interval sampling requires the random placement of a transect line, where a number of quadrats are placed along the transect line, according to either a random or predefined interval.

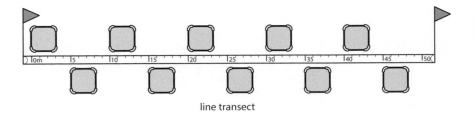

line transect

interval sampling simply lay out a transect line with regular divisions (such as a tape measure, as depicted in **Figure 3.7**) and place their quadrat(s) according to some predetermined rule. If the location and orientation of the transect is determined randomly, it is permissible to define a regular interval of measurement (such as one quadrat every 5 meters) without endangering the "randomness" of the survey. However, if the investigator is forced to orient the transect in a nonrandom manner (for example, according to some set ecological pattern or strata), the points of quadrat deployment along the transect line should be randomized instead (for example, one quadrat every 1 to 10 meters along the transect line, determined randomly each time).

For most benthic surveys, it is relatively simple to use a physical transect line resting on the **substrate**. However, three-dimensional surveys of aquatic systems typically require the use of a "virtual" transect line, usually defined by using standard map coordinates (such as latitude and longitude). Using this method, surface measurements are treated as though the transect line had been "placed" on the surface of the water. However, because of the strong vertical gradients present in all aquatic systems, it will be necessary to divide the vertical dimension into different strata if vertical profiles of the transect line are needed (**Figure 3.8**). It is also important to note that the measurements taken along a transect line must possess some areal or volumetric frame of reference (such as abundance per km^2, or concentration per m^3) in order for this to be considered a true plot method.

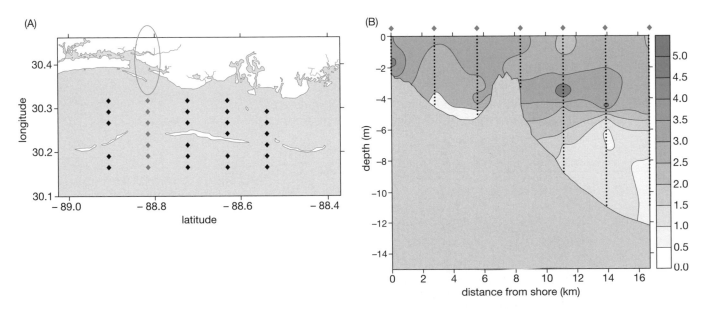

Figure 3.8 A series of regularly spaced station locations (using fixed latitute–longitude coordinates) can be easily oriented in a grid pattern, so each station location can serve as a point within the horizontal stratum, defined by the along-shelf and cross-shelf transects that compose the grid (A). At each station, the vertical dimension can also be subdivided into strata, typically by sampling at specific depth intervals. In this figure, measurements of *in situ* chlorophyll *a* (mg m^{-3}) along the transect highlighted in blue were pooled according to the vertical strata, at 0.25 m depth intervals (B).

Adaptive Sampling Is an Excellent Method for Detecting Rare Events

Unlike most plot methods, adaptive sampling is only quasi-random, since its primary purpose is the directed (nonrandom) survey of rare or elusive features. A typical method is to start with a simple random sampling method (nested quadrat or point contact), with a subsampling rule designed to concentrate effort only on those quadrats or grid points that successfully detect the feature under investigation (**Figure 3.9**). The subsampling rule (or "search pattern") can be just about anything and is limited only by the investigator's imagination. However, once the adaptive sampling rules are chosen, they should remain consistent throughout the duration of the survey.

Basic Plotless Methods

Although plot methods are generally considered to yield data of greater accuracy than plotless methods, plot surveys are usually more labor intensive and time consuming. If the features under investigation are assumed to be stationary and spatially distributed in a random manner (as depicted in Figure 3.3B), it may be preferable to use a plotless density estimator for ease of use in the field, or in those locations where it would otherwise be impractical to use a plot method. Even nonrandom distributions can be surveyed with a plotless method, provided that an appropriate sample size and methodology is chosen.

Plotless Density Estimators (PDEs) Do Not Require Exhaustive Surveys of an Area

There are a variety of methods used to estimate object densities by measuring the relative distances between objects rather than using counts per unit area. The simplest of these **plotless density estimators** (or PDEs) requires only a tape measure and a number of random starting points within the area of study. Although there are many different PDE formulations, they all require the investigator to begin from some random location in the study area and continue in a regular search pattern until the object of interest is encountered. By calculating the relative distances between a series of random starting points and each encountered object (**Figure 3.10**), the density of those objects in the study area can be estimated.

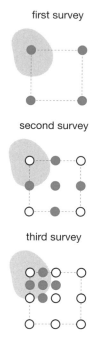

first survey

second survey

third survey

● new samples
○ samples reused from the previous survey

Figure 3.9 For adaptive sampling methods, a random point-contact grid is first established, with an adaptive sampling rule to halve the distance between contiguous grid points for every successful "hit." In the first survey, the rare event (shaded in grey) is detected in only one of four locations. In each successive survey, the investigation becomes increasingly focused on those regions where previous attempts were successful in detecting the event.

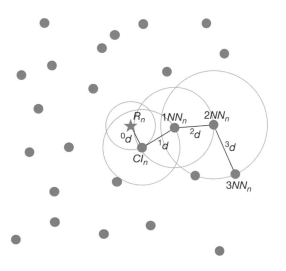

Figure 3.10 Schematic representation of how the simplest PDEs are implemented in the field. R_n is the nth random starting point; CI_n is the closest individual to R_n; $1NN_n$ is the first nearest neighbor to CI_n; $2NN_n$ is the second nearest neighbor to CI_n; $3NN_n$ is the third nearest neighbor to CI_n; ^0d is the shortest distance between R_n and CI_n; ^1d is the shortest distance between CI_n and $1NN_n$; and ^2d is the shortest distance between $1NN_n$ and $2NN_n$.

Basic Distance Estimators (BDEs)

The simplest of all common PDEs, the closest individual method (*CIM*) simply uses the mean distance $^{0}d_n$ between each random starting point R_n and the closest individual CI_n to R_n for N number of samples to estimate the object density:

$$D_{CIM} = \frac{1}{\left[4 \left(\dfrac{\sum {}^{0}d_n^2}{N} \right) \right]}$$ (3.13)

Nearest neighbor methods are a variation of the *CIM*, and comparative studies have shown that distance measurements $^{1}d_n$ between the closest individual CI_n and its first nearest neighbor $1NN_n$ generally yield more accurate object density estimates:

$$D_{1NN} = \frac{1}{2.778 \cdot \left(\dfrac{\sum_1 d_n^2}{N} \right)}$$ (3.14)

Often, density estimates can be further improved by using the distance measurement to the second nearest neighbor $2NN_n$ instead of $1NN_n$:

$$D_{2NN} = \frac{1}{2.778 \cdot \left(\dfrac{\sum_2 d_n^2}{N} \right)}$$ (3.15)

In a comprehensive review of PDE performance, an average of the results from each of the *CIM*, *1NN*, and *2NN* methods ($\bar{D}$) yielded the best general performance compared to all other PDEs, provided a minimum sample size N of 10:

$$\bar{D} = \frac{D_{CIM} + D_{1NN} + D_{2NN}}{3}$$ (3.16)

Ordered Distance Estimators (ODEs)

This method is very similar to the closest individual method (see Equation 3.13), but uses the distance between a randomly determined starting point R_n and the ith closest individual ($^{i}d_n$), which differs significantly from the nearest neighbor method of distance measurement (**Figure 3.11**). Although the second- and third-closest individual are most commonly used, researchers have found that for uniformly distributed objects, the formulation utilizing the third-closest individual ($i = 3$) performed best for sample sizes N in excess of 25:

$$D_{OD} = \frac{(3N - 1)}{\pi \cdot \sum {}^{3}d_n^2}$$ (3.17)

Angle-Order Estimators (AOEs)

This method is a variant of the ordered distance estimation method, except that the study area in proximity to a randomly determined starting point R_n is divided into four quadrants, and the distance to the ith closest individual in each of the

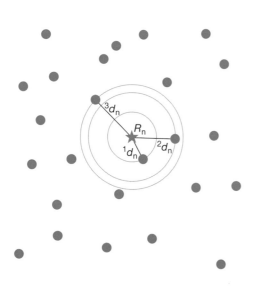

Figure 3.11 Schematic representation of the ordered distance estimator method, where the distance from the third-closest individual ($^{3}d_n$) to some random starting point (R_n, blue star) can be used as the best PDE for uniform distributions.

EXAMPLE BOX 3.1

Calculating the Basic Distance Estimators (BDEs)

Let us assume that Figure 3.10 represents an area of 1 m^2 (10,000 cm^2), and that the "true" density is 25 individuals per m^2. If we used 10 random starting points R_n in this area, and measured the distances to the closest individual (^0d_n), the first nearest neighbor (^1d_n), and the second nearest neighbor (^2d_n) for each R_n, our data might resemble Table 1.

Table 1 Measured distances (cm) to the closest individual (^0d_n), the first nearest neighbor (^1d_n), and the second nearest neighbor (^2d_n) using a variety of plotless methods

n	^0d_n	$^0d_n^2$	^1d_n	$^1d_n^2$	^2d_n	$^2d_n^2$
1	8.7	75.69	14.1	198.81	14.7	216.09
2	6.5	42.25	11.3	127.69	9.9	98.01
3	9.2	84.64	10.1	102.01	12.3	151.29
4	7.8	60.84	9.6	92.16	18.7	349.69
5	8.1	65.61	6.9	47.61	16.3	265.69
6	11.3	127.69	7.8	60.84	10.0	100.00
7	9.2	84.64	13.6	184.96	7.7	59.29
8	10.7	114.49	9.5	90.25	11.6	134.56
9	8.9	79.21	12.3	151.29	8.9	79.21
10	13.4	179.56	18.9	357.21	6.7	44.89
$N = 10$		$\Sigma = 914.6$		$\Sigma = 1412.83$		$\Sigma = 1498.72$

These data are then used to calculate the density estimators following the closest individual (D_{CIM}, Equation 3.13), first nearest neighbor (D_{1NN}, Equation 3.14), and second nearest neighbor (D_{2NN}, Equation 3.15) methods:

$$D_{CIM} = \left(\frac{1 \text{ individual}}{4 \cdot \left(\frac{914.62 \text{ cm}^2}{10} \right)} \right) \approx 27.3 \text{ ind m}^{-2} \qquad D_{1NN} = \left(\frac{1 \text{ individual}}{2.778 \cdot \left(\frac{1412.83 \text{ cm}^2}{10} \right)} \right) \approx 25.5 \text{ ind m}^{-2}$$

$$D_{2NN} = \left(\frac{1 \text{ individual}}{2.778 \cdot \left(\frac{1498.72 \text{ cm}^2}{10} \right)} \right) \approx 24.0 \text{ ind m}^{-2} \qquad \overline{D} = \left(\frac{27.3 + 25.5 + 24.0}{3} \right) \approx 25.6 \text{ ind m}^{-2}$$

Although the *1NN* analysis provided the most accurate estimate of the true population density in this particular example, the average of these three methods ($\overline{D}$), calculated using Equation 3.16, is generally the best choice among the basic distance estimators.

Figure 3.12 Schematic representation of the angle-order estimator method, where the study area is subdivided into four quadrants and the distance from the third-closest individual (^3d_n) is determined for each quadrant.

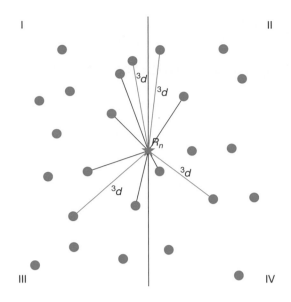

four quadrats (^id_n) is determined (**Figure 3.12**). Although this method is somewhat time consuming, it performs quite well for objects with an aggregated distribution pattern, especially when using the formulation that utilizes distance measurements to the third-closest individual ($i = 3$) for each quadrant:

$$D_{AO} = \frac{44N}{\pi \cdot \sum \left(1/^3d_n^2\right)} \tag{3.18}$$

Line Transect Density Estimators (LTDEs) Are the Best PDE for "Line-of-Sight" Search Patterns

As we have discussed, many PDEs require that the investigator follow a radial search pattern to determine the shortest distance between the survey starting point and the ith closest individual (or its nth nearest neighbor). However, the use of line transects may be preferable when the terrain would otherwise make it difficult to utilize a radial search pattern. Typically, **line transect density estimator** (or **LTDE**) methods simply measure the distance between the starting point of the transect and some encounter with the object or event along the transect line.

Variable Area Transect (VAT)

This method is not a true line transect method, as it uses a fixed-width, variable-length belt transect with a randomly selected starting point, and is extended in a random, straight direction until the object of study is encountered. Although this method can be used for "first" encounters, some researchers have found that the *VAT* method performed best when the distance l to the third encounter ($i = 3$) was used (**Figure 3.13**). Using a belt transect of uniform width w, the formulation is simply

$$D_{VAT} = \frac{(3N - 1)}{w \cdot \sum l} \tag{3.19}$$

Generally, the *VAT* is very easy to perform in the field and performs moderately well in most cases; however, the *VAT* performs best when sample sizes are relatively large ($N > 50$) and for those objects or events that exhibit a Poisson distribution.

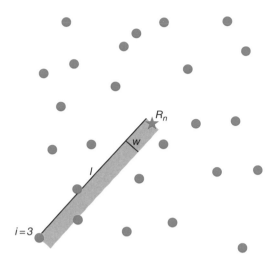

Figure 3.13 Schematic representation of the variable area transect (*VAT*) method, where a belt transect of fixed width *w* and variable length *l* is extended out from a random starting point R_n until the *i*th closest individual comes in contact with the belt transect. In most cases, the *VAT* method provides the most accurate results when the belt transect is extended out to the third encounter.

The Strong Method (SM)

The Strong method utilizes a true line transect and estimates object densities as a combined function of their overall abundance per unit area, as well as the objects' physical size and orientation (as these will also affect the probability of randomly encountering the object within the study area). For any transect of a fixed length *l*, every object encountered along the length of that transect is described instead by its maximum orthogonal width w_n (**Figure 3.14**), so the density can be estimated as

$$D_{SM} = \frac{A}{l} \times \sum \frac{1}{w_n} \qquad (3.20)$$

where *A* simply represents the unit area. Note that Equation 3.20 requires that *A*, *l*, and w_n all share consistent units.

Plotless Methods for Mobile Subjects Require a Strategy Different than Most PDEs

As we have seen from all the plotless methods discussed thus far, they each assume that the object or event being surveyed is somewhat fixed in place. In fact, it is that presumption of semipermanence that allows the investigator to measure distances between some theoretical starting point and the object of interest (either as the *i*th closest individual or as its *n*th nearest neighbor). But what are we to do if we wish to perform a plotless survey of objects that move freely (and often) throughout the study area? Unfortunately, most PDEs are insufficient for estimating population densities of mobile species; for them, you'll need a different approach.

Roaming Surveys

Fishes are notoriously difficult to assess using standard plotless methods, and are virtually impossible to assess using a plot method (unless the plot is inescapable, such as a net or cage). For rapidly or haphazardly swimming species, it is often advisable to estimate the fish density D_{RS} using the Southwood method (see Equation 3.21) for rapid visual assessments:

$$D_{RS} = \frac{Z}{2000rV} \qquad (3.21)$$

where *Z* is the number of encounters per hour, *r* is the effective viewing radius of the surveyor (in meters), and *V* is the average swimming speed of the organism (in kilometers per hour).

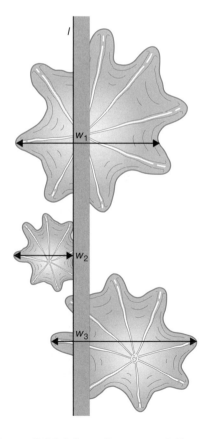

Figure 3.14 Schematic representation of the Strong method, which uses the maximum orthogonal width w_n for each object encountered along a line transect of fixed length *l*.

Mark and Recapture

These methods simply involve the capture and marking (or tagging) of individuals, which are subsequently released. At a later time, the investigator returns to capture more individuals and simply count those that were already marked from a previous capture. Of course, capture (and recapture) efficiencies are usually quite low, so these methods require a long-term commitment to the mark and recapture regime. However, such data can be extremely useful not only in establishing the population densities, but also in estimating birth, death, and emigration rates. These methods are discussed more thoroughly in Chapter 6.

Catch per Unit Effort (CPUE)

One of the simplest methods to perform in the field, measurements of CPUE can be used in virtually any biological context to estimate densities as a function of the collection device. For example, crabs harvested from a number of baited traps can be measured according to any number of relevant biometrics (size, quantity, biomass, etc.). As long as the collection device is used consistently, those biometrics can be related to the sampling effort. Although the CPUE method may not be the best choice for estimating instantaneous densities, CPUE data can be easily used for trend analyses in continuous, long-term data sets. These methods are discussed more thoroughly in Chapter 7.

References

Cochran WG (1977) Sampling Techniques, 3rd ed. John Wiley and Sons.

Dell RB, Holleran S, & Ramakrishnan R (2002) Sample size determination. *ILAR* 43(4):207–213.

Fleiss JL (1981) Statistical Methods for Rates and Proportions, 2nd ed. Wiley.

Krebs CJ (1999) Ecological Methodology. Harper & Row.

Math Open Reference Project (2011) http://www.mathopenref.com/polygonregulararea.html

Sanjerehei MM (2011) Determination of an appropriate quadrat size and shape for detecting association between plant species. *Ecological Modelling* 222:1790–1792.

Snedecor GW & Cochran WG (1989) Statistical Methods, 8th ed. Iowa State Press.

Southwood TRE (1978) Ecological Methods. Chapman and Hall.

Strong CW (1966) An improved method of obtaining density from line transect data. *Ecology* 47: 311–313.

U.S. Department of the Interior, National Park Service, Virgin Islands National Park (2001) *Coral Reef Monitoring Manual for the Caribbean and Western Atlantic.*

White NA, Engeman RM, Sugihara RT, & Krupa HW (2008) A comparison of plotless density estimators using Monte Carlo simulation on totally enumerated field data sets. *BioMed Central Ecology* 8:6. doi:10.1186/1472-6785-8-6.

Further Reading

Bakus GJ (2007) Quantitative Analysis of Marine Biological Communities: Field Biology and Environment. John Wiley and Sons.

Engeman R, Sugihara R, Pank L, & Dusenberry W (1994) A comparison of plotless density estimators using Monte Carlo simulation. *Ecology* 75:1769–1779.

Morisita M (1957) A new method for the estimation of density by spacing method applicable to nonrandomly distributed populations. *Physiology and Ecology Japan* 7:134–144.

Pollard J (1971) On distance estimators of density in randomly distributed forests. *Biometrics* 27:991–1002

Sutherland WJ (ed) (2006) Ecological Census Techniques: A Handbook, 2nd ed. Cambridge University Press.

Chapter 4

Oceanographic Variables

"Facts which at first seem improbable will, even on scant explanation, drop the cloak which has hidden them and stand forth in naked and simple beauty." – Galileo Galilei

Now that we have a firm understanding of how our experiment should be structured, it is time to turn our attention to the actual variables we shall be measuring in pursuit of that experimental design. But how do we determine what those variables should be? At the most basic level, we must ask ourselves—what must I measure to reveal the information most pertinent to my hypothesis?

Of course, there are no universal answers—that is a question you must answer for yourself. However, there are some fundamental oceanographic variables that offer the greatest bang for your buck and can at least get you started in the right direction.

Keep in mind that this chapter is not meant to serve as an exhaustive list of only those variables with which to concern yourself. Quite the contrary— it would be impossible to establish in an entire volume, let alone a single chapter, a definitive list of every oceanographic variable of critical import. Rather, this chapter is meant to outline what should be considered a minimalist's view of the most basic variables to consider when formulating your research plan.

Key Concepts

- The Lagrangian frame of reference describes ocean dynamics from the perspective of the fluid parcel itself as it flows through space and time.

- The Eulerian frame of reference describes ocean dynamics from the perspective of the fluid parcel as it moves through a fixed region of interest.

- The measurement of any oceanographic variable must be constrained by appropriate spatial and temporal boundaries, as defined by the investigator.

- There exist certain key oceanographic variables that can be considered essential to virtually all types of investigations of ocean dynamics according to their relevance in the biological, chemical, geological, and physical subdisciplines of oceanography.

Establishing the Frame of Reference

Before deciding on the specific oceanographic variables we will wish to measure in support of our field research, it is necessary that we first consider the appropriate frame of reference for those measurements. This is especially important in the context of aquatic systems because virtually all of our measurements will vary in the three spatial dimensions, as well as in time.

This presents us with one of the fundamental challenges of oceanographic research: Is it possible to simultaneously define four-dimensional variability? And if so, should our measurements be taken from the perspective of the fluid medium itself (within the flow), or should those measurements be taken from a fixed position (exterior to the flow)? What those measurements "mean" will be highly dependent on our frame of reference.

In the context of fluid dynamics, oceanographic observations are typically made from either of two different perspectives (**Figure 4.1**). The **Lagrangian** point of view (POV) seeks to describe fluid motion from the perspective of the fluid parcel itself as it flows through space and time. If we could shrink ourselves down and sit astride a rubber duck floating on the ocean surface, we would be observing the flow from a Lagrangian perspective. By contrast, the **Eulerian** POV demands that we measure the fluid as it moves through a fixed region of interest as time passes. So, if we were sitting on a pier and happened to observe a rubber duck floating by, we would be observing the flow from an Eulerian perspective.

As another example, let us consider an idealized loop current where the stream velocity within the current is much stronger than the flow outside the current in the surrounding water (**Figure 4.2**). With the passage of time, the loop current pinches off, forms a truncated loop, and spawns a small eddy. A drifter caught in the eddy would continue to measure the water properties within the flow; from its Lagrangian perspective, the drifter would not necessarily recognize any differences in the surrounding water properties, or that it was even "trapped" in the eddy. If we were observing these dynamics from a fixed location outside the flow, our Eulerian perspective would make it rather easy to recognize the spatial evolution of the flow with the passage of time. In this case, the data that describe the same oceanographic feature (eddy formation) could be analyzed in very different ways depending on the frame of reference.

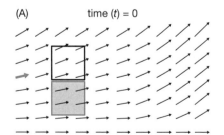
(A)   time (t) = 0

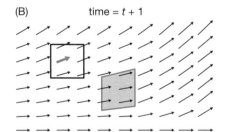
(B)   time = $t + 1$

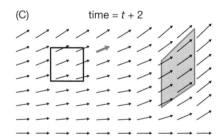
(C)   time = $t + 2$

Figure 4.1 A stylized view of water advection, where the length of the arrows represents the relative speed of the current, and the direction of the arrows represents the direction of water movement. In the Lagrangian view (*shaded box*), the region of interest (ROI) travels with the water and is monitored as it evolves over time. In the Eulerian view (*unshaded box*), the ROI remains fixed in a single location and the water is monitored as it flows into and then departs the ROI. Note that at time $t = 0$ (A), the Lagrangian and Eulerian views are indistinguishable from each other, and neither has experienced the discrete parcel of water located upstream from the flow (*blue arrow*). At some future time (B), the Eulerian view experiences the parcel of water (blue arrow) as it enters the ROI, while the area within the Lagrangian view is advected downstream. Even further into the future (C), the discrete water parcel (*blue arrow*) exits the Eulerian view and can no longer be measured within the ROI. Of course, the area within the Lagrangian view was never able to witness the passage of that particular water parcel, and has been deformed and advected further downstream by the flow. (Adapted from Janicke H & Scheuermann G, 1981)

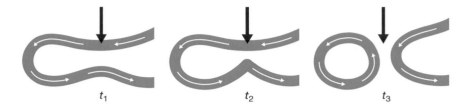

t_1　　　　　t_2　　　　　t_3

Figure 4.2 In the Lagrangian view, the properties of a water parcel *within the flow* (*white arrows*) will not seem to change from time t_1 to t_3. However, from the Eulerian perspective, the properties of the water parcel will seem quite different from t_1 to t_3, when viewed from a fixed location *outside the flow* (*black arrow*).

In practice, it is generally much easier to conduct oceanographic studies from an Eulerian perspective. Whether measurements are collected from a coastal landmark, a moored buoy, a research vessel, or from a satellite, the data can be easily related to a latitude–longitude–depth coordinate system—an Eulerian POV. To conduct a truly Lagrangian study, the instrument would be required to maintain its position within the evolving water mass, continuously recording data over the course of its travels. Although such drifters do exist, there are significant technical and financial challenges associated with their deployment, use, and recovery.

If the researcher is truly interested in the evolution of **discrete** water masses over time, it may be possible to perform a pseudo-Lagrangian study by simply adding more stations upstream or downstream of the main region of interest and increasing the sampling interval. As long as the same measurements are being recorded at all stations and at regular time intervals, the continuous evolution of those water masses will be captured in the "snap-shot" measurements recorded at each of the fixed locations over time.

Defining Spatial and Temporal Boundaries

Once you have established your preferred frame of reference, it will be necessary to define the boundaries of your study in space and time. In nearly all oceanographic applications, fixed positions on the Earth's surface are defined using the two-dimensional coordinate system that defines the **easting** direction in degrees longitude and the **northing** direction in degrees latitude (**Figure 4.3**). For ancient cartographers and navigators, determining one's exact position on the face of the Earth (with nothing but a compass in hand) was no easy task. With the advent of **GPS** technology, obtaining high-precision global positioning has now become routine.

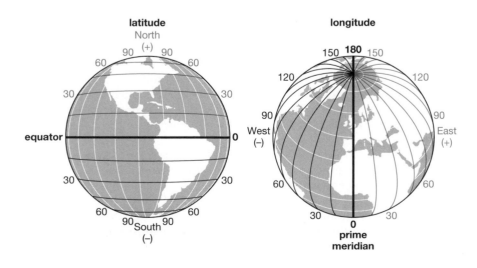

Figure 4.3 The precise location of any object on the Earth can be described using a Cartesian coordinate system, where latitude represents the *y*-coordinate (0 to 90° north or south of the equator) and the longitude represents the *x*-coordinate (0 to 180° east or west of the prime meridian).

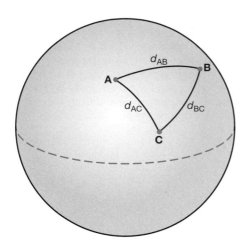

Figure 4.4 Although the latitude and longitude coordinates at points A, B, C can be easily determined using GPS technology, it may be more meaningful to relate these points to each other as a function of their relative distances (*d*) from each other. For distance calculations performed on a curved plane, we must use the Haversine functions.

Distance Measures are a Simple but Critical Way to Define Spatial Relationships

In some cases, it may be more useful to define the horizontal dimension not simply as latitude and longitude coordinates, but as distances from a fixed reference point. For example, if you wished to collect water samples at regular 20-m intervals offshore from the coastline, the latitude and longitude coordinates for all of your sampling sites would differ by only a few arc seconds. In this case, it might make more sense to define the latitude and longitude coordinates of the initial reference point at the coastline, and describe your sampling stations according to their relative distances from that point.

Depending on the sophistication of the surveying equipment being used, horizontal distances of 500 m or less can usually be measured directly. Over moderate distances (1 to 5 km), it becomes increasingly difficult to survey distances with sufficient accuracy and precision. When distances exceed 5 km, the curvature of the Earth becomes significant enough that we can no longer calculate distances according to planar geometry.

To calculate the distance between two points on a curved surface (**Figure 4.4**), we must use the **Haversine functions** (see Equations 4.1–4.4). Using a volumetric mean radius *r* of 6,371.0 km for the Earth, we can calculate the distance *d* between two points on the Earth's surface as

$$\Delta lon = lon_2 - lon_1 \tag{4.1}$$

$$\Delta lat = lat_2 - lat_1 \tag{4.2}$$

$$a = \frac{[1 - cos(\Delta lat)]}{2} + \left\{ cos(lat_1) \cdot cos(lat_2) \cdot \frac{[1 - cos(\Delta lon)]}{2} \right\} \tag{4.3}$$

$$d = r \cdot \left[2\,tan^{-1}\left(\frac{\sqrt{a}}{\sqrt{1-a}} \right) \right] \tag{4.4}$$

where all latitude and longitude coordinates must first be converted to **radians**, yielding the computed distance *d* in kilometers.

Depth Measures are Used to Extend Spatial Relationships into the Vertical Dimension

In aquatic systems, it is not sufficient to merely define our horizontal dimension; we must also define the vertical dimension as well. But measuring discrete ocean depths is far trickier in practice than it would at first seem. To accurately measure ocean depths, we must first establish a fixed reference point in the vertical to represent our "zero depth." This is where our troubles begin, because the surface of the ocean is never uniform—it is constantly being deformed by physical forces beyond our control (such as the tides, or swells caused by wind). So how do we accurately measure depths from the ocean surface if that surface is constantly changing?

One solution is to establish your frame of reference at the ocean floor, measuring heights from the bottom rather than depths from the surface. In many physical oceanography applications, this is the preferred method (especially if your instruments are designed to float at various heights, tethered to the bottom). If it is necessary to measure depths from the surface, the best solution is to use a submersible pressure sensor, such as a

A HAVERSINE EXAMPLE

If we wanted to calculate the distance between the latitude and longitude coordinates (+45.00, –100.00) and (+35.00, –98.00), our critical values for the Haversine functions are

$\Delta lon = 0.034907$

$\Delta lat = -0.174533$

$lat_1 = 0.785398$

$lon_1 = -1.745329$

$lat_2 = 0.610865$

$lon_2 = -1.710423$

$a = 0.007773$

$d = 1{,}125\,km$

Note that all latitude and longitude coordinates were converted from decimal degrees to radians using the conversion factor $\pi/180$ radians/degree.

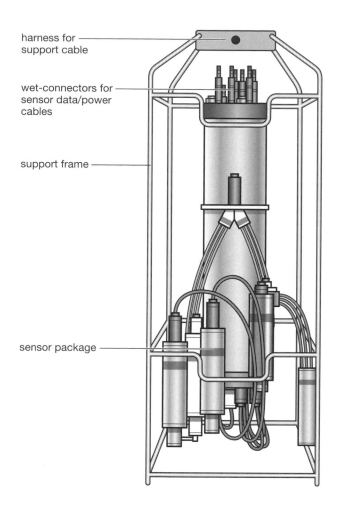

harness for
support cable

wet-connectors for
sensor data/power
cables

support frame

sensor package

Figure 4.5 An integrated conductivity-temperature-depth (CTD) Recorder (SBE 9 model). (Courtesy of Seabird Electronics, SBE 911plus model).)

conductivity–temperature–depth (CTD) recorder (**Figure 4.5**), to continuously record the **hydrostatic pressure** of the water column above the device as it is lowered. With such devices, any changes in hydrostatic pressure can be instantaneously correlated with the density and height of the overlying water column that is causing those pressure changes. In this way, the exact height of the water column (from the submerged sensor to the ocean surface) can be measured and defined as the *in situ* depth.

Although it is possible to measure water depths from the surface without a CTD, such low-tech methods usually require the use of a marked and weighted line that is dropped to the appropriate depth (an inherently imprecise solution). Such measurements can be improved by referencing the measurements against the **mean sea level** (MSL), but such efforts require the investigator to correct for any influences caused by currents, waves, and tides, as they will cause the local sea level to differ (sometimes significantly) from MSL.

Time is Required to Define the Succession of Events

With the three-dimensional spatial scale defined, any measurement taken at a specific location can only be represented as an instantaneous value, unless we are conducting those measurements at regular time intervals. Defining the temporal dimension of our variables is absolutely critical to success, as we must adhere to time scales that are relevant to the phenomena we wish to capture in our data. For example, if we were interested in studying the tidal cycle at various coastal locations, we might have a very difficult time making sense of our data if we decided to take our measurements at 15-hour intervals.

It is also important to consider how time should be scaled. For most long-term investigations, data are usually depicted first by the calendar year, and then by the calendar day (January 1 = Day 1, December 31 = day 365 for non-leap years). To compensate for leap years (or if the study has a significant seasonal component to it), it may be beneficial to convert the calendar day to 2π radians (360°), thereby relating the calendar day to the Earth's relative "completion" of its annual orbit around the Sun.

Shorter time intervals are typically denoted using the 24-hour military-style clock, as HH:MM:SS. Using this convention, the time of day is counted from midnight to midnight and is divided into 24 hours. As an example, 7:00 a.m. and 6:30 p.m. would be cited as 07:00:00 and 18:30:00, respectively (with midnight returning to 00:00:00).

Because time of day is relative to longitude, the global timekeeping convention usually considers 0° longitude (the **prime meridian**) as the global "time zero." Because Greenwich, England, is located on the prime meridian, the time designation used is sometimes called Greenwich Mean Time (GMT). In more recent years, the GMT designation has been replaced by Universal Time Coordinated (UTC), but they are synonymous.

Using the UTC convention, the various time zones across the Earth are calculated as positive or negative offsets from UTC (**Figure 4.6**). Based on your specific longitude, it would be necessary for you to determine the appropriate UTC offset if you wished to cite your time-stamped data relative to UTC. For example, the UTC offset for Puerto Rico is –4.0 hours. If we were lucky enough to be performing field research off Puerto Rico and took our measurement at 11:34:27 local time, that time-stamp is –4.0 hours from UTC. Therefore, we would simply have to add 4.0 hours to cite our data consistent with UTC (collected at 15:34:27 UTC, in fact).

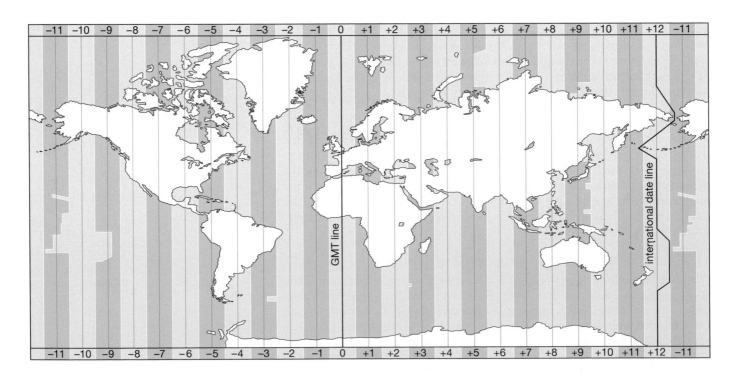

Figure 4.6 UTC time zones of the world.

Physical Variables for Consideration and the Essential Measures of Ocean Physics

When we speak in terms of physical oceanography, we are essentially attempting to describe the physical structure of the ocean and how the ocean obeys the laws of motion, the laws of thermodynamics, and the laws of conservation of mass and energy. When these fundamental concepts of physics are considered within the context of ocean science, we are able to elucidate the variables that more clearly define those phenomena of special interest, such as fluid dynamics (like waves, currents, and tides) as well as the propagation of light and sound within the ocean.

The physical properties of the ocean that are most important to fluid dynamics and ocean structure are the measurements of a water parcel's mass m, temperature T, and pressure P. With regard to seawater, mass cannot be considered solely from a physical perspective; in fact, the mass of seawater is intimately related to its salinity S—a chemical property. Remarkably, measurements of water mass, temperature, and pressure can be used by a clever oceanographer to derive (or estimate) many other physical variables of relevance, as we will soon discuss.

Mass is the Most Important Way to Quantify Existence

When we think of an object's mass, most of us (mistakenly) think in terms of the object's weight; as if the two were perfectly synonymous. Technically, an object's mass is a measure of the quantity of **matter** that comprises that object. One of the most fundamental principles in the physical sciences is that matter can be neither created nor destroyed. In other words, the amount of matter within an object must either remain unchanged, or must be 100% accounted for—this is the law of conservation of mass. Thus, the amount of matter will be absolute, whereas the "weight" of that matter will always be relative to the balance of forces exerted on that mass while it sits on the scale. Imagine if we stepped onto a scale while we were shoulder deep in a swimming pool—the buoyancy of our submerged bodies would counteract the pull of Earth's gravity, and the scale would read as if we were much lighter than we really are (but our true mass would remain unchanged).

In most cases, measuring the mass of a solid object is so elementary that we don't even think about it. After all, it is not difficult to draw the boundary that separates an 8-kg bowling ball from the rest of the natural world. It's an entirely different challenge for us to define the physical boundaries that separate an 8-kg mass of fluid from the rest of the ocean. How do we conceptually "draw the line" between 8 kg of ocean, or perhaps 8 kg of atmospheric gas?

When dealing with the masses of liquids or gases, it is necessary for us to also define the three-dimensional space that "contains" the fluid (**Figure 4.7**). This is why we use **density** (ρ, in units of g cm^{-3}, or kg m^{-3}) as the preferred expression of fluid mass in oceanography. Thus, the measure of a fluid's density is simply the quantity of matter (its mass) that simultaneously occupies a certain amount of space (its volume). Density (ρ) can be easily determined by calculating the mass-to-volume ratio; or we can express the density of a solid or fluid by the **specific volume** α it occupies:

$$\rho = \frac{m}{V}; \quad \alpha = \frac{1}{\rho} \qquad (4.5)$$

For example, a fluid sample with a density ρ of 1012.35 kg m^{-3} would also have a specific volume α of 9.87801×10^{-4} m^3 kg^{-1}.

Figure 4.7 The density of a solid (or fluid) is defined by the amount of mass contained within a specific volume. However, it is important to note that density is affected either by changes to the amount of mass or by changes to the volume it occupies. For example, if more mass is packed into the same volume (*lower left*), density will increase. Likewise, if the same amount of mass is squeezed into a smaller volume (*lower right*), density will increase as well.

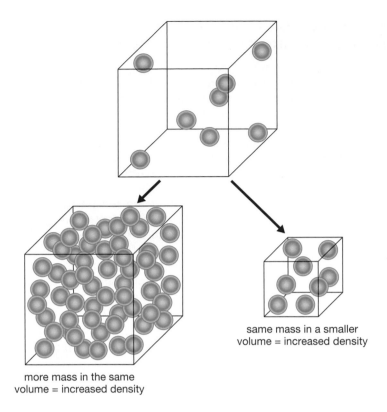

more mass in the same
volume = increased density

same mass in a smaller
volume = increased density

Because it is necessary for us to define a fluid's mass in terms of the volume it occupies, we must consider the total mass contained within that volume. That makes things a bit more difficult for us because we can rarely expect to encounter a pure fluid in nature. In the "simple" case of ocean water, we know that any volume of fluid we collect will contain a certain mass of water, several different species of dissolved salts, and several different species of dissolved gases (all with different atomic masses and in various concentrations).

Since the density of a fluid is defined by both its mass and volume, changes to either of those will result in a different measure of overall density. In the context of oceanography, changes in salinity can have a dramatic effect on the overall mass of the water parcel, thereby changing its density. Changes in temperature and pressure can have considerable effects instead on the volume of the water parcel, also resulting in a modified measurement of density.

As you can see, this makes an accurate measurement of density completely dependent on our ability to assess the mass m, temperature T, and pressure P simultaneously. But before we get too ahead of ourselves, let's take a closer look at the effects of temperature and pressure—we'll rejoin our discussion of density a bit later.

Temperature Affects the Physical Properties of All Matter

Most of us are so used to measuring temperatures that we rarely give it much thought. In the context of oceanography, the simple task of measuring the ambient temperature (whether at the surface or at discrete depths) is really a measurement of the *in situ* temperature T. When we are dealing with relatively shallow water depths, it is usually sufficient to measure T.

However, because seawater is slightly compressible, any water parcel at great depth will be compressed into a smaller volume. If that same water parcel

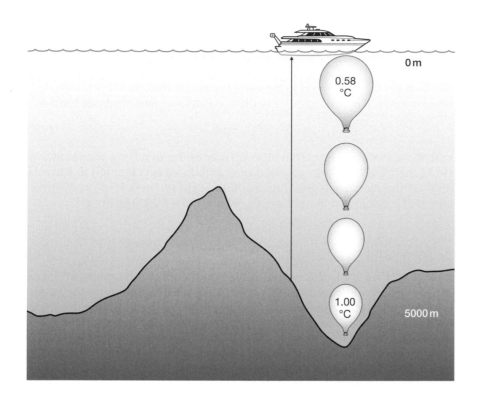

Figure 4.8 The *in situ* temperature of a parcel of water 5000 m deep may be 1.00°C at depth, but if this parcel were raised to sea level, the relief in hydrostatic pressure would result in the adiabatic expansion of the water parcel, causing it to cool to 0.58°C (its potential temperature, or θ).

were raised to the surface without gaining or losing heat to the surrounding water, the alleviation of pressure would cause the water parcel to slightly expand and therefore cool (**Figure 4.8**). If we were to measure the theoretical temperature of this deep-ocean water parcel after raising it to the surface, we would measure it as the parcel's potential temperature (θ). Potential temperatures are extremely useful measures when comparing water samples from different depths, or when water masses are transported over significant vertical distances, because θ compensates for any and all compressibility effects (whereas measures of *in situ* temperature do not).

Regardless of whether temperatures are measured *in situ* (T) or as θ, the preferred scale of measurement is in decimal degrees Celsius (°C). In oceanography, seldom are temperatures measured in degrees Fahrenheit (°F); however, the two are easily interchangeable using

$$^{\circ}F = \left(\frac{9}{5} \cdot {^{\circ}C} \right) + 32 \tag{4.6}$$

$$^{\circ}C = \left(\frac{5}{9} \right) \cdot \left(^{\circ}F - 32 \right) \tag{4.7}$$

Occasionally, it may be necessary to convert to kelvins (K), a measure of absolute temperature. The Kelvin scale is simply defined as

$$K = {^{\circ}C} + 273.15 \tag{4.8}$$

Hydrostatic Pressure is Used to Describe the Potential Energy within a Fluid

In most oceanographic applications, pressure (P) is measured *in situ* and refers only to hydrostatic pressure and excludes the contributions of atmospheric pressure. In most cases, the variations in hydrostatic pressure far exceed the subtle variations in atmospheric pressure, so we can assume that standard atmospheric pressure is essentially a constant (1 atm = 1013.25 mbar, or 101.325 kPa). Thus, by removing the effects of atmospheric pressure, $P = 0$ kPa at the ocean surface.

The best course of action is to measure the actual hydrostatic pressure P_h *in situ*. However, if it is necessary to estimate P, its simplest algebraic form is

$$P_h = \rho g d \qquad (4.9)$$

where ρ specifically represents the density of the overlying water column (in $kg\ m^{-3}$), g is the mean constant of gravitational acceleration ($9.8\ m\ s^{-2}$), and d is the discrete depth of measurement (in meters). Hydrostatic pressure (P_h) is calculated in units of kPa, where 1 kPa = $1000\ kg\ m^{-1}\ s^{-2}$.

Density Describes the Relationship between Matter and the Space it Occupies

Because density is a highly variable function of salinity S, temperature T, and pressure P, *in situ* measurements are usually preferred because they automatically reflect the true effects of the (S, T, P) conditions on the density of the measured water parcel. However, this also means that we must cite ρ as a function of its (S, T, P) conditions where the data were collected. Since the density of pure, fresh water is $1000\ kg\ m^{-3}$ and the density of seawater rarely exceeds $1060\ kg\ m^{-3}$, the total variability in ρ, for all the world's oceans, will occur only in the last two figures. To focus our attention where it really belongs (on those last two figures), many oceanographers convert ρ (S, T, P) to **sigma** (σ):

$$\sigma\left(S,T,P\right) = \left[\rho\left(S,T,P\right) - 1000\ \text{kg}\,\text{m}^{-3} \right] \qquad (4.10)$$

If we cannot measure *in situ* density, we must instead estimate ρ according to its unique (S, T, P) conditions. For most applications where the accuracy of ρ can be $\pm 0.5\ kg\ m^{-3}$, the simple linear equation of state should be sufficient:

$$\rho\left(S,T,P\right) = \left[\bar{a}\left(T - T_0\right) + \bar{b}\left(S - S_0\right) + \bar{k}P \right] + \rho_0 \qquad (4.11)$$

where $\bar{a} = -0.15\ kg\ m^{-3}$ per °C, $T_0 = 10$°C, $\bar{b} = 0.78\ kg\ m^{-3}$ per ‰ salinity, $S_0 = 35$‰ salinity, $\bar{k} = 4.5 \times 10^{-3}\ kg\ m^{-3}$ per decibar, and $\rho_0 = 1027\ kg\ m^{-3}$. When a more precise estimate of $\rho(S, T, P)$ is needed, it is preferable to use the 1980 International Equation of State (IES 80) of Seawater (**Technical Box 4.1**).

Supplemental Measures of Ocean Physics

Within the context of physical oceanography, measures of *in situ* temperature, density, and hydrostatic pressure are so critical, in so many different applications, that they should be collected as part of any research effort. Keep in mind that as important as these measurements are, it may be necessary to expand your list of physical parameters to include those variables that have a more direct influence on the fluid dynamics of the ocean.

Flow Fields (and Currents) are Used to Describe both the Speed and Direction of Water Movement

Because fluids are free to flow in three dimensions, it is often necessary to determine the speed and direction of these flows (as distinct **currents**) in three dimensions. In practice, this is done using a directional current meter (**Figure 4.9**), a submersible device that can measure current speeds in three different directions simultaneously: the easting (u), the northing (v), and the

TECHNICAL BOX 4.1

1980 International Equation of State (IES 80) of Seawater

The most recent (and most precise) accepted method for estimating the density of sea water at any given salinity S (in ‰), temperature T (in °C), and hydrostatic pressure P (in bars) is defined by the IES 80 as the function

$$\rho(S,T,P) = \frac{\rho(S,T,0)}{\left\langle 1 - \left[\dfrac{P}{K(S,T,P)}\right]\right\rangle} \tag{4.12}$$

where $\rho(S, T, 0)$ represents the density of seawater at the surface $(P = 0)$, such that

$$
\begin{aligned}
\rho(S,T,0) = {}& 999.842594 \\
& + \left(T \cdot 6.793952 \times 10^{-2}\right) - \left(T^2 \cdot 9.095290 \times 10^{-3}\right) \\
& + \left(T^3 \cdot 1.001685 \times 10^{-4}\right) - \left(T^4 \cdot 1.120083 \times 10^{-6}\right) \\
& + \left(T^5 \cdot 6.536332 \times 10^{-9}\right) + \left(S \cdot 8.24493 \times 10^{-1}\right) \\
& - \left(T \cdot S \cdot 4.0899 \times 10^{-3}\right) + \left(T^2 \cdot S \cdot 7.6438 \times 10^{-5}\right) \\
& - \left(T^3 \cdot S \cdot 8.2467 \times 10^{-7}\right) + \left(T^4 \cdot S \cdot 5.3875 \times 10^{-9}\right) \\
& - \left(S^{3/2} \cdot 5.72466 \times 10^{-3}\right) + \left(T \cdot S^{3/2} \cdot 1.0227 \times 10^{-4}\right) \\
& - \left(T^2 \cdot S^{3/2} \cdot 1.6546 \times 10^{-6}\right) + \left(S^2 \cdot 4.8314 \times 10^{-4}\right)
\end{aligned}
\tag{4.13}
$$

$K(S, T, P)$ is a complex regression equation, determined experimentally, that represents the secant bulk modulus, mathematically defined by the function

$$
\begin{aligned}
K(S,T,P) = {}& 19{,}652.21 \\
& + (T \cdot 148.4206) - (T^2 \cdot 2.327105) \\
& + (T^3 \cdot 1.360477 \times 10^{-2}) - (T^4 \cdot 5.155288 \times 10^{-5}) \\
& + (P \cdot 3.239908) + (T \cdot P \cdot 1.43713 \times 10^{-3}) \\
& + (T^2 \cdot P \cdot 1.16092 \times 10^{-4}) - (T^3 \cdot P \cdot 5.77905 \times 10^{-7}) \\
& + (P^2 \cdot 8.50935 \times 10^{-5}) - (T \cdot P^2 \cdot 6.12293 \times 10^{-6}) \\
& + (T^2 \cdot P^2 \cdot 5.2787 \times 10^{-8}) + (S \cdot 54.6746) \\
& - (T \cdot S \cdot 6.03459 \times 10^{-1}) + (T^2 \cdot S \cdot 1.09987 \times 10^{-2}) \\
& - (T^3 \cdot S \cdot 6.1670 \times 10^{-5}) + (S^{3/2} \cdot 7.944 \times 10^{-2}) \\
& + (T \cdot S^{3/2} \cdot 1.6483 \times 10^{-2}) - (T^2 \cdot S^{3/2} \cdot 5.3009 \times 10^{-4}) \\
& + (P \cdot S \cdot 2.2838 \times 10^{-3}) - (T \cdot P \cdot S \cdot 1.0981 \times 10^{-5}) \\
& - (T^2 \cdot P \cdot S \cdot 1.6078 \times 10^{-6}) + (P \cdot S^{3/2} \cdot 1.91075 \times 10^{-4}) \\
& - (P^2 \cdot S \cdot 9.9348 \times 10^{-7}) + (T \cdot P^2 \cdot S \cdot 2.0816 \times 10^{-8}) \\
& + (T^2 \cdot P^2 \cdot S \cdot 9.1697 \times 10^{-10})
\end{aligned}
\tag{4.14}
$$

Note that ρ is given in units of kg m^{-3} and the secant bulk modulus K is given in units of bars.

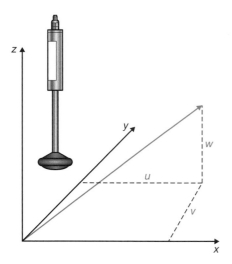

Figure 4.9 An example of a submersible three-dimensional current meter, which can simultaneously measure the (u, v, w) vector components of any flow.

vertical (w) vector components of flow. Since the movement of water within the ocean is described in terms of both the speed and direction of any given current, it is often critical to measure the evolution of three-dimensional flow fields as they might affect particle **advection**, mixing, and the overall stability of the water column.

Wave Geometry Is Used to Discern Wave Patterns in a Complex Ocean

For most of us, it is impossible to think about the ocean without also thinking about ocean waves. Whether we observe them as breaking waves on the shore, or as rising and falling swells causing our vessel to bob like a cork on the ocean surface, ocean waves are an integral part of water movement. When we seek to define wave dynamics, it is most helpful for us to consider the ideal case of an individual wave, propagating in a specific direction, exhibiting the classic sinusoidal wave geometry (**Figure 4.10**).

If we were fortunate enough to encounter something so simple as a single wave in the ocean, it would be a relatively easy task to measure its length λ or its height h. If we had a stopwatch, we could also measure the wave's frequency f and period T with very little difficulty. Unfortunately, the real ocean is never so agreeable—in fact, what we observe in the ocean is a complex mixture of many different waves, of many different wave geometries, that all "arrive" at our location according to their different wave periods and frequencies. Because of this complexity, it is often necessary to describe the sea state according to the **significant wave height** H_S, which is defined as the average wave height of the highest third of all wave heights (**Figure 4.11**).

Tides Represent the Influence of Heavenly Bodies on Ocean Dynamics

Among the most significant of all waves within the ocean are our tides, generated primarily by the gravitational forces between the Earth, Sun, and Moon. Because of the stability of the Earth–Sun and Earth–Moon orbits, our tides produce tidal amplitudes (tidal wave heights) and currents that ebb and flow with remarkable predictability. Because of this regularity, the Earth's tides exert a tremendous influence on virtually every aspect of ocean science. As a practical matter, field researchers must always consider how these tidal influences might complicate the design of their field experiments and/or the acquisition of field data.

Secchi Depth Is Used as a Simple Way to Define the Optical Properties of Water

Although the penetration of light in the ocean is most relevant in the context of photosynthesis (see the section "Biological Variables"), the transmission and attenuation of light is largely a physical phenomenon. In most cases, the effective depth of penetration for natural sunlight can be estimated quite easily using a Secchi disk (**Figure 4.12**). Using this simple methodology, a weighted disk is attached to a marked line and lowered until the disk first

Figure 4.10 An idealized ocean wave, indicating the defining elements of its sinusoidal geometry, including the wave crest, trough, length (λ,) and height (h). The amount of time required for the wave crest to travel from point A to point B (that is, a single wavelength) is defined as the wave period T. The number of wave crests passing point A every second is defined as the wave frequency f.

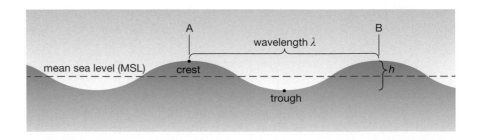

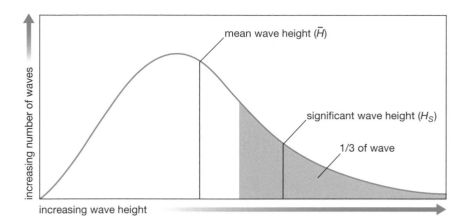

increasing number of waves

increasing wave height

mean wave height ($\bar{H}$)

significant wave height (H_S)

1/3 of wave

Figure 4.11 A theoretical spectrum of measured wave heights, where the average wave height of the highest 1/3 of all waves is defined as the significant wave height, or H_S.

disappears from view. The distance from the disk to the water line is noted. Then, the disk is slowly raised until it reappears and the new distance (from the disk to the water line) is recorded. The average between the two measurements is called the Secchi depth *SD*. Measures of Secchi depth loosely represent the maximum depth through which visible light can penetrate, with sufficient spectral quality and intensity, to support photosynthesis.

Chemical Variables

Because water is such an effective solvent, it is capable of dissolving nearly any molecule to which it is exposed. Since the world ocean has been in existence for somewhere between 3.5 and 3.8 billion years, it is no surprise that the oceans would contain such a complex mixture of chemicals, not the least of which are the salts that define our global marine waters.

Although the water is free to evaporate from the ocean and fall back to Earth as precipitation, the chemicals that are dissolved in the ocean cannot evaporate—they are "trapped" in the ocean basins and will remain there for untold epochs, unless removed by some other natural phenomenon. And so the primary challenge posed to our ocean chemists is the task of determining

⚠ FINDING HARMONY WITH THE TIDES

Periodic tidal currents can have a significant impact on the net direction and speed of oceanic currents. Tides will also have a significant impact on the overall depth of the water column, which is especially significant in shelf or coastal waters.

Luckily, the hydrodynamic effects of the tides are also monitored regularly at coastal observation stations across the globe. Typically, tide table data are cited for a particular location as a $\pm\Delta$ (in meters) from mean sea level (MSL), measured as a function of time *t*.

Tide tables can be used to predict the time and date of arrival for high and low tides ($\pm\Delta$ MSL) so field researchers can schedule their sampling efforts to compensate for tidal effects. They can also use the $\pm\Delta$ MSL predictions (or observations) to estimate the tidal current speed v_T as

$$v_T = \frac{(\Delta_2 - \Delta_1)}{(t_2 - t_1)}$$

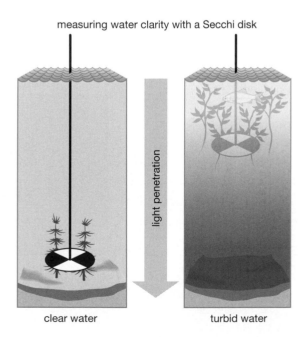

measuring water clarity with a Secchi disk

light penetration

clear water

turbid water

Figure 4.12 The relative turbidity of the water column will determine the depth to which sunlight can penetrate. The depth to which a lowered Secchi disk disappears from view (defined as the Secchi depth) can be used quite effectively to make comparisons of water clarity between water bodies. (Courtesy of the Minnesota Pollution Control Agency.)

both the presence and relative concentration of particular chemical species, without confusion or interference from any of the other chemicals within the same volume of seawater.

The primary constituents of seawater (other than the water itself) are the multitude of salts dissolved within. Not only do these salts contribute to the overall chemical environment of the ocean, but they also contribute to the overall density of the water. But beyond the basic measurement of the ocean's salt content (or **salinity**), there are many other chemicals present within seawater that have tremendous significance and must therefore be quantified.

Salinity Is the Fundamental Measure of Salt Content in the Ocean

Measuring the overall salt content within a sample of seawater is not as easy as one would think. Certainly, if seawater were a simple mixture of water and a single variety of salt (like NaCl), we would simply be required to measure the concentration of either the sodium ions (Na^+) or the chloride ions (Cl^-) in solution. Unfortunately, the major constituents of seawater include several ions as well, such as sulfate (SO_4^{2-}), magnesium (Mg^{2+}), calcium (Ca^{2+}), potassium (K^+), and many others. From a chemist's point of view, this is the worst-case scenario, as this would require a different chemical test to determine the concentration of each and every ion in solution.

Fortunately, the salt ions within the world ocean are relatively **inert** and their concentrations always remain in constant proportion to each other (**Figure 4.13**). That means that no matter where a sample of seawater is taken, and no matter how salty the water is, the proportion of salt ions will always be the same: 55.5% Cl^-, 30.6% Na^+, 7.7% SO_4^{2-}, 3.7% Mg^{2+}, and so on. Because all seawater follows this **principle of constant proportions**, ocean chemists only need to determine the concentration of one major salt ion in order to derive all the rest.

Historically, the concentration of the most dominant salt ion (Cl^-) was first determined and ultimately related to the overall salinity of the water sample. When using chemical tests for chloride concentration (or **chlorinity**), salinity ($S‰$) can be determined according to the experimental relationship defined in Equation 4.15, where

$$S‰= 1.80655 \ Cl^-‰ \tag{4.15}$$

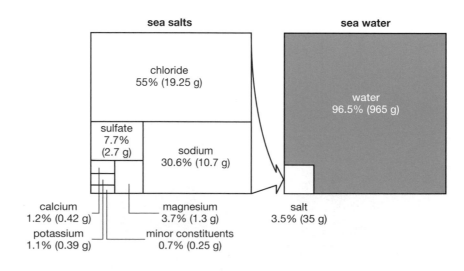

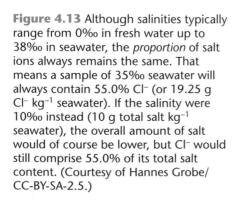

Figure 4.13 Although salinities typically range from 0‰ in fresh water up to 38‰ in seawater, the *proportion* of salt ions always remains the same. That means a sample of 35‰ seawater will always contain 55.0% Cl^- (or 19.25 g Cl^- kg^{-1} seawater). If the salinity were 10‰ instead (10 g total salt kg^{-1} seawater), the overall amount of salt would of course be lower, but Cl^- would still comprise 55.0% of its total salt content. (Courtesy of Hannes Grobe/ CC-BY-SA-2.5.)

Because there are so many different salt species present within seawater, it is more correct to consider salinity as the sum of all salt ions present within a known quantity of water. By convention, oceanographers typically use "parts per thousand" (‰) as a unit of measure for salinity, based on the mass ratio between solute (salt) and solvent (water). Thus, a 1-kg (1000-g) sample of seawater that contains a total of 35 g of dissolved salt ions has 35 g salt per 1000 g total, or 35‰.

As a practical matter, salinity is almost never measured by determining the weight of dissolved salts. Instead, most oceanographers measure the **conductivity** of sea water and ultimately relate those measurements to the overall salt content of the water sample. This practice led to the adoption of the **Practical Salinity Scale** (PSS) in 1978, when the global oceanographic community agreed to redefine the method for measuring salinity by comparing the conductivity of seawater samples against a standard salt solution of potassium chloride (KCl). Salinity measurements that are determined according to the PSS are in fact unitless; however, some oceanographers cite these salinity measures in terms of practical salinity units (PSU).

Conductivity Represents an Indirect Way to Measure Salinity Using Electricity

Because seawater contains so many dissolved salt ions, it behaves as an **electrolyte** capable of conducting electricity. Since the conductance of seawater is directly related to the amount of salt dissolved in solution, oceanographers are able to determine the salinity of seawater based on the flow of electricity through the sample. Although the best laboratory practices can usually detect salinities with 1/8000‰ precision, conductivity measurements are far superior and can detect the most subtle differences in salinity, up to 1/40,000‰! As a result, most modern methods of measuring salinity rely on conductivity meters, like those integrated in CTD recorders (see Figure 4.5).

Metabolic Nutrients Are Dissolved Molecules which are Necessary for Life in the Ocean

Beyond the inert salts that define the salinity (and density) of seawater, there are several other chemical species found in seawater, in much smaller concentrations, that have significant implications on the **biogeochemistry** of the ocean. The most important of these are the inorganic nitrogen- and phosphorus-based nutrients that are extremely **labile** and heavily utilized by living organisms to metabolize or accumulate biomass. As most biologically important nutrients are found only in trace amounts, it is always advisable to use the most sensitive tests available to determine their *in situ* concentrations.

Among the several different varieties of nitrogen-based nutrients present in the ocean, the most important chemical species are the dissolved ammonium (NH_4^+) and nitrate (NO_3^-) ions. The growth of all living organisms is largely dependent on the availability of NH_4^+ and/or NO_3^- in the ocean; therefore, any oceanographic study that investigates some aspect of biological production in the ocean would require an assessment of *in situ* NH_4^+ and NO_3^- concentrations.

With regard to phosphorus-based nutrients, the orthophosphate ion (PO_4^{3-}) is utilized almost exclusively of any other form of phosphorus. Like NH_4^+ and/or NO_3^-, orthophosphate is absolutely necessary for biological growth (making it a vital nutrient for study).

Supplemental Measures of Ocean Chemistry

Structural Nutrients Are Dissolved Molecules which Can Form Solid Mineral Deposits

Dissolved species of silicon (Si) and calcium (Ca) are absolutely necessary for the secretion of protective coatings or "shells" for some of the most important microorganisms in the ocean (which form the base of the entire oceanic food web). Dissolved calcium (as Ca^{2+}) is available as a major constituent of seawater and is used by organisms most commonly in the formation of calcium carbonate ($CaCO_3$) shells or skeletons. Soluble forms of silicon, such as $Si(OH)_4$ (silicon hydroxide), are used by certain marine microorganisms in the same proportions as NH_4^+ or NO_3^- (a testament to the importance of dissolved silica as a structural nutrient). Because of these relationships, investigations that focus primarily on the capacity for biological production will benefit from the inclusion of structural nutrients as well as the metabolic nutrients.

Dissolved O_2 Is One of the Most Influential Molecules that Affects the Biogeochemistry of the Ocean

Although it is perhaps easier to conceptualize how salts become dissolved in seawater, we must not forget that gases can also dissolve in seawater. The availability of dissolved oxygen (DO) gas is a critical concern for many biological investigations, particularly as it relates to the ability of seawater to sustain animal life. DO concentrations can be used to estimate gross respiration and/or photosynthetic rates (as they relate to biological processes) as well as gas exchange rates with the atmosphere (as a function of turbulent mixing, water column stability, and a variety of other physical phenomena). The availability of DO also determines the type of chemical reactions which can occur within the ocean, which in turn has a huge influence on all aspects of marine geochemistry.

Dissolved CO_2 Can Be Used to Investigate Gas Exchange and Ocean Acidification Dynamics

Carbon dioxide represents the flip side of the dissolved O_2 picture: as O_2 is consumed in the water column, it is largely replaced by CO_2 during respiration. During photosynthesis, CO_2 is removed from solution and replaced (more or less) with O_2. By monitoring the changes in these dissolved gases, it is often possible to quantify the relative biological contributions of both the consumers and the producers in the ecosystem.

CO_2 concentrations also have a critical impact on the contributions of several different inorganic forms of carbon within the ocean, according to the carbonate equilibrium:

$$CO_2 + H_2O \leftrightarrow H_2CO_3 \leftrightarrow HCO_3^- + H^+ \leftrightarrow CO_3^{2-} + 2H^+ \qquad (4.16)$$

The most direct consequence of rising CO_2 concentrations is the liberation of H^+, which ultimately affects the acidity (pH) of seawater, potentially resulting in ocean acidification.

pH and alkalinity: The pH of a solution refers to the molar concentration of the acidic hydronium ion, $[H^+]$, where

$$pH = -log_{10} [H^+] \qquad (4.17)$$

Although changes in pH can cause fluctuations in the carbonate equilibrium (see Equation 4.17), they can also affect the rates and outcomes of several important **hydrolysis** and **redox** reactions within the ocean. The natural ability of seawater to negate the effects of rising [H⁺], sometimes called its "**alkalinity**," can become diminished if there is too much H⁺ in solution. When this happens, slight increases in [H⁺] can have a disproportionately strong effect on ocean pH, exacerbating the effects of ocean acidification.

Biological Variables

Of all the oceanographic variables to consider, those of biological significance are perhaps the most difficult to quantify. Living organisms are adaptive by nature and exhibit behaviors that can complicate our efforts to collect those data we find most pertinent. Biology is anything but passive, and rarely do we have the luxury of performing a particular measurement that doesn't affect the organism and induce a change in the very parameter we were trying to nail down. That being said, it is possible to reduce the most basic biological investigations down to just a few of the essentials.

Regardless of whether you are interested in the biology of the oceans from the zoological perspective of a single species or the ecological perspective of entire communities of organisms, the essential measures are consistent. In very general terms, we are most commonly interested in three basic biological questions: 1) who is there, 2) how many are there, and 3) what are the indicators of organism or ecosystem function? The first two are easy—it's that last one that poses the most significant challenges.

Species Identification Is the Only Way to Describe the Diversity of Life in the Ocean

In the context of natural selection and evolution, if more than one species exist in a given habitat, we must assume that they possess enough differences in their ecological functionality to have evolved into different species in the first place. Why evolve into different species if you utilize all the same resources, occupy all the same habitats, and serve all the same functions? Therefore, the simultaneous presence of multiple species must be indicative of their biological differences. By simply identifying what species are present, we are much better equipped to investigate what those differences might be.

The simplest method of species identification is to assess **species richness**: a simple count of all the different species collected (or found) in the area of interest. Although it is always best to identify organisms to **genus** and **species**, such efforts can be time consuming and require a certain degree of expertise in **taxonomy**. A collection of organisms can still be categorized into different "species," even if the true taxonomic names are unknown. For example, you may not know the scientific names of an Atlantic flounder and a striped anchovy, but you could very easily determine that they are indeed two very different species of fish.

Species Abundance and Biomass Measures Are Used to Quantify the Amount of Biologic Matter

When assessing species abundance, there are two very basic strategies: either we can choose to perform counts of whole individuals or we can measure their **biomass** (and if the resources are available to do so, we might measure both). There is no universal rule as to when each strategy should be employed; however, the chosen expression of biological abundance should adhere to the context of the study. For instance, the results of a study focused on the disease rates of brown pelicans would be far more meaningful if we cited our results in terms of the number of diseased individuals per 1000 pelicans. However, if that same study sought to associate disease rates as a function of pelican body weight, we would be wise to collect the biomass data from those diseased and healthy pelicans.

When it is necessary to take counts of individual organisms, those data are most meaningful when the counts are related to the area (or volume) from which they were taken. For benthic organisms, areal abundances are the norm (number of individuals per square meter). For organisms surveyed within the water column, such counts are typically reported as volumetric abundances (number of individuals per square meter, or number of individuals per litre).

When it is necessary to gather biomass data, it is also important to consider whether such information will be collected from a **wet weight** versus **dry weight** perspective. The determination of wet weight is by far the simplest: the organism is simply blotted dry (or the surrounding water is eliminated by vacuum filtration) and weighed. Another advantage of assessing wet weight is that the organism can survive the procedure (if the investigator is both quick and gentle about it). Unfortunately, many aquatic organisms can retain a significant amount of fluid within their bodies, and the amount of fluid can be highly variable among different individuals of the same species; therefore, wet weights can be a misleading measure of the specimen's true biomass. A more accurate method of measuring true biomass is to first **desiccate** the specimen (in a 70°C drying oven) and then measure the organism's dry weight biomass.

Occasionally, it may be preferable to determine the dry organic weight of a specimen (as a measure of the specimen's total organic content). This is done by first determining the specimen's dry weight biomass, and then placing the specimen in a 500°C furnace to thoroughly combust all organic carbon to CO_2. The remaining ash is allowed to cool and then weighed, so that the difference between the pre- and post-combustion mass can be determined as the specimen's dry organic weight (or **ash-free dry weight**).

Species Biometrics Represent Those Measures that Suggest a Specific Biological Function

For most organismal studies, it is not enough to simply account for species richness and species abundance. For those investigations that seek to measure an organism's function (either as an individual or within the larger context of its role in the ecosystem), it is necessary that we decide on the appropriate **biometric** that represents the best anatomical or biochemical indicator(s) of that function. A fundamental assumption in biology is that all organisms exhibit adaptations that define their overall fitness for survival. If we seek to explore their unique function (that is, their unique fitness for competition and survival), we should be able to find some measurable feature that is indicative of their capacity for success within the ecosystem.

It is impossible to offer blanket guidance on an "appropriate" biometric for all investigations. The broad diversity of marine organisms, from bacteria to blue whales, compels us to consider a much larger panoply of biometrics with which to describe organism function. As a general rule, anatomical biometrics typically include measures of body or cell size and mass. Although these measures are largely regarded as "must-haves" for any investigation, they are also very general and may not be truly indicative of the specific function being investigated.

From a molecular perspective, it is even more difficult to suggest a universal biometric. However, one of the most fundamental molecular measurements, performed routinely in most oceanographic studies, is the assessment of *in situ* concentrations of **chlorophyll *a*** (chl *a*). Since chl *a* is the primary biochemical pigment that enables carbon fixation, it is found in every single organism capable of performing photosynthesis. Because of that universality, measures of chl *a* concentration are directly indicative of the photosynthetic biomass and the overall potential to add new biomass to the ecosystem, at the very base of the aquatic food web.

Supplemental Measures of Ocean Biology

With the immense diversity of life present in the world's oceans, it is impossible to issue a "one-size-fits-all" declaration of the important biological variables to consider. However, in addition to the essential measurements suggested previously in this section, the following can be used to build on those foundations, often as a simple augmentation to the field efforts in which we are already engaged.

Accessory Pigments Can Be Used To Identify Different Species of Plants and Algae

In addition to chl *a*, photosynthetic organisms also possess a tremendously diverse assortment of other photosynthetic and photoprotective pigments (such as chlorophyll *b*, beta-carotene, zeanthin, phycocyanin, etc.). Since many taxonomic groups can be identified by comparing the type (and cellular abundance) of these different pigments, it is possible to use pigment analysis to identify the taxonomic groups present and their relative abundances. Since photosynthetic organisms also possess the ability to alter their internal pigment ratios in response to the available light, detailed analyses of accessory pigments can also be used to document the photosynthetic status of primary producers.

Organic C-N-P Ratios Define the Nutritional Status of the Living Biomass

Although chl *a* and accessory pigments are only found in photosynthetic organisms, all the world's life-forms are made from the elements of carbon, nitrogen, and phosphorus. Organisms growing at peak nutritional efficiency will typically accumulate organic material with a C–N–P ratio of 106:16:1, also known as the **Redfield ratio**. When organisms are surviving in nutrient-stressed conditions, they will accumulate C–N–P in very different ratios. By assessing the organic C–N–P content of a particular organism (or group of organisms), it is possible to determine the nutritional conditions under which they have been living—information that can have rather critical implications for the transfer of those nutrients (and food energy) to higher-level consumers in the food web.

Age Structure and Size Class Measures Establish the Demographics within a Specific Population

Many of the ocean's creatures begin their lives as microscopic larvae and grow, sometimes to remarkable size, according to very regular developmental stages. Therefore, it is often possible to determine the breeding histories of individual species based on the relative abundance of specimens collected in each size class (**Figure 4.14**). For many species, the time needed to develop into each particular life stage is well-known, so the size classes can be easily correlated with the age structure of the population. Such information can be particularly valuable for stock assessments of ecologically and commercially important species.

Geological Variables

The pertinent spatial and temporal scales in geological oceanography can run the gamut and easily stretch to the extremes. After all, many geological investigations deal with entire ocean basins or entire continental margins, and for many geologic processes, we measure the passage of time in millions

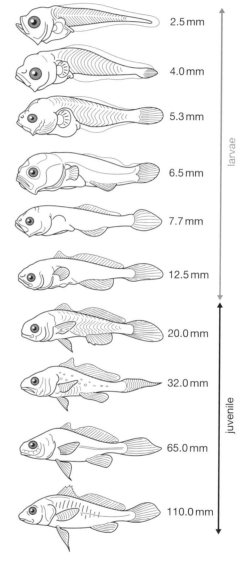

Figure 4.14 Developmental stages and size classes of the Atlantic croaker, *Micropogonias undulatus*. Since many marine species possess multiple larval and juvenile stages of growth that are highly time dependent, this information can be used to back-calculate breeding periods and to estimate maturation rates (age structure) of a given population, based on life-stage collections performed in the field. (Adapted from, Johnson GD [1978] U.S. Department of the Interior).

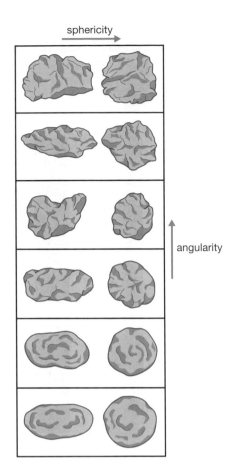

sphericity

angularity

Figure 4.17 Sketches of microscopic analysis of both the angularity and sphericity of sediment grains within each layer of a sediment core can yield a tremendous amount of information about the age of the sediments, as well as the erosional intensity prior to deposition. (Adapted from Powers, MC [1953] *Journal of Sedimentary Petrology*, 23(2):118.)

Barometric Pressure Can Affect the Hydrostatic Pressure of Ocean Water Beneath It

Like wind speed and direction, barometric pressures are routinely measured at weather stations located at sea and along the coast in virtually every location on the globe. If high-precision calculations of hydrostatic pressure are required, it will be necessary to use barometric pressure data to correct for any departures from the "standard atmospheric pressure" of 101.325 kPa, which is assumed to be constant to establish a hydrostatic pressure of $P = 0$ for the Equation of State of Seawater (IES 80; see Technical Box 4.1)

Air Temperature Can Provide a Delayed Influence on Water Temperature

Like wind velocity and barometric pressure, air temperatures are also routine measures at coastal and oceanic weather stations. Because seawater is slow to gain and lose heat, the effects of air temperature on ocean dynamics can largely be ignored if water temperatures are known. However, changes in air temperature can have a significant effect at the air–sea interface and lead to heating or cooling of the uppermost layers of the ocean (which can lead to increased mixing and overturn). Certainly, extreme seasonal variations (or transient temperature fronts) can lead to significant heating or cooling of surface waters.

Precipitation Will Affect Both the Temperature and Salinity of Surface Water in the Ocean

Although the quantity (and type) of precipitation is routinely monitored at land-based weather stations, it is seldom monitored at offshore stations. Depending on the nature of the study, precipitation can have a significant impact on surface salinities and may even increase the surface concentrations of certain nutrients and salts that are scoured from the atmosphere and delivered to the ocean surface during precipitation events.

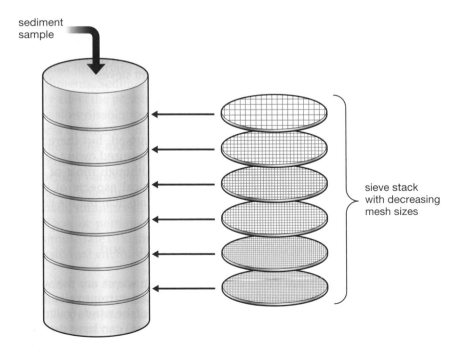

sediment sample

sieve stack with decreasing mesh sizes

Figure 4.18 A series of sieves with progressively smaller mesh sizes, commonly used to sort sediments according to their size class.

References

Bakus GJ (2007) Quantitative Analysis of Marine Biological Communities: Field Biology and Environment. John Wiley and Sons.

Knauss JA (1997) Introduction to Physical Oceanography, 2nd ed. Prentice-Hall.

Omori M & Ikeda T (1992) Methods in Marine Zooplankton Ecology. Kreiger Publishing Company.

Parsons TR, Takahashi M, & Hargrave B (1984) Biological Oceanographic Processes, 3rd ed. Pergamon Press.

Pilson MEQ (1998) An Introduction to the Chemistry of the Sea. Prentice-Hall.

Pond S & Pickard GL (1983) Introductory Dynamical Oceanography, 2nd ed. Pergamon Press,.

Further Reading

Diemer FP, Vernberg FJ, & Mirkes DZ (eds) (1980) Advanced Concepts in Ocean Measurements for Marine Biology. University of South Carolina Press.

Dyer KR (ed) (1979) Estuarine Hydrography and Sedimentation: A Handbook. Cambridge University Press.

Head PC (2011) Practical Estuarine Chemistry: A Handbook. Cambridge University Press.

Kennish MJ (2000) Practical Handbook of Marine Science, 3rd ed. CRC Press

Knödel K, Lange G, & Voigt H-J (2007) Environmental Geology: Handbook of Field Methods and Case Studies. Springer-Verlag.

Mudroch A & MacKnight SD (eds) (1994) Handbook of Techniques for Aquatic Sediments Sampling, 2nd ed. CRC Press.

Wangersky PJ (2000) Marine Chemistry (The Handbook of Environmental Chemistry). Springer-Verlag.

Chapter 5

Common Hydrologic Census Methods

"Water is the driving force of all nature." – Leonardo Da Vinci

Within the aquatic sciences, it is nearly impossible (in practice or in concept) to separate natural waters from the myriad of dissolved and suspended constituents therein. Therefore, the aquatic sciences are as much about the study of water as everything else. So it should come as no surprise that our consideration of the fluid medium—the study of water itself—should be the foundation on which we must build our field research program.

But there is far more to this task than simply following a prescribed analytical methodology. The very best methods, like the very best equipment, can be rendered impotent if our sampling procedures were inherently flawed, or if our samples were inexpertly collected. In truth, the careful planning and execution of our sampling plan is perhaps the most critical aspect of our analyses—more so than the latest gizmos or the most advanced laboratory techniques.

So when it comes to hydrologic census techniques, deciding when, where, and how to collect the most pertinent hydrologic data from the field will go a long way to guaranteeing solid experimental results. After all, little can be wrought from a poor sampling plan, except poor-quality data.

Key Concepts

- Hydrology is different from hydrography in that it focuses exclusively on the intrinsic qualities of natural waters.
- The periodicity of hydrologic data collection will determine what analyses are possible and which time-sensitive dynamics are captured in the dataset.
- The number, location, and orientation of the chosen sampling sites will define the horizontal and vertical gradients of all hydrologic parameters.
- The methods chosen for measuring the dissolved and suspended constituents of natural waters will require a combination of colorimetric, titrimetric, and electronic analyses.

The Difference Between Hydrology and Hydrography

As an integrative discipline, the marine sciences require that we know a little something about both the hydrology and hydrography of the ocean. **Hydrology** is quite literally the "study of water" that focuses on the movement and distribution of water, as well as the composition of the aquatic medium and all its constituents. In natural waters, hydrology includes much more than just the water molecules themselves—it also includes the quantitative and qualitative measurement of all the other materials suspended or dissolved in that same water (more generally regarded as **water quality**). For this reason, hydrologic parameters (especially those that affect water quality) are considered to be **intrinsic** to the water; in other words, they are qualities that "belong" to the water. For example, the salinity of seawater is an excellent example of a quantitative hydrologic measurement, because it defines the total dissolved salt concentration within the water.

This is different from **hydrography**, which focuses instead on the geophysical parameters that define a particular body of water, usually from the perspective of global positioning, navigability, and shipping safety (such as coastline geometry, bathymetry, tides, and currents). These sorts of parameters are not a natural part of the water; they are actually defined by factors external to the water. For example, if we wished to define the depth of the water column in a particular location, we could not accomplish this without also defining our position and perhaps the tides and currents affecting the height of the water column at the instant we wished to make that measurement. All of these features provide an external influence on our measurements; thus, hydrographic parameters such as these are considered to be **extrinsic** to the water.

As you can well imagine, hydrography and hydrology are both critical to the study of marine science, but they are not necessarily equal in their importance. Typically, hydrographic parameters are measured in order to define the dynamic range and context of the pertinent external influences on our efforts in the field. However, the hydrologic parameters are most often viewed as the most powerful and most informative analyses that can be conducted in marine science. This is because the hydrologic properties of water are inherent to the water, so they come with the water and will not change, even if we remove the water from its original place. Hydrographic properties are just the opposite. Since they are dependent on external forces, they will change (or be impossible to measure) when the water is removed from its natural place of origin.

When to Measure Hydrologic Data

Deciding just how often you must collect samples is much more difficult than you would think. From a theoretical perspective, the more often you collect a water sample (and perform a measurement), the better your chances of capturing rare events and time-sensitive dynamics. Of course, the more often you collect a water sample, the quicker you exhaust your supplies, storage space, handling time, and crew morale! No matter what, **periodicity** of sampling swiftly becomes an issue with which you must deal.

In some cases, you might be able to get away with storing your samples for months on end, but if you must perform your analyses immediately (or with a 24- to 48-hour turnaround, which is much more typical), you will absolutely need to determine the best strategy to not only collect but also analyze all of your samples. If you have access to automated and/or electronic equipment that can make rapid measurements and store the results in a digital format, **continuous sampling** is certainly the way to go. However, if your measurement and analysis methods are decidedly more low tech, you will be forced to adopt a **periodic sampling** strategy.

Continuous Sampling Methods Are Best When Measuring Rapidly Changing Hydrologic Variables

Owing to critical technological advancements in the past two decades of ocean research, many of the most commonly measured hydrologic parameters (salinity, temperature, pH, dissolved oxygen, and chlorophyll *a* (chl *a*), for example) can now be measured using differential optical and/or electrical resistivity. Since these electrical responses occur, and can be accurately measured, in mere microseconds, what used to take minutes to measure can now be measured several times each and every second! This has led to the development of near-continuous oceanographic sensors that are able to take hundreds of nearly instantaneous measurements and transmit those data to a laptop or flash memory device for easy data storage.

Because of the tremendously swift response time of these sensors, the data they collect are only useful if they are exposed to different water masses over the duration of their deployment. After all, why take thousands upon thousands of measurements of an unchanging water mass when just a few, redundant measurements would suffice? Thus, hydrologic sensors are best used on a platform that is guaranteed to be exposed to different water masses, like on a moving vessel or on a buoy affected by changing currents and the tides. Although expensive, these systems are the backbone of modern oceanographic research and are typically employed either in an **underway system** or a **moored system** for continuous sampling.

Underway Systems Provide Continuous Sampling During Transit Between Multiple Locations

These are typically integrated into a seagoing vessel, often as a self-contained plumbing system, external to the ships systems (to avoid sample contamination). An intake port is usually oriented at the bow of the vessel, whereby ocean water is forced into the intake port by the vessel's forward progress. As the water is forced through the plumbing system (**Figure 5.1**), its hydrologic parameters are continuously measured by an oceanographic sensor

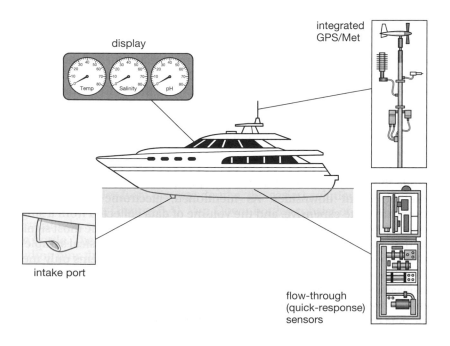

Figure 5.1 An underway system, which allows rapid measurement of important hydrologic parameters on a near-continuous basis. As water is forced into the intake vent at the ship's bow, it travels downpipe to integrated oceanographic sensors that make rapid measurements and store the data prior to the water exiting through the excurrent vent at the ship's stern. Underway data are easily displayed to the ship's captain or research team for real-time visualization of critical oceanographic variables.

Figure 5.4 A Van Dorn bottle, often used to collect water samples from discrete depths when only a few samples are needed (or when field conditions will not allow the use of powered oceanographic sensors). The Van Dorn bottle is first loaded with the end seals held open with a tension rod, and then lowered to the desired depth. A heavy brass weight (called a "messenger") is sent down the line to trip the release mechanism, which causes the end seals to snap shut and form a watertight seal, capturing a fixed volume of water within the Van Dorn bottle. When it is hauled back aboard the ship, the drain valve is opened to decant the water sample collected at depth.

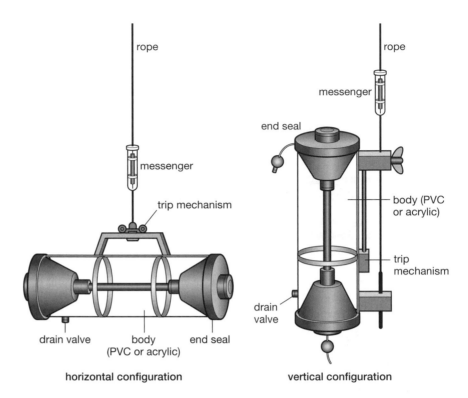

horizontal configuration   vertical configuration

messenger will trip the bottle closed and seal the water sample inside the bottle (which can then be hauled up and placed into storage bottles for preservation or testing).

To collect a large number of water samples, it is often necessary to use powered oceanographic sensors, which can be integrated with a **rosette** of several large collection vessels (**Figure 5.5**). The principle of collecting water samples from discrete depths is the same; however, for most rosette systems,

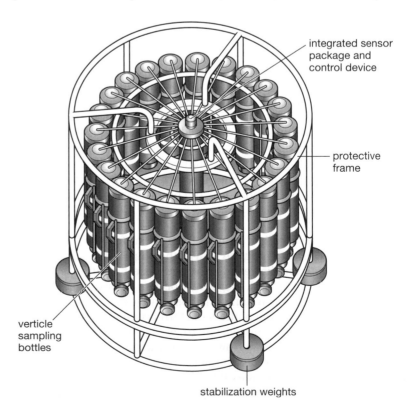

Figure 5.5 A rosette of several large sample bottles, integrated with a CTD sensor package, and housed within a protective metal cage to prevent damage to the sensor package. Each vertical bottle within the rosette can be triggered separately via an electrical current that is delivered through a conducting cable, allowing the collection of several different water samples (from several different depths) in a single deployment.

the bottles are triggered remotely by radio transmission or by electrical signal, sent through a conducting cable. Such systems are typically deployed with a depth sensor that can relay the actual depth of the rosette back to a shipboard observer who lowers the rosette to the desired depth. When the rosette is in position, a simple press of a button will send an electrical signal to the rosette, causing a specific bottle to "fire."

Because the collection of water samples takes a significant amount of time, it is impossible to sample continuously in this fashion. Instead, the physical collection of samples must be conducted periodically. Even if oceanographic sensors are continuously recording measurements while the sampling apparatus is being lowered, only those measurements that were recorded at the moment the bottle was "tripped" can be associated with the water sample.

Whether water samples are first collected with a bucket, from a hand-lowered Van Dorn bottle, or from an electronically controlled rosette, it is critical that all water samples are handled correctly, stored in an appropriate container, and properly preserved to minimize contamination and to protect against sample loss. In most cases, the handling and storage of water samples should adhere to the following general methodology:

1. Sample collection and storage bottles should be made from rugged, nonreactant materials (for example, borosilicate glass, HDPE, or Nalgene™).

2. All collection bottles and lids should be soaked in an acid bath (10% HCl) for a minimum of 15 minutes, thoroughly rinsed with distilled or deionized water, and protected from contamination until use.

3. Each collection bottle should be rinsed twice with the sample water before collecting the final sample.

4. All collected samples should be stored at 4°C and out of direct sunlight. Light-sensitive samples should be stored in opaque bottles in complete darkness.

5. It is best to perform analyses within 12–24 hours post-collection.

Of course, the nature of the research project will dictate the specific method of collection, storage, and analysis for your water samples. But regardless of whether you have chosen to collect sensor-derived measurements on a continuous basis, or instead to collect periodic discrete samples, it is not enough to simply consider the timing of your collections and be done with it—you must also determine where you wish to collect your samples.

Where to Collect Hydrologic Data

Depending on the focus of your field research, you should already have a basic idea of the local or regional areas where you shall focus your attention. Thus, in a very general sense, deciding on your general area of interest should be relatively straightforward. However, when it comes down to pinpointing a number of very specific locations within that general area of study, you will be required to invest a considerable amount of forethought into the process.

Proper Site Selection is the Key to Success for Any Sampling Strategy

For all aquatic field studies, there are a few general considerations to keep in mind whenever you engage in the process of selecting your study sites. Although the goals of your specific research project should drive the decision-making process, you should always keep the following at the forefront of your mind when deciding where your study sites should be located:

- Greatest likelihood to meet research objectives
- Indicative of control versus experimental conditions
- Easily accessed for repeat sampling
- Broadly representative of the water body

Whatever the goals of your field study, the sites you select should offer the greatest likelihood that you will be able to collect the data you require to meet your research objectives. To that end, it is important to select a variety of sites that can serve as internal comparisons: some of which can be deemed "experimental" sites, while others can serve as baseline or "control" sites. It is also critical that you are able to easily access each of your study sites whenever your research plan requires a field measurement.

Your sites should also be located where the measurements being conducted will yield data that are representative of the water body where those sites are located. For example, if you wished to establish a coastal study site to represent a "typical" undeveloped coastal area, it would be unwise to locate your site near the outfall of a storm sewer.

With these points in mind, you must now take the time to consider the specific aquatic environment you wish to study, and how the unique features of that water body will inform your decisions. The proper placement of your study site(s) is critical to the success of your field research, so don't just wing it—take the time to get it right.

Fluvial Study Sites Are Those Located Along Streams and Rivers

The site selection process for **fluvial** (riverine) studies is among the simplest of aquatic field studies because of the very obvious connectivity and flow pattern among streams and rivers within the **watershed**. Despite the winding paths of the many tributaries within your general study area, it is relatively easy to map out the **stream order** in an effort to explicitly define how all of the streams and rivers are connected to each other in a hierarchical pattern (**Figure 5.6**).

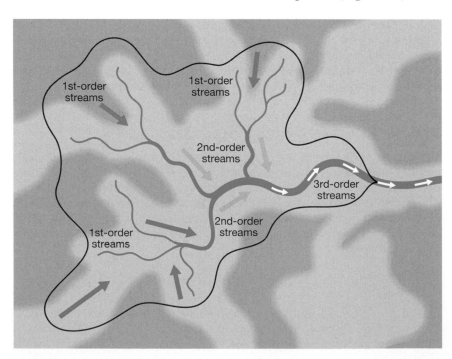

Figure 5.6 An idealized watershed, indicating the connectivity of all associated streams and rivers, as well as the streamflow. This connectivity can be further defined by using a "ranking" process, or stream order, to elucidate exactly how the all the tributaries are connected to each other.

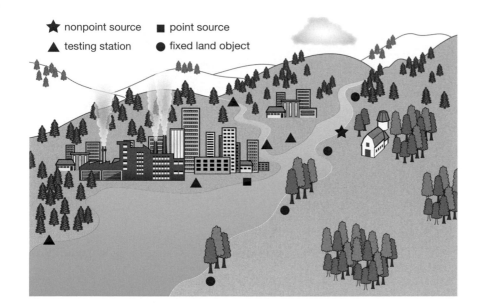

Figure 5.7 An example of a fluvial study region, appropriately stream-ordered and with testing stations located up- and downstream in each tributary to capture the potential effects of various point and nonpoint sources of pollution. Although it is always preferable to use GPS coordinates to pinpoint the location of each study site, the low-tech method of using fixed land objects for positioning is a viable backup (Adapted from Campbell G & Wildberger S [2001] The Monitor's Handbook, 2nd ed. LaMotte Company.)

If the streams are relatively shallow, or if the rivers are well mixed, it should make no difference whether we collect our data from the surface or the bottom of the stream; nor should it make a difference whether we collect our data from either shore, or even from the middle of the tributary. If this is the case, you can treat each tributary as a one-dimensional waterway, locating your study sites according to stream order and relative to each other as being some distance up- or downstream from a neighboring site (**Figure 5.7**).

For most fluvial studies, it is always a good idea to locate a site at the "mouth" of each tributary, with at least one additional site some distance upstream. If there are other points of interest along the route of the tributary (for example, sites of **point** and **nonpoint sources** of pollution outfall), you should sandwich it between a pair of up- and downstream sites as well.

Lacustrine and Estuarine Study Sites Are Those Located in Bays and Inland Coastal Areas

Performing field studies within a large body of water, such as a lake or coastal embayment, can be much more complicated than an idealized fluvial study. Freshwater lake (or **lacustrine**) studies are similar to those involving coastal bays in that they both exhibit complex, three-dimensional variability and are more of a logistical challenge simply due to their size. Choosing study sites within bay (or **estuarine**) studies can be even more complicated than lacustrine studies because bays are fed not only by freshwater tributaries, but also by marine water encroachment due to tidal influences.

Because of these factors, it can be quite difficult to affirmatively establish which locations within a lake or bay will yield measurements that are most representative of the entire system. This will make it necessary to establish several different study sites within each lake or bay, taking great care to provide ample geographic coverage (**Figure 5.8**).

As a general rule, a study site should be located at each major inlet and outlet. If the inlets and outlets are themselves quite large, it may be necessary to establish both a nearshore and mid-water station at each inlet and outlet. Likewise, a mixture of nearshore and mid-water stations should be located throughout the larger body of the lake or bay. If possible, stations should be oriented according to some kind of geometric pattern that can be used to establish regular transects along or across the major dimension of the water

Figure 5.8 An example of a lacustrine or estuarine study region, showing a variety of sampling stations that would provide excellent geographic coverage of an irregularly shaped lake or bay. Note that the spatial coverage is augmented by including a number of nearshore and mid-water stations throughout the entire system.

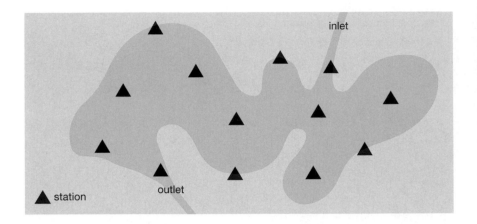

body. Transects can be extremely useful in establishing ecological **gradients** within complex systems.

Marine Study Sites Are Those Located Along the Coast or Out to Sea

In some ways, deciding the location of marine study sites is an easier task because marine waters are less constrained by land features. In fact, the only geographic feature to consider in nearshore studies is the coastline; for open ocean studies, there are no land masses to contend with. Although this may grant you the ultimate flexibility in positioning your study sites, it also presents a unique problem: How do you choose a location that will maintain its "representativeness" if the water at each station is free to flow in or out, in any direction, at any time?

If you are attempting a Lagrangian study, it is not necessary to define a fixed study site. By definition, all of the measurements in a Lagrangian study are taken from within the same water mass, no matter where that water mass happens to travel—the ultimate in "representativeness." But for Eulerian studies, it is necessary to choose fixed positions for repeat sampling (typically done using high-precision GPS coordinates). In such studies, the sites are not meant to be representative of the ocean per se; they are meant to be representative of the dynamics experienced at those locations over time.

In most marine studies, study sites are located along specific transect lines according to some fixed geographic pattern (for example, along the same latitudinal or longitudinal lines, separated from each other by some equal distance). The number and orientation of those transect lines depend largely on the objectives of the research. For those studies that anticipate strong coastal or shelf influences, the transect lines are usually oriented perpendicular to the coast in order to determine **cross-shelf** gradients. If multiple transects are used, data from stations located along the same **isobath** can be used to determine **along-shelf** gradients as well (**Figure 5.9**). For those studies that anticipate a dilution or "spreading" effect (like the simultaneous propagation of some event, extending in multiple directions from the location of the event), transect lines are oriented in a radial pattern instead.

Don't Forget the Vertical Dimension When Planning Your Study Sites

With the exception of most fluvial studies, it will be important to define the vertical dimension of your sampling strategy at each study site. Although each study site is chosen largely on the basis of geographic information, the three-dimensional structure of the aquatic medium often requires that we take samples from multiple depths at each study site—only then can we hope to describe the three-dimensional variability that makes aquatic field research such a challenging endeavor.

Although the primary objectives of your field research will dictate your vertical sampling regime, standard practice within the aquatic sciences is to collect measurements from both the surface and bottom (or within 0.5 m of the bottom—the closer, the better). Surface and bottom water sampling should be regarded as the bare minimum; very rarely is it justified to collect measurements from a single depth (unless you are certain the water column is well mixed).

Of course, increasing the number of samples taken in the vertical dimension will enable a much more realistic description of the true vertical structure of the water column. If the bathymetry is known at each station, it is advisable to take a mid-depth sample at each station as well. If the main objective of the research is to define the vertical structure of the water column, it is not unusual to collect data from several discrete depths at each station. If this is done, it shall be important to define the "rules" for sampling at multiple depths. For example, you may choose to collect measurements at every 10-m depth increment. Another strategy often utilized is to sample at depth intervals that are proportional to the overall depth at each station. For example, if you chose to collect water samples at 0/25/50/75/100% depth intervals, a station with an overall depth of 30 m would indicate that samples be taken at 0 m, 7.5 m, 15 m, 22.5 m, and 30 m. Breaking the vertical dimension into different depth layers and collecting samples from each of those layers is a type of **stratified sampling** (Figure 5.10). In the aquatic sciences, this method is absolutely essential for describing the hydrologic features of discrete water masses in the vertical dimension.

(A)

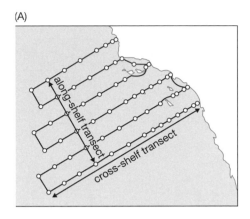

(B)

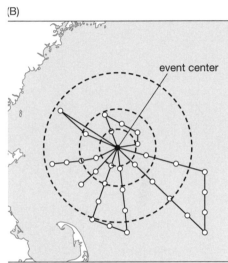

Figure 5.9 (A) For marine studies that focus on coastal influences, it is typical to establish a number of study sites along transects that are oriented both parallel (along-shelf) and perpendicular (cross-shelf) to the coastline. (B) For marine studies that focus instead on the influence of a localized and unique event (such as an oil spill), the transects are oriented in a radial pattern, extending outward from the "epicenter" of the event.

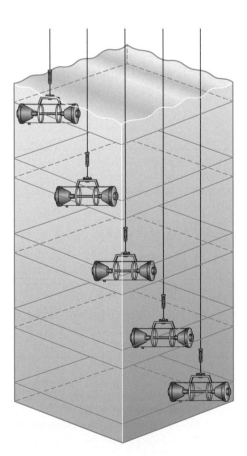

Figure 5.10 Any water column can be subdivided into different depth layers. Water samples collected from each of these layers (using a Van Dorn sampler, or similar device) can then be analyzed separately to define the hydrologic parameters for each of the discrete water masses collected.

How to Measure Hydrologic Data

For the most common hydrologic parameters, the equipment and methods used to collect those measurements fall into one of three main categories: colorimetric, titrimetric, and electronic analysis. If a particular hydrologic parameter can be measured by different methodologies, you should choose the method that provides the greatest precision and accuracy, whenever possible.

Colorimetric Analysis Uses Color as a Method to Quantify Hydrologic Parameters

Occasionally, there are dissolved or suspended constituents in seawater that can be analyzed quantitatively by treating those constituents with specific chemicals that cause a unique reaction, thereby producing a colored solution. Since all colors can be defined by their unique wavelength, the development of different colors can be measured by instruments that are extremely sensitive to different wavelengths of light. Not only can these instruments (called spectrophotometers) measure the specific wavelength of each unique color, they can also determine the relative intensity of those colors. In **colorimetric** analyses, the color and intensity can be used to identify specific hydrologic parameters and determine their relative abundance (concentration) within the sample (**Figure 5.11**).

Some constituents of seawater (like chlorophyll *a*) exhibit **fluorescence**, a phenomenon that causes the spontaneous emission of light of a specific intensity and wavelength when doused with higher-energy light at a special "excitation" wavelength. If the colorimetric device being used is sensitive to fluorescence, the intensity of fluorescence can be used to determine the concentration of the fluorescing constituents in seawater. In fact, this colorimetric method is one of the most important and heavily utilized methods in the aquatic sciences. Since photosynthesis requires chl *a* it is possible to directly measure chl *a* concentrations by simply measuring the intensity of chl *a* fluorescence in a water sample.

Although not as sensitive as a spectrophotometer, the human eye is also capable of discriminating between color tone and intensity. Thus, simple

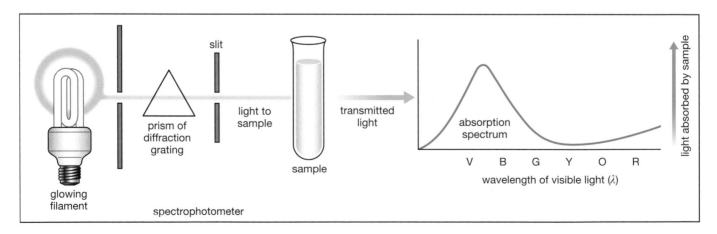

Figure 5.11 The most common application of colorimetric analysis uses a light-sensitive device, called a spectrophotometer. Light of a specific wavelength and intensity is generated by the lamp and passed through the water sample. As different wavelengths λ of light are transmitted through (or absorbed by) the constituents in the water sample, the unique color (and intensity of color) can be measured as an absorption spectrum, which can then be used for quantitative analysis of many different hydrologic parameters.

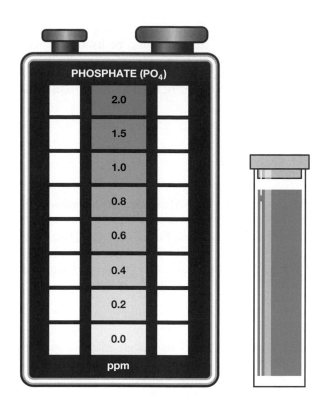

Figure 5.12 Colorimetric test kits require either a few drops of water placed upon a test strip, or a small volume of water in a sealed test tube, combined with colorimetric reagents. When a positive reaction occurs, the intensity of color will indicate the relative concentration of the hydrologic parameter being tested. Most test kits measure chemical concentrations in parts per million (ppm), which is equivalent to mg/kg.

colorimetry test kits can be used in the field for rapid detection and quantification of seawater constituents (**Figure 5.12**).

Titrimetric Analysis Quantifies Hydrologic Parameters by Inducing a Specific Chemical Reaction

A variation of the colorimetric method is the **titrimetric** method of analysis, which involves the addition of a chemical solution of known concentration to induce a chemical reaction with a specific constituent of seawater. If the constituent is absent, the reaction does not take place and there are no measureable results. If the constituent is present, it will react in direct proportion to its concentration in the water sample being tested.

Depending on the titrimetric method being used, a colored product may develop (in which case, a colorimetric method could be used to determine its concentration). Occasionally, solid crystals may form as a result of the reaction. If those crystals are collected and weighed (a type of **gravimetric** analysis), the data can be used to back-calculate the concentration of the constituent in seawater that caused the crystals to develop.

Electronic Analysis Measures Small Changes in Conductance to Quantify Hydrologic Parameters

By far, the preferred analytical method for most hydrologic parameters involves the use of highly sensitive electronic sensors that have been engineered to respond only to a very specific constituent in water. If that particular constituent is present, it will induce an electrical impulse in the device that is proportional to its concentration in seawater: the stronger the concentration, the stronger the impulse. As these instruments can detect electrical changes, measured as a thousandth of a volt in just a fraction of a second, they provide the most rapid, accurate, and precise measurements obtainable by modern methods.

Of course, such instruments can be quite expensive and require both a power source and a data storage solution. Many of these systems can be powered by an onboard battery pack, or by a ship's generator and an attached electrical cable. The same electrical cable can also be used to transfer data from the sensors, uploaded to a weatherproof laptop computer or hard drive connected to the cable. To ensure against data loss, many sensors also have self-contained data storage capability, usually as a simple flash drive or memory card.

References

American Public Health Association (APHA), American Water Works Association (AWWA), and Water Pollution Control Federation (WPCF) (1989) Standard Methods for the Examination of Water and Wastewater. American Public Health Association.

Campbell G & Wildberger S (2001) The Monitor's Handbook, 2nd ed. LaMotte Company.

Cross FL Jr (1974) Handbook on Environmental Monitoring. Technomic Publishing Company.

Grasshoff K, Ehrhardt M, & Kremling K (eds) (1983) Methods of Seawater Analysis: 2nd rev & ext ed. Verlag-Chemie.

Lind OT (1979) Handbook of Common Methods in Limnology, 2nd ed. C.V. Mosby Company.

Further Reading

de Jong CD, Lachapelle G, Elema IA, & Skone S (2006) Hydrography. Delft Academic Press (VSSD).

Dyer KR (ed) (1979) Estuarine Hydrography and Sedimentation: A Handbook. Cambridge University Press.

Hailwood EA & Kidd R (eds) (2013) Marine Geological Surveying and Sampling. Kluwer Academic Publishers.

McCuen RH (2004) Hydrologic Analysis and Design, 3rd ed. Prentice Hall.

Renn CE (1970) Investigating Water Problems: A Water Analysis Manual. LaMotte Company.

Strickland JDH & Parsons TR (1965) A Manual of Sea Water Analysis. Fisheries Research Board of Canada.

Chapter 6

Census Methods for Benthic Organisms

"Consider what each soil will bear, and what each refuses." – Virgil

Unlike hydrologic data, the information we seek to gather from living organisms is far more ephemeral. After all, we expect organisms to exhibit variable features, even among members of the same species. Virtually every conceivable measure of an organism's features—its body shape, its coloration, its biomass, or even its gender—will be different from its conspecific fellows. And while variety may be the spice of life, it certainly makes it more difficult for us to assess what is a "typical" biometric for that species.

Fortunately for us, a wide variety of census techniques are available for the assessment of highly variable biological communities. And among the biotic habits of organisms found in the sea, those that make their home on or in the bottom are the easiest to measure, largely because we can ignore the vast volumes of the ocean and restrict our attention to the two-dimensional world of the bottom-dwellers. That is good news, because it is always easier to analyze complexity in two dimensions rather than three. So, as long as our study subjects stay on the bottom we can use the classic terrestrial census techniques to assess population density, frequency of occurrence, percent occupation of the habitat, or virtually any other biometric we may wish to quantify.

Key Concepts

- The census methods used for benthic aquatic organisms are similar to those used in classic terrestrial ecology assessments.

- The biotic habit of the target species will define which census method is most appropriate for use.

- If a particular plot or plotless method is indicated for sessile benthic species, the same method can be used to assess motile benthic species as well; however, the corollary may not be true.

- Benthic species that can freely migrate into or out from a study area are said to be from "open" populations and may require highly specialized census techniques to estimate their true population densities.

Choosing the Right Method for Each Biotic Habit

When deciding on the most appropriate census method for a particular biological community, it is critical that you first consider the **biotic habit** of the marine organism(s) you wish to study (**Figure 6.1**). Since a biotic habit is essentially an organism's preferred lifestyle, it only makes sense that we would have better success studying an organism "where it lives."

If an organism in the ocean is powerless against the currents and will float wherever the ocean takes it, it is considered to be **plankton**. Although plankton can be found floating and/or drifting anywhere in the water column, those plankton that can only be found at the surface belong to a special class of organisms called **neuston**. If an organism is powerful enough to swim against the currents of the ocean, it is classified as **nekton**. Because the plankton and nekton can be found anywhere (and everywhere) within the water column, we are usually forced to adopt more complex, three-dimensional census methods when surveying these kinds of **pelagic** organisms (which we shall discuss in the following chapter).

Of course, there are many marine creatures that neither swim nor float. Those organisms that are associated with the bottom are considered to be **benthic** in their biotic habit; that is, they are part of the **benthos**. Within the larger context of the benthos, the **epifauna** are those animals that make their living upon the substrate, while the **infauna** are those that live within the substrate itself.

Although all of the pelagic organisms can be assumed to be **motile** (either carried by the currents or swimming on their own), this is not necessarily the case among benthic organisms. Certainly, there are many benthic organisms that are also motile (that is, free to move about and within the substrate). However, one of the most significant consequences of the benthic habit is that it allows certain species to remain stationery, either attached to or buried within the substrate. Hence, those creatures that move extremely slowly (or not at all) are considered instead to be nonmotile, or **sessile**.

In general, benthic organisms are much more easily surveyed because we know exactly where they can be found: somewhere on (or in) the bottom. That means we can utilize the more traditional, two-dimensional census

Figure 6.1 The classic paradigm of benthic and pelagic habits is depicted here. Within the pelagic environment, organisms that are powerful enough to swim against the currents are categorized as nekton, while those powerless to resist water movement are categorized as plankton (with the neuston constrained to the surface). Within the benthic environment, animals that live upon the substrate are categorized as the epifauna, while those that live buried (or encapsulated) within the substrate are called the infauna.

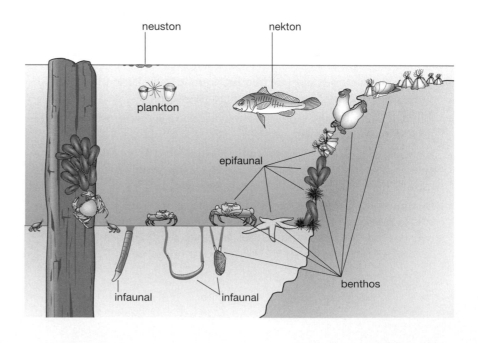

methods used in terrestrial ecology—adapted for use in an aquatic environment. But unlike the pelagic organisms (which are all motile), you must take great care in choosing the most appropriate census method for benthic organisms, particularly if the community you wish to investigate has a complicated mixture of both motile and sessile forms.

Starting Simple Is the Key to Every Biological Census

As a field researcher, you should never feel limited by what others have done in the past, or by what is considered to be a "typical" census method. By the same token, it is often useful to consider such things and to seek guidance from other investigations as inspiration for your own. In most cases, biological surveys seek to accomplish one or more of the following assessments in an ecological context:

1. Population density (number of individuals per unit area)
2. Frequency of occurrence (relative abundance)
3. Live coverage (area occupied per unit area of available habitat)
4. Living biomass (or other relevant biometric, such as body length)

Although these sorts of data may seem a bit routine or perhaps simplistic, it would not be possible to perform any of the higher-order assessments of community dynamics without them. So for now, let us focus on the fundamentals—we will have ample time to explore how these data can be used in the analyses of single- and multivariate dynamics in later chapters.

Take Note Whether the Census Includes Open or Closed Populations

Because benthic populations can have a mixture of motile and sessile forms, it is often a challenge to account for population gains and losses over the period of investigation. Not surprisingly, an organism's capability for reproduction may cause the population size within the area of study to increase specifically as a function of **natality** and/or larval **recruitment**. Conversely, the population size may decrease as a result of **mortality**. Each of these represent an inherent change to the overall population size, in the sense that reproductive additions to the population size (or growth additions to the population biomass) are determined by an organism's intrinsic life cycle and can only be "undone" by the organism's death.

However, there can also be changes to the population size (or biomass) that are caused by transient migration events. Any organism migrating into the study area from another location will of course represent an increase to the population size due to **immigration**; likewise, any organism migrating out of the study area will cause the population size to decrease as a result of **emigration**. In either case, the true population size is not affected at all, but there will be an apparent change to the population size within the area of study depending on the migration rates into or out of the confines of your census area (**Figure 6.2**).

Although one might expect a natural population to always fluctuate as a result of reproduction, growth, and mortality, field researchers must consider whether motile species are free to immigrate into (or emigrate from) the study site—a truly "open" population. Sessile species will also be affected by reproduction, growth, and mortality, but they can usually be considered to belong to a "closed" population because their numbers are largely unaffected by migration. That being said, it is important to keep in mind that there are very few populations that are truly closed, because most aquatic species spend the early phases of their life cycle

Figure 6.2 Changes in population size will typically increase as a result of reproductive gains (as natality or recruitment) or immigration into a fixed study area. Decreases in population size will result from mortality losses and/or emigration out of the study area. Although all organisms are affected by natality and mortality, only the motile species are affected by migration.

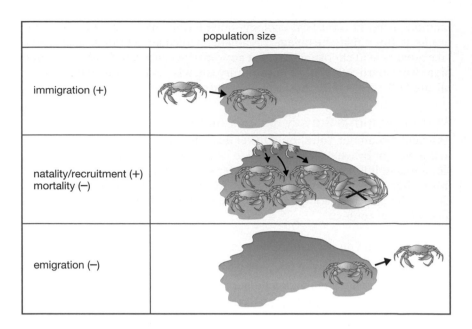

as planktonic larvae (so there is always some capacity for drifting larvae to migrate).

The duration of the field study can also have an influence on whether the population(s) are considered to be open or closed. If a field survey is conducted over a long enough time period that the organisms under investigation become subject to changing climactic and/or breeding cycles, the population must be assumed to be open. Similarly, slow-moving motile species can be considered to be closed populations, so long as the field study is conducted over such a brief period of time that migration effects are eliminated. Although there are ways to compensate for these effects, it is nonetheless important for the field researcher to consider whether the population is open or closed when deciding what census technique to use.

Benthic Organisms Require a Two-Dimensional Census Method

Because benthic organisms must be located on (or in) the substrate, it is almost always appropriate to think in terms of population density, live coverage, or biomass relative to some unit area rather than by volume. In a very practical sense, it is far easier to conduct the general plot and plotless methods discussed earlier when you're dealing with two spatial dimensions rather than three.

Although the bottom is never truly "flat," in areas where the bottom has very limited topographic relief, it is perfectly acceptable to assume planar geometries when surveying the epifauna, which can easily be accomplished with a plot method utilizing rigid quadrats. However, for those epifauna living upon highly irregular surfaces, it is more appropriate to use a survey method utilizing a line or belt transect that can be draped over an irregular bottom topography. This method has the added benefit of providing an objective measure of substrate **rugosity**, whereby biological data obtained from the plot method can be related to (1) the true surface area of the substrate (A^0), (2) the idealized surface area of the substrate (A'), and/or (3) the overall rugosity of the substrate (**Figure 6.3**). Of course, one might opt to abandon the plot methods altogether in favor of a plotless one.

For assessments of infauna, it may be more appropriate to utilize a stratified sampling technique, where infauna can be assessed at discrete depth intervals in the substrate, which are treated as successive, two-dimensional layers extending down into the substrate. For example, let us presume that the organisms depicted in **Figure 6.4** are each found in a different "depth zone" in the sediment, but were each collected with a box core with a cross-sectional area of 0.50 m^2, 25 cm deep into the sediment. If Layer 4 is explicitly defined as inclusive of sediment depths 7–12 cm, all of the organisms collected within Layer 4 (9 cm mean depth) can ultimately be cited relative to the 0.50 m^2 sampling area of the box core.

Census Methods for the Sessile Benthos

Among the most versatile of all census methods in marine science are the "nearest neighbor" plotless methods and the "line transect" and "quadrat" plot methods discussed earlier. With very little modification, these census methods are ideal for use in surveys of sessile benthic organisms. However, there are several other options that may have more specific utility, depending on the habit of the organisms that are at the center of your field investigation.

Rare Species Are Best Surveyed Using the T-Square Plotless Method

For those sessile organisms that occur very rarely, or are sparsely distributed, it is very likely that most of the investigator's time in the field will be wasted conducting multiple plot assessments in which the species of interest is simply not recorded. In such cases, it is usually advisable to conduct a plotless "T-square" survey instead. Similar to the nearest neighbor method (*1NN*, see Equation 3.14), the T-square method employs a series of random starting points (P_i) within the study area, as illustrated in **Figure 6.5**. For each randomly determined starting point P_i in the study area, the nearest individual of the target species (O_i) is sought out and its shortest distance to P_i is measured as x_i. A conceptual line is then drawn perpendicular to the

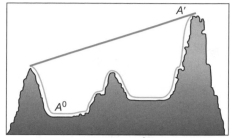

Rugosity = A^0/A'

Figure 6.3 Rugosity as a measure of small-scale variations in height across an irregular surface height. Mathematically, it is simply defined as a ratio of the actual (true) length or surface area A^0 in relation to the idealized (geometric) length or surface area A'.

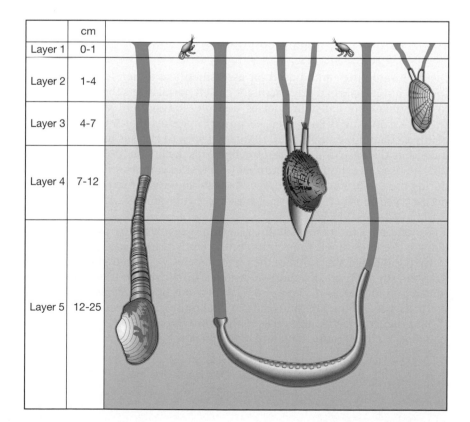

	cm
Layer 1	0-1
Layer 2	1-4
Layer 3	4-7
Layer 4	7-12
Layer 5	12-25

Figure 6.4 Despite the vertical differences of infauna diversity and biomass, if several discrete vertical layers are chosen and each is surveyed as a single "section," the data collected from each section may be reported as biomass per unit volume (for example, mg cm^{-3}), performed for each layer).

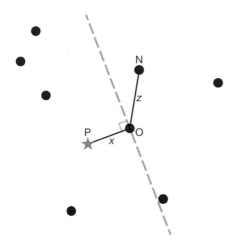

Figure 6.5 The T-square method. The black circles represent individual sessile organisms belonging to the target species. O is the nearest individual to the random starting point P. N is O's first nearest neighbor on the opposite side of the line perpendicular to OP.

line between P_i and O_i, and the distance to the first nearest neighbor (N_i) to O_i is then measured as z_i (as long as N_i is located on the opposite side of the line, relative to P_i).

The population density D (the number of individuals per unit area) can then be estimated using Equation 6.1:

$$D = \frac{\left(\sum i\right)^2}{2.828\left(\sum x_i \cdot \sum z_i\right)} \tag{6.1}$$

where i is simply the number of random starting points used in the assessment. In order to test whether the organisms are distributed randomly throughout the study area, you may use Equation 6.2 to calculate the associated test statistic t':

$$t' = \left(\sum i \left[\frac{x_i^2}{\left\langle\left(x_i^2 + z_i^2\right)/2\right\rangle}\right] - \frac{i}{2}\right) \cdot \sqrt{\frac{12}{i}} \tag{6.2}$$

Determining the distribution pattern of a particular species can often yield information about whether the individuals in that population are associative and therefore exhibit some kind of density-dependent relationship. In virtually all cases, the distribution patterns can be categorized either as uniform, random, or aggregated (see Figure 3.3). Using the test statistic calculated in Equation 6.2, the distribution pattern can be determined as

$t' > +2$	Uniform
$-2 < t' < +2$	Random
$t' < -2$	Aggregated

The Distance Sampling Plot Method Is Best for Closed Populations

Distance sampling can be used for assessments of either sessile or motile species, but it is particularly useful for closed populations that are assumed to have relatively constant species densities throughout the study area. Distance sampling is so named because it assumes that the investigator's ability to detect members of the target species decreases as a function of increasing distance from the transect line. This decline in "detectability" can thus be used to estimate the number of specimens that were not recorded during the execution of the survey. As with all survey methods, it is often necessary to conduct several replicate surveys to minimize errors.

This method is remarkably simple, as it simply involves an observer moving along a predetermined route through the study area (usually along a line transect). When a specimen of the target species is spotted at any distance from the survey route, its shortest distance from the line transect is determined and logged as d_i (**Figure 6.6**). When the observer has traveled the entire length L of the line transect, the total number of specimens detected (i) and their respective distances d_i from the line transect can be used in Equation 6.3 to estimate the population density D:

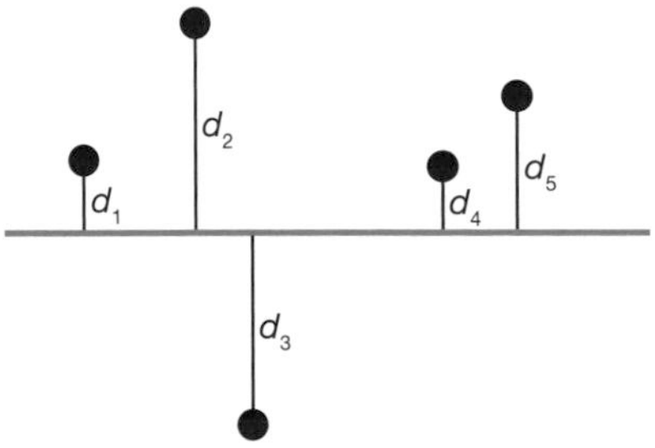

Figure 6.6 The distance sampling method. The black circles represent individual sessile organisms belonging to the target species; their shortest distances from the line transect are measured as d_i.

$$D = \frac{i\sqrt{\left(2i/\sum\left\langle d_i^2\right\rangle\right)}}{2L} \tag{6.3}$$

The Line Transect Plot Method Is Among the Simplest to Use

The line transect method simply involves a line (or chain) of a specified length randomly laid out within a study site. The observer then transits the line transect, recording biometric data for the organisms of interest that are encountered (touching the line) at various points along the length of the transect (**Figure 6.7**). If the observer wishes to conduct a complete count, everything in contact with the transect line is surveyed. However, if the data volume generated from a complete count is too burdensome, the investigator may instead survey only those organisms encountered at regular distance intervals along the line (such as every 1 m along a 100-m line). Once interval data are collected, they can be used to provide an estimate of species densities, based on the encounter rate of each species and the relative distances between encounters (**Example Box 6.1**).

In low-energy environments, the use of a tethered nylon rope or tape measure is an excellent choice. For interval sampling along the line, knots can be tied in the line (or marked with permanent ink) according to the chosen sampling interval. Of course, the use of a nylon tape measure is the most straightforward and has the added benefit of allowing the investigator to monitor "along-transect" distances at a glance. In higher energy environments, it may be necessary to instead use a heavy brass chain to prevent the line transect from becoming dislodged from its original position. Individual links in the chain can easily be marked (or tagged) according to the chosen sampling interval.

Line transects have the added benefit of deformability; that is, either the line can be stretched taut and surveyed as a straight geometric line, or it can be draped over the substrate or along horizontal contours. Typically, line transects are deployed either parallel to shore to investigate "along-shelf" gradients, perpendicular to the shore to investigate "across-shelf" gradients, or along depth contours for **isobathyal** surveys. Their ease of placement makes them an ideal choice as a plot survey method on structurally complex substrates.

Recall that when plot methods were first introduced in Chapter 3, it was apparent that any plot method must be chosen with careful consideration when determining (1) sufficient sampling effort, (2) the shape and size of the plot method, and (3) the randomness of plot placement. Although line

Figure 6.7 A line transect draped over the substrate to be surveyed. For complete counts, everything touching the white transect line is surveyed. For interval sampling, only those species touching the line at predetermined intervals (black hash marks) would be surveyed.

EXAMPLE BOX 6.1

Estimating Population Densities from Line Transect Data

Because line transects are inherently one-dimensional, it is not possible to directly determine population densities from line transect data. However, since species are detected along the line transect as a distance relationship, we can easily determine the mean distances between members of the same species:

Distance interval	Coral species
0.00 m	PPOR
0.25 m	AAGA
0.50 m	MANN
0.75 m	MCAV
1.00 m	SSID
1.25 m	PPOR
1.50 m	SSID
1.75 m	MANN
2.00 m	SSID
2.25 m	MANN
2.50 m	PPOR
2.75 m	MCAV
3.00 m	SSID

According to these line transect data, the coral species *Montastraea annularis* (MANN) was encountered at 0.5 m, 1.75 m, and 2.25 m. The distance between $MANN_1$ and $MANN_2$ was measured as 1.25 m, while the distance between $MANN_2$ and $MANN_3$ was 0.50 m; thus, the mean distance between MANN colonies is calculated as 0.87 m. This means we can expect to encounter 1 MANN colony in each 0.87 m × 0.87 m (0.76 m²) area. Thus, we can estimate our population density as 1 MANN per 0.76 m² (or 1.32 MANN m⁻²). Although this example demonstrates how easy it is use line transect data to estimate population densities for each species encountered, we would obviously want to collect much more data than are presented here.

transects are easy to deploy in just about any environment, they are a simple linear survey with limited power to resolve the spatial heterogeneity of the study site. The addition of several replicate line transects can ameliorate this, but the quadrat method is generally considered to be superior to the line transect method for most ecological assessments.

True Area Surveys Are Accomplished Using Quadrat Plot Methods

The term "quadrat" generally refers to a square frame of a predetermined size used as a measuring unit by placing the frame over (or upon) the substrate and subsequently documenting the data for all objects or organisms contained within the boundary of the quadrat. As we first discussed in Chapter 3, the use of a circular "quadrat" is preferable to a square frame in order to reduce the sampling bias at the edges of the quadrat boundary. However, the use of a square frame quadrat is perfectly acceptable, so long as the size of the quadrat is appropriate for the size of organisms (or objects) that are the focus of the research.

An appropriate quadrat size can be determined analytically by selecting the quadrat size with the lowest variance-to-mean ratio (VMR) according to the "3 & 3 rule" (see Chapter 3). However, densely populated areas and/or small organisms are usually sampled using 0.25 m² quadrats; more sparsely populated areas and larger organisms are sampled using either 0.5 m² or 1.0 m² quadrats. The use of quadrats larger than 1.0 m² may certainly be indicated for large field investigations; however, this usually requires that a plot of land be staked off and defined using a line or light chain stretched taut around the perimeter of the study area.

Gridded quadrats offer the greatest flexibility in the field (**Figure 6.8**), as the observer can use the same device to either (1) perform complete counts over the entire quadrat, (2) randomly determine some number of nested sub-quadrats in which to perform complete counts, or (3) perform counts only at the intersection points within the gridded quadrat (which is sometimes called a "point-intercept" method). Quadrats can also be used to improve the line transect method by performing the quadrat surveys at predetermined intervals along the line transect.

Because all of the objects surveyed within a quadrat are automatically assessed as a function of spatial coverage, the use of quadrats makes it incredibly easy to determine population densities, frequencies, and areas of live coverage in relation to the area contained within the quadrat. For example, if you conducted a complete count and found 20 individuals within your 1.0 m² quadrat, your final count represents your estimate for the actual population density per unit area (that is, 20 individuals m⁻²). Frequencies f are calculated in Equation 6.4 as

$$f = \frac{n}{N} \qquad (6.4)$$

where n is the total number of quadrats in which the species was observed and N is the total number of quadrats performed in the entire survey.

Figure 6.8 A gridded quadrat placed over the substrate to be surveyed. For complete counts, everything contained within the main body of the quadrat is surveyed. Otherwise, a set number of smaller sub-quadrats can be selected randomly for complete counts. The simplest analysis requires the observer to count only those organisms found beneath each of the intersection points of the grid. (Courtesy of the Flower Garden Banks National Marine Sanctuary; Hickerson/FGBNMS.)

Calculations of percent live coverage are typically used to assess distribution patterns of benthic species (like corals, algae, seagrass, or barnacles), but can also be used to determine changes in the availability of habitable space on the substrate. Since measures of percent live coverage are unitless, we can use data either from line transects or from quadrats. For point-intercept data from line transects or gridded quadrats, percent live cover (%) is calculated in Equation 6.5 as

$$\% = \frac{n_P}{N_P} \times 100 \qquad (6.5)$$

where n_p is the number of points along the line transect (or within the gridded quadrat) where the living species was encountered and N_p is the total number of points along the line transect (or within the gridded quadrat).

For gridded quadrats, another easy method for assessing percent live cover is to simply determine the number of sub-quadrats where the species in question occupies more than 50% of that sub-quadrat (n_{50}). When related to the total number of sub-quadrats (N_T) within the "parent" quadrat, percent live cover (%) can be imprecisely estimated by Equation 6.6:

$$\% = \frac{n_{50}}{N_T} \times 100 \qquad (6.6)$$

Of course, the most precise estimates of percent live cover require that the observer define the physical dimensions of the living organism in relation to the total area within the quadrat. In the field, this may be too time consuming for practical application. However, if it is possible to map all of the quadrats within the survey (usually done by taking digital photographs), the contents of each quadrat can be analyzed at a later date when more time can be dedicated to the analysis.

The use of "photo-quadrats" is a practice well used in oceanography, because digital images can be analyzed in several different color spectra and at several different spatial resolutions. As long as each photo-quadrat depicts a ruler (or some other size standard), the observer can later convert the real dimensions of the photo-quadrat into digital pixels and thereby "measure" the complex geometries of irregularly shaped organisms by tracing their shapes and then calculating the total number of pixels contained within that tracing (**Figure 6.9**). These measurements are then back-calculated to meaningful dimensions, from which precise calculations of percent live cover (or any other spatial assessments) can be conducted.

Long-Term Studies May Require a Fixed Quadrat Plot Method

For long-term monitoring programs, it is sometimes advisable to establish a series of semipermanent (fixed) quadrats that are repeatedly surveyed according to a predetermined time interval. These are especially useful in monitoring the larval settlement, growth, and mortality of certain species over time. This method is also extremely useful when conducting traditional before-after-control-impact (BACI) studies, where several semipermanent quadrats are established in an area before an ecological perturbation, and then surveyed again after the impacts of the perturbation are realized. So long as several quadrats are located in similar but unaffected (controlled) habitats, the observer can use data from the BACI investigation to monitor ecosystem resilience and recovery.

Fundamentally, there is no difference between the standard quadrat plot method and the fixed quadrat method, save that the fixed quadrats are not

Figure 6.9 An example of a photo-quadrat used in a coral reef survey. Note that the device depicted in the photograph contains a 10.0 cm strip of silver duct tape to serve as a size standard. Using planigraphic software, the physical dimensions of each coral colony can be traced as a complex polygon, and the number of pixels contained within that polygon can be easily determined. By simply measuring the number of pixels along the 10.0 cm size standard, pixel information can be used to determine the total area surveyed within the photo-quadrat (for example, 54.55 cm × 36.36 cm = 1,983 cm²). Then each coral colony polygon can be expressed in terms of percent cover (total pixels of polygon ÷ total pixels of quadrat) or by its estimated area (in cm²) of live coverage.

moved once they are initially placed in the area of interest, and they are repeat-sampled. It is important that the investigator consider the appropriate periodicity of repeat sampling within fixed quadrats, as many species exhibit different sensitivities to diurnal, tidal, breeding, and seasonal cycles (which can all significantly impact the results from repeat-sampled study areas).

Census Methods for the Motile Benthos

All the census methods discussed earlier for the sessile benthos are perfectly suited for motile species as well. This section is instead devoted to the few, specialized census techniques that are uniquely suited for bottom-dwelling organisms that are capable of migration—a special circumstance that we have ignored until now.

Because motile species are capable of migration into or out of the study area, we can use these behaviors as an integral part of the census methods used to assess their numbers. This is typically done in one of two general ways: either we can physically collect the migrators and assess their number as a function of trapping success, or we can devise some clever labeling system to mark the organisms we have already surveyed and assess trapping success by keeping track of how many individuals are captured in successive collections (using "mark-and-recapture" methods).

Trapping and Removal Methods Are the Most Common Census Strategy for Motile Benthic Specimens

All trapping methods essentially estimate population densities as a function of catch per unit effort (CPUE), a concept first discussed in Chapter 3. Among the benefits of the various trapping methods, the most significant is the fact that the organisms collected within the traps may be analyzed for virtually any biometric of interest to the investigator. Species identification and enumeration are the most basic data collected from traps, but with a little additional effort, the field researcher can also assess size and biomass data from each collected specimen. If specimens are collected alive, they can be easily transported to the laboratory for behavioral experiments or sacrificed for dissection, tissue analysis, or preservation and long-term archival.

It should be noted that trapping efficiency will never be consistent across study sites and among species. As a result, there are significant biases in data collected from traps. Although the investigator can compensate for this inherent bias by increasing the quantity of traps used in a particular study, the use of passive traps will always carry the risk that the specimens captured are not truly representative of the larger population from where they came. Despite these imperfections, comparisons can still be made between CPUE and trapping data that follow the same methodology, for the same species and areas of interest. So long as we maintain consistency in our chosen sampling method, the biases inherent to those methods can be constrained.

It is also important to recognize that a haphazard network of traps serves as a plotless census method. As a result, it is usually quite difficult to assess population densities because there is no way to preserve the "spatial history" of each collected organism. After all, how would you know if two specimens collected in a trap were originally located within 5 cm of each other, or within 0.5 km of each other, prior to their journey to the same trap? In an effort to minimize this fundamental weakness among trapping methods, it is useful to adopt a geometric pattern of trap placement, either as a rectilinear trapping grid or as a radial trapping web.

Trapping Grids/Webs: Trapping grids are used when it is assumed that a motile species can be readily trapped, and that the density (and mobility) of the organisms is the same throughout the study area. The simple geometry of a symmetrical, rectilinear grid (**Figure 6.10A**) can be used in Equation 6.7 to estimate the population density D as

$$D = \frac{N_{MR}}{\left(L^2 + 4LW + \pi W^2\right)} \tag{6.7}$$

where the total number of organisms estimated to be present in the study area (N_{MR}) must be determined by a mark-and-recapture method (discussed later in this chapter). Although the overall length of the grid (L) is easily defined, the width of the outermost strip of habitat (W) from which organisms may migrate into the main grid and become trapped is unknown. However, if multiple trapping grids of different sizes (ΔL) are used, W can be estimated using Equation 6.7 as long as we assume N_{MR} and D do not vary.

Trapping webs simply utilize a uniform number of traps arranged in concentric circles (**Figure 6.10B**). Such webs are also used to introduce a gradient of trapping efficiency, whereby organisms located near the center of the web are more likely to encounter a trap and become ensnared compared to those located near the outermost ring of traps.

Figure 6.10 Typical geometric patterns used to define (A) a rectilinear trapping grid and (B) a radial trapping web. Trapping grids can be used to establish population densities as a function of some unknown width (W) from whence "outside" specimens will immigrate into the grid, relative to the overall size of the trapping grid (L). Trapping webs are typically used with an attractant (or deterrent) located in the center of the grid or web.

(A)

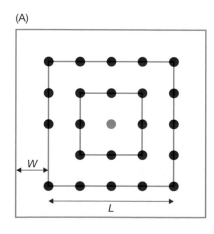

(B)

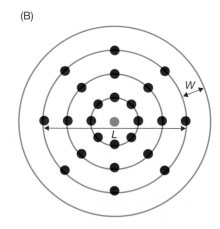

Because of this gradient in trapping efficiency, successive surveys can be used to determine migration patterns into (or out of) the center of the study area. Since it is assumed that organisms in the center of the study area are more likely to be caught in the innermost ring of traps, capture rates in the interior of the web should decline over time. By comparison, capture rates in the outermost rings will more than likely be due to "new" organisms entering the study area. The placement of an attractant (or deterrent) at the center of the trapping web can also be used to investigate the differences in random versus induced migration.

Beyond the investigator's choice of trapping pattern, it is also important to choose the appropriate style of trap for the organism you wish to capture. Although there are a wide variety of traps available, there are four general types: (1) pitfall, (2) attractant, (3) emergence, and (4) enclosure and exclosure traps. Take care that you consider the motility and migration behavior of the species you wish to capture when deciding which kind of trap to use.

Pitfall Traps Are Best for Capturing Small, Motile Species Above Sea Level

Pitfall traps can be constructed from simple plastic cups, buried just beneath the surface of the substrate, and protected by an overhanging platform to prevent unwanted debris from falling into the trap (**Figure 6.11**). The space between the substrate and the overhanging platform can be modified to exclude larger organisms by simply not giving them enough space to crawl under the platform. To prevent trapped organisms from crawling out, it is best to use tall, sheer-walled containers (hard plastic or glass). Of course, the traps can also be baited in order to increase capture rates; just be sure to use a bait consistent with your target organism's diet.

In most cases, you should check your traps daily, especially if you wish to collect live specimens. If your traps are located in a coastal area that will not be inundated with water, you may wish to place a small amount of formalin in the bottom of your traps to ensure that any captured organisms are killed and preserved before they have a chance to escape. Of course, it is not advisable to place any toxic fixatives or anaesthetizing agents in submerged traps, as these chemicals may be "flushed out" by tides or currents and pollute nearby habitats.

Attractant Traps Are Excellent for Capturing Forager and Scavenger Species

Attractant traps are those that are placed in plain view on the substrate and are baited with some kind of attractant. A classic type of attractant trap is the crab trap, usually with one or two entry points, oriented horizontally to

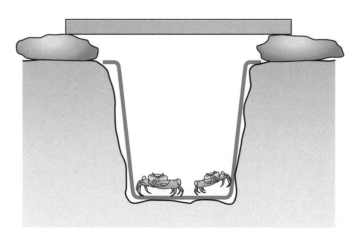

Figure 6.11 A pitfall trap, buried in the substrate so that the lip of the trap is just below grade. The use of an overhanging platform will help prevent unwanted debris from cluttering your trap. If the trap is baited, a cover will also keep unwanted scavengers from stealing the bait.

Figure 6.12 A wire-mesh crab trap, baited and used as a classic attractant trap.

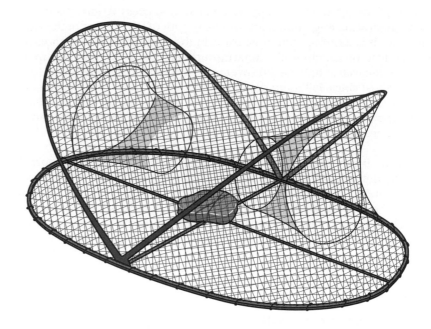

the substrate, that functions as one-way "valves" to allow entry and prevent escape. Many trap varieties simply use an "inverted funnel" design rather than a physical barrier to prevent escape, but either type is suitable (**Figure 6.12**).

Most attractant traps will open easily to allow the investigator to change bait and/or collect the captured animals. Traps can either be tethered to fixed objects (such as rocks or wharf pilings) or be anchored in soft sediment and marked with a floating buoy. Many sizes and varieties of these traps are easily available, but take care that you select a size most appropriate for your species of interest. It is not uncommon for your larger prisoners to feast on the smaller ones, so it is best to limit the size of the trap openings to discourage larger carnivores from entering your traps, and be sure to check your traps often. To catch very small organisms, you may wish to use glass mason jars fitted with an inverted funnel before screwing on the collar. Otherwise, the smallest organisms can easily sneak out of the mesh traps.

Emergence Traps Are Best Used in the Capture of Cryptic Species

Emergence traps are very similar to attractant traps, but they are typically oriented so that the funnel mouths of the traps are perpendicular to the substrate in order to capture motile species emerging from the substrate (**Figure 6.13**). These traps are particularly effective at trapping cryptic

Figure 6.13 Emergence traps can be placed directly upon the substrate (*left*) or anchored and allowed to float a set distance above the substrate (*right*), with the collecting funnel oriented with the broad opening facing downward in order to trap organisms emerging from the substrate and swimming/floating upwards.

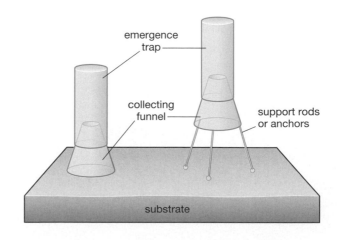

and infaunal species, which spend their daylight hours hidden within the substrate. After dark, when there is no danger of them being seen by predators, they will leave the relative safety of their hiding places and move or swim upwards. Like attractant traps, emergence traps can also be baited.

Enclosure and Exclosure Traps Are Best Suited for Mesocosm Studies

Traps that are constructed to enclose (or exclude) a specific organism are not designed to "collect" specimens; they are more commonly used to perform controlled ecological experiments in a natural environment, whereby the subjects of the study are prevented from leaving (an enclosure system), or the subjects of the study are protected from marauding outsiders (an exclosure system). The most common type of enclosure/exclosure trap is a wire cage with a uniform mesh size that is smaller than the shortest body dimension of the organism of interest. Wire cages have the benefit of allowing free exchange of water into and out of the trap while preventing the migration of the target species. However, if you wish to specifically disallow water exchange in the trap to create a truly closed aquatic **mesocosm**, it may be preferable to use a large, solid-walled tube or box that can be firmly fixed to the substrate (**Figure 6.14**).

Mark-and-Recapture Methods Are a "Trapless" Solution to Census Motile Species

Once a suitable trap design is chosen, the investigator may choose an appropriate "mark-and-recapture" method in order to make sense of the data collected from those traps without sacrificing the caught specimens. Although the methods we have discussed earlier require that specimens be removed from consideration once collected, other methods require that the collected specimens be tagged and allowed to return to the population unharmed, where they may be subsequently collected again (and again).

For mark-and-recapture methods to be useful, we must assume that, in a well-mixed population, the proportion of marked specimens in any collection will be the same as the proportion of marked specimens in the population from which they were collected. If n_1 represents the number of animals marked and released in the first outing, then n_2 can be used to represent the total number of specimens collected in the second outing. Among the n_2 organisms collected, the number that bear the markings placed on them from our first outing can be represented as m_2. Thus, we should be able to

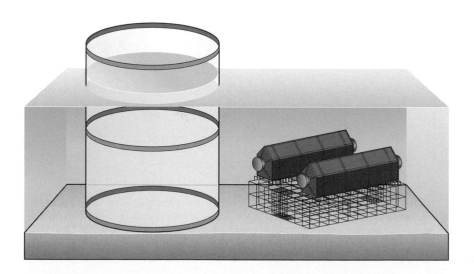

Figure 6.14 Enclosure and exclosure traps are typically fashioned as nets or wire-mesh cages designed to function as a flow-through system (*right*) or as solid-walled traps to literally trap the water along with the organisms enclosed therein (*left*). By their nature, these traps are generally not used for migration studies; instead, they are more commonly used to study closed-system dynamics.

use Equation 6.8 to estimate the total population size N_0 according to the mathematical relationship

$$N_0 = \frac{(n_1 + 1)(n_2 + 1)}{(m_2 + 1)} - 1 \qquad (6.8)$$

Although this represents the simplest of all mark-and-recapture relationships (that is, a two-sample method), it is the mathematical foundation on which all mark-and-recapture methods are built.

The Pseudo-removal Method Can be Used to Estimate the Size of a Closed Population by Simply Marking Specimens

In a closed population, it is easy to see that as animals are continually trapped, marked, and released, the number of unmarked animals captured in successive collections should decline. At some point, every member of the population will be captured and marked, and at that point, the total number of individuals in the population will be known.

Of course, it is not practical to mark and recapture specimens until every member of the population is collected. However, by plotting the number of unmarked animals collected (on the y-axis) against the cumulative number of marked animals from all previous collections (on the x-axis), a least-squares regression of the resultant line ($y = mx + b$) can be used to estimate the total population size x' when $y = 0$ (**Figure 6.15**). If animals are simply removed when trapped, this same method can be applied to the results by plotting the number of organisms caught and removed (on the y-axis) against the cumulative catch (on the x-axis). As more and more members of the closed population are caught and removed, $y \to 0$.

The Wileyto Removal Method for Closed Populations Uses a Combination of Marked and Removed Specimens to Assess Population Size

Another simple method for closed populations is the Wileyto method, which relies on two different sorts of traps deployed simultaneously. The first is a simple trap that catches and permanently removes animals from the population. The second trap is identical to the first, except that animals entering it are marked and immediately set free. Since the capture trap will collect

collection series	no. unmarked specimens per collection series (y)	cumulative catch (x)
1	47	0
2	52	47
3	25	99
4	33	123
5	11	156
6	8	167
7	8	175
8	10	184

x = 223 total individuals when y = 0

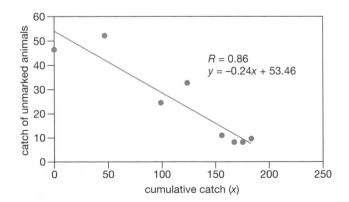

$R = 0.86$
$y = -0.24x + 53.46$

Figure 6.15 An example of the pseudo-removal method. By using the cumulative catch as the independent variable x and the number of unmarked specimens per collection as the dependent variable y, a least-squares linear regression ($y = mx + b$) can be used to estimate the total number of members in the population (x) when the number of unmarked specimens falls to zero (that is, when $y = 0$). In this example, the regression equation is rearranged to solve for x when $y = 0$, yielding the solution $x =$ 233 as the estimated size of the total population.

both marked and unmarked specimens, their proportions can be used in Equation 6.9 to provide an estimate of the total population size (N_0) according to the formula

$$N_0 = \frac{(U + M)^2}{2(M + 1)} \tag{6.9}$$

where U is the number of unmarked animals caught in the "permanent removal" trap and M is the number of marked animals caught in the same trap. From these simple results, a 95% confidence interval can be estimated using Equation 6.10:

$$N_0 \pm \frac{2\left(\sqrt{N_0 \langle U(U + M) + M(U - M)\rangle}\right)}{M} \tag{6.10}$$

The Multiple Mark-and-Recapture (Schnabel) Method for Closed Populations is among the Most Rigorous Estimates of Population Size

Also known as the Schnabel method, this particular method requires that specimens are marked only when they are first captured; successive captures do not merit successive markings. Assuming that the collected organisms do not vary in their "trapability" over time, estimates of population size can be calculated from the proportion of marked and unmarked specimens captured in successive collections. Although the Schnabel method is similar to the pseudo-removal method discussed earlier, it utilizes more information when estimating population size. Generally speaking, this makes the Schnabel method the preferred method, but it also requires a more rigorous mathematical treatment of the data.

The fundamental variables for use in the Schnabel method are defined as follows:

i = Number of collections
n_i = Number of total specimens in the ith collection
m_i = Number of marked specimens in the ith collection
u_i = Number of unmarked specimens in the ith collection
M_i = Cumulative number of specimens marked prior to the ith collection

From these data, Equation 6.11 is used to define the following relationships:

$$A = \sum_i n_i M_i^2; \quad B = \sum_i m_i M_i \tag{6.11}$$

As an example, let us use the data in **Table 6.1** for a mark-and-recapture study of lightning whelk snails (**Figure 6.16**) collected from a South Texas mud flat.

According to the Schnabel method, the final estimate of population size (N_0) is calculated in Equation 6.12 simply as

$$N_0 = \frac{A}{B} \tag{6.12}$$

Using the data from Table 6.1, the quantity (A/B) is calculated to be 64.7627; rounded to the nearest whole number, the Schnabel method estimates that the total population size of lightning whelk snails in our example study area is 65 individuals.

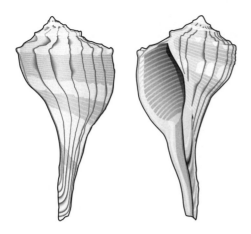

Figure 6.16 The lightning whelk (*Busycon contrarium*) is a large, predatory snail commonly found on sandy and muddy substrates in the Gulf of Mexico and the Caribbean.

Table 6.1 Data for Lightning Whelk (*Busycon contrarium*)

i	n_i	m_i	u_i	M_i	$n_iM_i^2$	m_iM_i
1	23	0	23	0	0	0
2	39	13	26	23	20631	299
3	27	23	4	49	64827	1127
4	44	34	10	53	123596	1802
SUM					$A = 209054$	$B = 3228$

Specimens collected using the Schnabel method, indicating the collection number (i), the total number of specimens in each collection (n_i), the number of marked (m_i) and unmarked (u_i) specimens in each collection, and the cumulative number of specimens collected in previous collections (M_i).

Open Populations are More Challenging to Assess and Require Complex Multiple Mark-and-Recapture Methods

For populations that are subject to immigration and emigration of specimens, the method used to estimate N_0 requires that the investigator conduct a minimum of three mark-and-recapture collections and requires that the specimens are marked with unique symbols or colors each time they are collected (thereby preserving their "capture histories").

The method most commonly used for multiple-recapture data from open populations is the Jolly–Seber method. Although this method is an excellent choice for estimating the total number of individuals within an open population, it is also quite rigorous in its execution and in the mathematical computations required for the manipulation of collected data. For further information on the Jolly-Seber method, please refer to the Further Reading section.

References

Bakus GJ (2007) Quantitative Analysis of Marine Biological Communities: Field Biology and Environment. John Wiley and Sons.

Cochran WG (1977) Sampling Techniques, 3rd ed. John Wiley and Sons.

Coyer J & Witman J (1990) The Underwater Catalog: A Guide to Methods in Underwater Research. Cornell University, Shoals Marine Laboratory.

Krebs CJ (1999) Ecological Methodology. Harper & Row.

Rogers CS, Garrison G, Grober R, Hillis Z-M, & Franke MA (1994) Coral Reef Monitoring Manual for the Caribbean and Western Atlantic. National Park Service, US Virgin Islands.

Southwood TRE (1978) Ecological Methods. Chapman and Hall.

Sutherland WJ (ed) (1996) Ecological Census Techniques: A Handbook, 2nd ed. Cambridge University Press.

Further Reading

Anderson DR, Burnham KP, White GC, & Otis DL (1983) Density estimation of small mammal populations using trapping web and distance sampling methods. *Ecology* 64:674–680.

Burnham KP & Overton WS (1978) Estimation of the size of a closed population when capture probabilities vary among animals. *Biometrika* 65:625–633.

Jolly GM (1965) Explicit estimates from capture-recapture data with both death and immigration: A stochastic model. *Biometrika* 52:225–247.

Manly BFJ (1984) Obtaining confidence limits on parameters of the Jolly-Seber model for capture-recapture data. *Biometrics* 40:749–758.

Schwartz CJ & Arnason AN (1996) A general methodology for the analysis of capture-recapture experiments in open populations. *Biometrics* 52:860–873.

Seber GAF (1982) The Estimation of Animal Abundance and Related Parameters. Charles Griffin Publishing.

Wileyto EP, Ewens WJ, & Mullen MA (1994) Mark-recapture population estimates: A tool for improving interpretation of trapping experiments. *Ecology* 75:1109–1117.

Zippin C (1956) An evaluation of the removal method of estimating animal populations. *Biometrics* 12:163–169.

Chapter 7

Census Methods for Pelagic Organisms

"Chance is always powerful. Let your hook always be cast; in the pool where you least expect it, there will be fish." – Ovid

In the previous chapter, we dealt exclusively with the methods for collecting the aquatic organisms that make their home on (or in) the benthos: the creepers, the crawlers, the slitherers, and the burrowers. In this chapter, we shall see that the addition of a third dimension (getting ourselves off the "floor" and into the water column) requires very different strategies when it comes to choosing the proper census technique.

Herein, we consider the dynamic, three-dimensional realm of the pelagic organism. And by "pelagic," we need not limit ourselves to the consideration of the deep waters of the open ocean. It is more correct to think of pelagic organisms as those creatures that occupy the water column, no matter its depth: the floaters, the sinkers, and the swimmers.

The ocean is a mighty big place, and it's also a mighty deep place. As you can well imagine, performing specimen collections in the vastness of the ocean is the ultimate three-dimensional experience. And it's not like our critters have the common decency to sit still for us to come along and collect them at our leisure. No, they are constantly moving: sinking into the gloom, floating towards the light, swimming against the currents or swept away by them. Far be it for life (or science) to be easy!

Key Concepts

- The census methods used for pelagic aquatic organisms require the inclusion of a vertical component in the sampling technique.
- The depth preference, body size, and swimming speed of the target species all define which pelagic census method is most appropriate for use.
- Virtually all pelagic census methods require the use of a net, used to filter the water and act as an ensnarement device for specimen collection.
- Physical collections often yield the largest volume and best quality specimen data; unfortunately, they are also the most lethal to nontarget species.
- Virtual collections are far less destructive than physical collections, but are limited by technical constraints that also limit the breadth of data that can be collected.

Life in a Fluid Medium

It is an inherently difficult task to survey pelagic organisms, particularly since we are dealing with species that are free to float or swim, with varied direction and speed, in open water. Unlike the traditional census techniques discussed in Chapter 6, pelagic organisms are not constrained to a two-dimensional substrate, so many of those methods are woefully inadequate for surveying populations that are uniquely adapted for life in a fluid medium.

Fortunately, we are not forced to completely re-invent the wheel. Benthic census methods are, by definition, designed to survey the two-dimensional benthos; that means they are quite effective at surveying the horizontal dimensions of a particular area. Whether we think of these horizontal dimensions in terms of length and width, or latitude and longitude, we need only to include the vertical dimension (the depth of the water column) to devise a three-dimensional survey method appropriate for pelagic organisms. So it should come as no surprise that we must carefully consider the bathymetry of the water body when defining the vertical dimension of our investigations.

Pelagic Surveys Require a Definition of the Vertical Dimension

Marine ecologists typically define the vertical dimension of the ocean according to "depth zones," traditionally as a function of light penetration relative to the overall depth of the water column (**Figure 7.1**). In the clear waters of the open ocean, sunlight is able to penetrate as deep as 200 m, and with sufficient intensity, to support **primary production**. Called the **euphotic zone**, this particular depth zone is the exclusive home of the ocean's

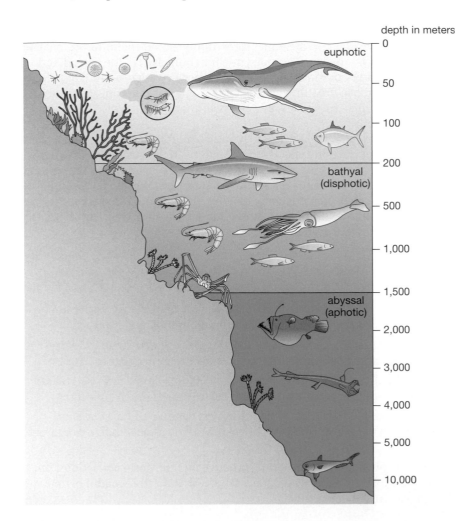

Figure 7.1 The traditional "depth zones" of the world ocean, including the euphotic zone (0–200 m), the disphotic (or bathyal) zone (200–1500 m), and the aphotic (or abyssal) zone (1500+ m). Although the continental shelf typically extends out to the 200-m isobath, it is incorrect to assume that all neritic or littoral waters are truly euphotic. In fact, due to the heavy influence of coastal runoff and primary production, it is certainly possible to have disphotic and aphotic zones, even in nearshore waters, constrained to very shallow depths.

primary producers (aquatic plants and algae). Since this is also the zone where all of the ocean's "new" food is produced, the euphotic zone is also the most heavily populated depth zone in the world ocean.

Because the euphotic zone only reaches down to the relatively shallow depth of 200 m, it is easy to deploy most oceanographic sampling equipment from modest shipborne platforms. Since all estuarine and continental shelf waters are <200 m deep, coastal access to the euphotic zone is also quite easy. Thus, it is plain to see why the euphotic zone is the most heavily studied, and best understood, ocean depth zone.

In the deeper reaches of the ocean where very faint sunlight can penetrate (<1% of the surface intensity), primary production is unsustainable. As a result, no plants or algae are able to survive for long in the **disphotic** (or **bathyal**) **zone**. Despite the small amount of ambient light, there are several species of nekton (mostly fishes) that prefer to live in this "twilight zone." Some creatures live here to avoid being seen by predators, while others prefer to lurk in the gloom in order to ambush their prey.

Although the sunlight is extremely faint in the disphotic zone, the creatures that live here are still quite aware of the day/night cycle and will frequently migrate to shallower waters at night (**Figure 7.2**), into the bottom of the euphotic zone, where food is far more plentiful. Just before sunrise, they will return once again to the dim waters of the disphotic zone, where they await their next opportunity to forage in the euphotic zone when night falls. These daily vertical migrations (DVMs) are so prevalent in the disphotic zone, even among microscopic organisms, that they are more the rule than the exception.

Logistically, pelagic surveys within the disphotic zone can be quite challenging. Although some regions with very narrow continental shelves may exhibit ocean depths 200–1500 m within a few kilometers or so from the coast (and are therefore quite easily accessible by ship), standard oceanographic equipment may not be operational at such extreme depths. Even the "simple" deployment and recovery of a collection vessel might require more than a kilometer of steel cable to be paid out on a heavy-duty cable drum and hydraulic winch.

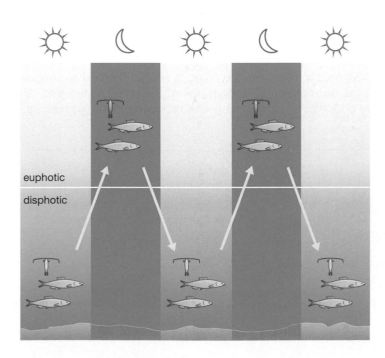

Figure 7.2 Nekton and plankton alike often prefer to remain in darker waters during the day to avoid visual predators. As sunset approaches, many oceanic species will begin their migration into the base of the euphotic zone to feed in the productive euphotic waters at night, only to return to their home in the disphotic (bathyal) zone before sunrise. Organisms that perform these daily vertical migrations often traverse significant vertical distances in just a few hours.

Below 1500 m, the **aphotic** (or **abyssal**) **zone** is shrouded in complete darkness, and has been ever since the formation of the world ocean about 3.8 billion years ago. With rare exception, population densities within the aphotic zone are very sparse and are dominated by scavengers in search of sinking carcasses and other detritus. As depths within the aphotic zone can reach 6000 m or more (the deepest being nearly 11,000 m), biological surveys within the aphotic zone require significant resources due to the technical challenges of reaching such extreme depths. However, with the proper equipment and budget, such research efforts are invaluable, as very little is known about the creatures that inhabit the depths of the bathyal and abyssal zones.

It is important to keep in mind that the traditional depth boundaries for the euphotic, disphotic, and aphotic (EDA) zones are relevant only to open ocean waters. In shallow estuarine and shelf waters, the significant influence of coastal runoff and enhanced primary production can dramatically reduce water clarity. In some extreme cases, the euphotic zone can be reduced to just the top few meters of the water column, where the water becomes aphotic at depths of just 20–30 m (or perhaps even shallower than that, depending on the turbidity of the water). Thus, when considering the vertical dimension of your pelagic surveys, don't rely on a rigid view of the traditional EDA cutoff depths; it is far more important that you consider the realistic depth zones in your particular study area.

Also keep in mind that many organisms, even those inhabiting estuaries and shallow coastal waters, can exhibit DVM behaviors. If your survey techniques rely heavily on dividing the water column into EDA "light layers" and surveying each layer discretely, the time of day of your surveys will have a dramatic influence on which species are found where. And since most field studies are conducted only during daylight hours (usually for the convenience of the researcher), nightly migration behaviors are easily excluded from biological surveys and may constitute a tremendous source of bias in the data.

Generic, "one-size-fits-all" sampling methodologies usually define the vertical dimension of the water column by dividing it geometrically between the surface and bottom. Regardless of the bathymetry of a particular site, there will always be an easily defined "surface" depth (0 m) and an easily defined "bottom" depth; the median depth can then be easily defined as well (Figure 7.3). While this generic sampling method is certainly not perfect, it should at least represent a starting point from which you can devise a more appropriate method for defining the vertical dimension within your own study.

When designing a pelagic survey plan, it is often helpful to consider both the vertical and horizontal dimensions of the study in tandem. Although there are no "perfect" survey plans, the recommendations given by Ramon

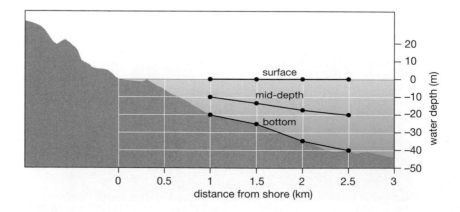

Figure 7.3 A simple example of the "mid-water sampling strategy," where samples are taken at the surface, at the bottom, and at the calculated median (mid-water) depth for all stations, relative to the changing bathymetry along the transect.

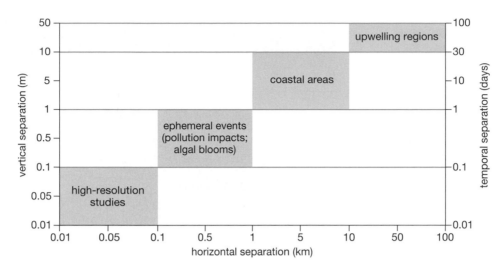

Figure 7.4 Recommended spatial and temporal separation between oceanic sampling stations located within upwelling regions or coastal areas. More ephemeral events (such as pollution impacts or algal blooms) require shorter distances between samples, and more frequent sampling, in order to capture the transient nature of those dynamics. High-resolution sampling should be reserved only for those events that are very small in size or occur over very rapid time scales. (Adapted from Ramon Margalef [1978] *Monographs on Oceanographic Methodology,* Volume 6.)

Margalef, one of the most influential ecologists in modern history, are always a good place to start (**Figure 7.4**).

Size Matters When Choosing the Proper Gear and Method for Pelagic Surveys

Another key consideration in pelagic census methods is the physical size of the organism(s) intended for survey. Although it is true that aquatic organisms span a broad range of body sizes, benthic organisms are generally large enough to "settle" on the benthos, so even the smallest creatures are relatively easy to find (at least we know where to look for them). In contrast, pelagic species can be found just about anywhere, moving horizontally and vertically, within the fluid medium.

Of course, the size of the organism will generally dictate whether it is classified as plankton or nekton (**Figure 7.5**). As a general rule, **microscopic**

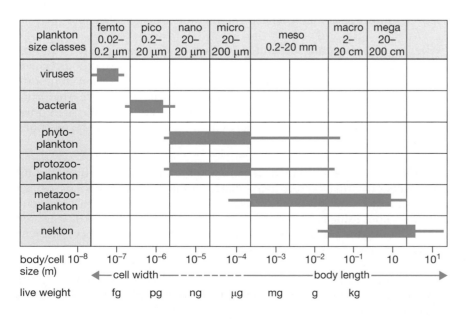

Figure 7.5 Size chart of aquatic organisms, including several subcategories based on longest body dimension (body or cell size) and/or biomass (live weight). Within the plankton, only the femto- to nanoplankton size classes (0.02–20 μm) are truly planktonic. Many species within the other plankton size classes possess weak swimming abilities and can exert some minor influence on their position in the water column. (Adapted from Sieburth JM, Smetacek V, & Lenz J [1978] *Limnology & Oceanography* 23: 1256–1263.)

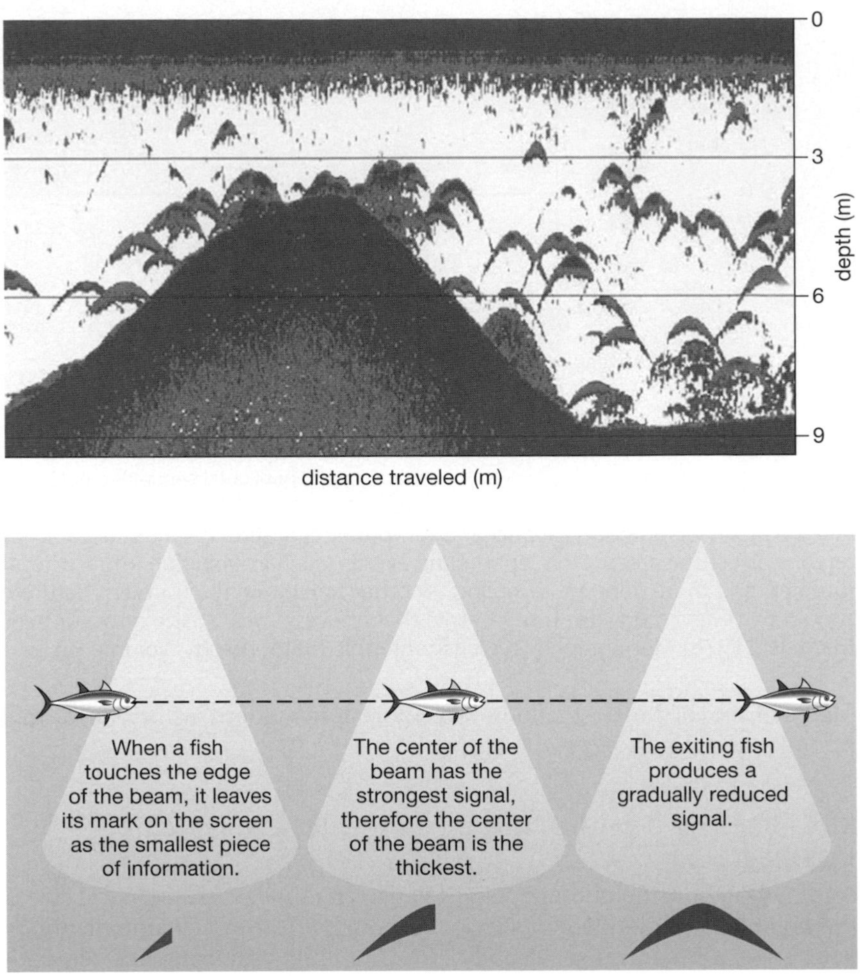

Figure 7.6 The visual display from a "Fish Finder" acoustic imaging device. Sound waves are emitted below the hull of a boat, which propagate through the water and return to the sensor when the sound waves are reflected off any submerged object (including fish, which are represented by the multitude of crescents in this image). Data from a device such as this could be used as "virtual capture," to help estimate the population density of fish in the survey area.

distance traveled (m)

When a fish touches the edge of the beam, it leaves its mark on the screen as the smallest piece of information.

The center of the beam has the strongest signal, therefore the center of the beam is the thickest.

The exiting fish produces a gradually reduced signal.

smallest size class collected	common mesh/ net sizes
nano (>13 µm)	10 µm
nano (>27 µm)	20 µm
micro (>40 µm)	30 µm
micro (>67 µm)	50 µm
micro (>133 µm)	100 µm
micro (>200 µm)	150 µm
micro (>267 µm)	200 µm
micro (>400 µm)	300 µm
meso (>400 µm)	400 µm
meso (>667 µm)	500 µm
meso & nekton	1/8″ (3175 µm)
meso & nekton	3/16″ (4762.5 µm)
meso & nekton	1/4″ (6350 µm)
meso & nekton	3/8″ (9525 µm)
meso & nekton	1″ (25,400 µm)

Figure 7.7 Common sizes of plankton nets and mesh filters for the capture of various size classes of microorganisms. Typical towing speeds for plankton collection should be rather slow (0.5–1.0 m s^{-1}) to avoid damaging the delicate organisms caught in the net. Larger mesh sizes (shaded in blue) should be towed at a quicker pace (1.0–2.0 m s^{-1}) to prevent the large mesoplankton and nekton from escaping the approaching net. (Adapted from Makoto Omori & Tsutomu Ikeda, [1992] Methods in Marine Zooplankton Ecology.)

organisms are exclusively planktonic, although some of the larger forms do exhibit the ability to migrate vertically. By contrast, **macroscopic** organisms are largely nektonic, and are frequently capable of swimming at great speeds and over long distances. Information about body size can also be used to make informed assumptions about several other important biometrics, such as the organism's biomass. These details can then be used to help inform the most appropriate sampling device and method for data collection.

Whether organisms are suspended in the water, or are actively swimming through it, pelagic census techniques usually require some method of specimen capture, either physical or virtual. Physical capture is most commonly accomplished by "sifting" the water using some type of net or mesh. Virtual capture instead collects evidence of the organism without requiring its physical capture, usually through the use of light- or acoustic-imaging technologies (**Figure 7.6**).

Although advances in modern technology are dramatically advancing the capabilities of virtual capture (particularly for sampling nekton), most pelagic survey methods still rely on the physical capture of organisms. This is accomplished either using bottle samples (capturing discrete volumes of seawater) or by filtering the water with a net in order to capture those organisms with body sizes larger than the openings in the chosen mesh (**Figure 7.7**).

Because of their extremely small size, the femto- and picoplankton (10^{-7}–10^{-6} m) are best collected using bottle samples. In contrast, the largest and fastest nekton (10^0–10^1 m) may be able to avoid capture, even with a net. However, all of the size classes ranging from the nano- to the megaplankton (10^{-5}–10^{-1} m), as well as small nekton, are captured quite effectively using nets.

Census Methods for Viruses, Bacteria, and Small Phytoplankton

For all plankton with cell sizes <10 μm, the most practical collection method requires that discrete water samples be taken using a bottle. Surface samples can be collected by gently scooping water into an appropriate container, even using something so simple as a bucket tied to a line and cast into the water. However, the collection of samples from anywhere beneath the surface will necessitate the use of Niskin or Van Dorn bottles, which can be lowered to the desired depth and "tripped" with a weighted messenger (**Figure 7.8**). Once the sample is collected at depth, the bottle can be hauled up and the water is easily decanted for processing, analysis, and/or preservation.

Typically, Niskin and Van Dorn bottles are constructed from clear acrylic or opaque PVC. Regardless of the type of material from which the bottle is made, designs that prevent the collected water sample from coming in contact with the bottle's metal components are always recommended. To insure against contamination from previous collections, bottles should be prerinsed three times in 5% hydrochloric (HCl) acid, and then thoroughly rinsed with copious amounts of distilled water prior to deployment. It is usually not necessary to acid rinse bottles between samples during multiple, same-day deployments of a singular sampling bottle (but bottles should be rinsed thoroughly between collections).

If **virioplankton** or **bacterioplankton** samples are the primary target, the use of several sterilized collection bottles is ideal. If this is not feasible in the field, water from the collection bottle(s) should be immediately transferred into smaller, sterilized sample bottles and sealed against potential contamination. For **phytoplankton** collections, sterilization procedures are usually not necessary.

Ideally, water samples should be processed immediately. If it is necessary to store water samples before they can be processed, they should be stored in the dark at 4°C for the shortest possible duration before processing (6–12 hours

PUTTING THE SQUEEZE ON PLANKTON

Water pressure can sometimes force organisms through a mesh that would otherwise keep them captured; therefore, it is often advisable to use a mesh size that is 75% of the target organism's smallest body dimension. This is particularly true among the larger plankton, as well as most nekton, whose body widths can be an order of magnitude smaller than their body lengths.

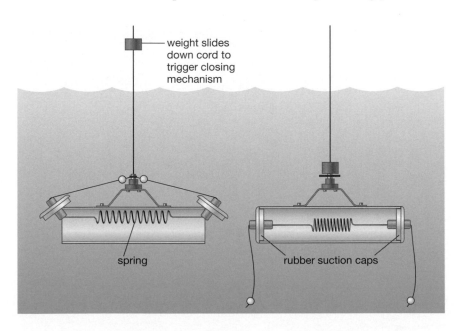

Figure 7.8 Bottle samplers (Niskin or Van Dorn) are lowered to a specific depth in the open position. When the bottle has reached the desired depth, the closing mechanism is triggered, sealing the water sample within the bottle for later analysis.

is fine, but it is wise to avoid refrigerated storage times that exceed 24 hours). If there is no way to process the samples within 24 hours of collection, it may be necessary to preserve your samples either by freezing them or by using a chemical preservative.

Processing the samples typically involves filtering the collected water through a filter with a pore size that will retain the cell sizes that are consistent with the target organisms. After filtration, the filter is used for biological analyses, while the filtered water can be used for other important physical and chemical data (such as measurements of temperature, salinity, pH, nutrient concentration, etc.).

Virio- and Bacterioplankton Census Methods Require Specialized Equipment

The general census method for virio- and bacterioplankton requires that the collected water samples be divided into 2–10 mL **aliquots** and filtered through 0.02 μm Al_2O_3 Anodisc filters for recovery of virio- and bacterioplankton. Since bacterial cell sizes are much larger than viruses, the use of a 0.2 μm filter will retain bacterioplankton only (thereby allowing the viruses to pass through the filter). Methods requiring live cells for RNA/DNA sequencing, staining, or culturing must be performed immediately after filtration; otherwise, cells can be fixed in sterilized and 0.02-μm-filtered 1% formalin for later processing without fear of organic degradation.

Stained samples can be placed on a slide and counted using standard light microscopy; alternatively, samples stained with fluorescent dyes (such as SYBR Green or SYBR Gold) are counted using epifluorescence microscopy (**Figure 7.9**). Live bacterioplankton samples that are filtered, deposited on plate nutrient media, and cultured in sterile incubation will eventually grow colonies visible to the naked eye, which can be easily counted or sized.

As most census methods for virio- and bacterioplankton require a significant amount of technical equipment and expertise, the reader is encouraged to seek additional guidance from reference materials for visible or fluorescent staining and microbiological culture methods. This is especially true for advanced genetic labeling and sequencing techniques.

Figure 7.9 A water sample from marine waters, prepared with a fluorescent stain. The large bright particles are counted as bacteria (large arrow), and the smaller, more numerous particles are counted as viruses (small arrow). The nucleus of a large diatom (80 μm long) is brightly stained. As long as the volume of water in the sample is known, a simple count of glowing bacterioplankton cells can be used to calculate the population density of bacteria. A variation of the same method can be used to quantify viral densities as well. (Courtesy of Rachel Noble & Jed Fuhrman, [1998] originally published in *Aquatic Microbial Ecology* 14:113–118.)

Small (<10 μm) Phytoplankton Are Most Effectively Collected by Bottle

Methods for the quantitation of very small (<10 μm) phytoplankton cells do not differ significantly from those methods discussed for phytoplankton in general. The only real difference is the initial method of collection: bottle versus net. As a practical matter, it is often difficult to find plankton net manufacturers who carry mesh sizes smaller than 10 μm. For those forms with cell sizes 0.2–10 μm, even the smallest mesh would be incapable of retaining them for collection.

Thus, for phytoplankton collections that seek to include all the potential phytoplankton size classes, it is often necessary to use a certain number of bottle samples for quantification of the phytoplankton biomass and species diversity, especially if it is suspected that picoplankton (0.2–2.0 μm) constitute a significant portion of the overall phytoplankton community. Because of their size, small phytoplankton are most effectively sampled by using a large-capacity bottle (or several smaller-capacity bottles deployed in replicate). A good rule of thumb is to collect at least 1 L of water from each site and 4 L if taken from **oligotrophic** or open ocean waters (in fact, surveys of very rare picoplankton can require sample sizes up to 20 L taken at each sampling location).

Census Methods for Large Phytoplankton, Zooplankton, and Larvae

One of the most effective and widely used methods for collecting plankton is the towed plankton net. Although designs and deployment strategies can vary, all plankton nets are made from strong, lightweight mesh (usually nylon) that is specifically engineered so that the holes in the mesh are a uniform size. Mesh sizes are then selected to best match the size of the organism(s) the researcher wishes to collect (see Figure 7.5). The mesh, when pulled through the water, acts as a selective screen, effectively catching all of the suspended particles that are larger than the holes in the mesh.

This is a simple but important concept to keep in mind, as the use of a small-mesh, or "catch everything," approach to field sampling may not always be the best strategy. For comprehensive surveys of the plankton community as a whole, it would obviously be wise to use a small mesh size to capture not only the phytoplankton, but also the much larger **zooplankton** and **larvae**. But if collecting **ichthyoplankton** and other larvae were the primary goal of the research, the use of a higher mesh size would more efficiently capture ichthyoplankton while allowing the other unwanted plankton groups to slip through the mesh.

The Geometric Dimensions of a Plankton Net Determine Its Ideal Function

Plankton nets are most commonly designed with the geometry of a cone, with a fixed opening at the "mouth" of the net, usually constructed from a rigid, stainless-steel hoop with a three-point bridle so that a rope or cable can be easily attached for deployment (**Figure 7.10**). The mesh extends behind the mouth of the net and tapers to the collecting cod end, which is designed to be easily detachable from the net assembly, so the plankton sample can be handled more easily.

The first and most important consideration of any plankton net is the mesh size, which should be strategically matched to the assumed body or cell size of the plankton to be collected (see Figure 7.5). The overall dimensions of the net are also important, as they will have a huge influence on the hydrodynamics of the water flowing into (and out of) the plankton net. Most

Figure 7.10 A conical plankton net of typical design, showing the rigid collar, three-point bridle, and the nylon mesh tapering to the collecting cod end, where the plankton sample is retained.

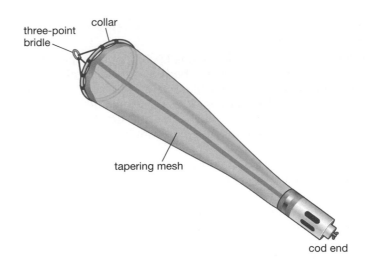

Figure 7.11 Typical plankton net proportions, where the ratio of the mouth diameter *d* to net length *l* is typically 1:3 (*left*) or 1:4 (*right*).

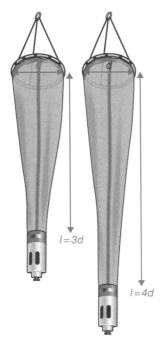

"off-the-shelf" conical nets have mouth diameters of 15, 30, or 60 cm. In estuarine or coastal waters where plankton densities are typically quite high, 15-cm or 30-cm nets are usually sufficient. For offshore deployments where plankton densities are expected to be relatively low, it is often an advantage to use nets that filter the maximum amount of water in each pass. For this reason, 60-cm (or larger) nets are often used in the open ocean.

The overall length of the net is also an important consideration. Typically, plankton nets are available with diameter-to-length ratios of 1:3 or 1:4 (**Figure 7.11**). As a practical matter, longer nets are more difficult to handle, but they also have better filtering efficiencies. Since water can escape through large meshes quite easily, mesh sizes 100+ μm usually perform quite well with 1:3 proportions. For very small mesh sizes (<30 μm), it is usually best to opt for 1:4 proportions.

The cod end of the plankton net is where the sample is ultimately collected, so choosing the right collection "bucket" is also an important consideration. Above all, the bucket should be securely fastened to the cod end of the net, be designed to retain the delicate plankton without damaging them, and still be easily detachable from the cod-end assembly to make it very simple to transfer the collected plankton into a sample bottle.

The best cod-end assemblies are made from a thick PVC collar, which is either threaded (so the bucket can be easily screwed on and off the cod end), or has stainless-steel clasps to securely fasten the bucket to the cod end (**Figure 7.12**). The bucket itself should be machined with 2–5 mesh "windows" near the top of

Figure 7.12 PVC cod-end assembly, showing the threaded endpiece that is affixed to the tapered end of the plankton net (*right*), as well as the collecting bucket (*left*). To facilitate swift drainage of the cod-end assembly, the collecting bucket may contain several mesh windows for easy flow-through design.

the bucket to allow water to escape from the bucket, thereby avoiding any water compression or backflow in the bucket and preventing damage to the collected plankton. Of course, it is critical that the mesh windows of the bucket are the same mesh size as the net itself.

Choosing the Appropriate Type of Plankton Tow Requires Some Knowledge of the Target Organisms' Vertical Distribution in the Ocean

Once a decision has been reached regarding the size and type of plankton net to be used, it will be necessary to consider the various towing paths as the chosen collection strategy. As it turns out, the horizontal scale of plankton patchiness is typically from 200 m to 10 km; in contrast, the vertical scale of plankton patchiness is only 1–20 m. In other words, plankton diversity and abundance tend to be more sensitive to the vertical dimension than to the horizontal dimension. Essentially, that means plankton patches are far more likely to be organized in relatively flat "cloud layers" at discrete depths.

So, if plankton are typically found in layers, it stands to reason that if we knew at what depth to sample, a simple horizontal tow (right in the thick of the plankton layer) would yield the greatest abundance of collected plankton. But what if you miscalculate and deploy the horizontal tow at a depth just above (or just below) that optimal layer? And what if the multitude of different plankton species, all occupying the same water column, were actually organized in several different depth layers? These are the types of considerations that should inform the development of your chosen method for plankton collection.

In practice, plankton tows typically follow one of three basic strategies: (1) the horizontal tow, (2) the vertical tow, or (3) the double-oblique tow. Each has its own strengths and weaknesses, but none are perfect—it is a decision that must be made by the investigator, based on the primary research goals and in consideration of the plankton species targeted for collection.

The Horizontal Tow Is Best for Sampling at a Single, Discrete Depth Layer

Surface horizontal tows (**Figure 7.13A**) are among the simplest to execute in the field and are well suited for phytoplankton collection. For most tows, standard conical plankton nets can be used, taking care to adjust the towing speed to keep the mouth of the net completely submerged at all times, just below the surface of the water. For neuston collection, special wide-mouthed rectangular nets (appropriately called "neuston nets") are used, which are specially designed to literally skim the surface of the water, collecting only those plankton in the upper 0.1 m of the water column.

Deep horizontal tows (**Figure 7.13B**) may be possible with the appropriate use of weights on the bridle and/or cod end (to keep the net submerged) and an opening–closing mechanism installed behind the mouth of the net (**Figure 7.14**). The closed-mouth net is lowered a bit beyond the desired depth and towed at an appropriate speed so that the water resistance (when the net is opened) drives the net up to the desired depth. When the tow is complete, the mouth of the net is closed again, and hauled to the surface for sample processing.

Deep horizontal tows are something of an art, and will require a fair bit of trial and error to perfect. Devising a system by which weights can be easily added to (or subtracted from) the net assembly is a tremendous asset when troubleshooting in the field. The use of a marked line (or metered cable) is also quite helpful, as net behavior can be influenced heavily by the amount

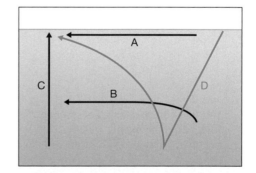

Figure 7.13 The towing paths most commonly used for plankton collection: (A) surface horizontal tow, (B) deep horizontal tow (for nets with an opening-closing mechanism), (C) vertical tow, and (D) double-oblique tow. (Adapted from, Omori M & Ikeda T [1992] Methods in Marine Zooplankton Ecology. With permission from Kreiger Publishing Company.)

Figure 7.14 Conical plankton net with a "choke collar" installed as an opening–closing mechanism, to control the depth at which the collection is taken.

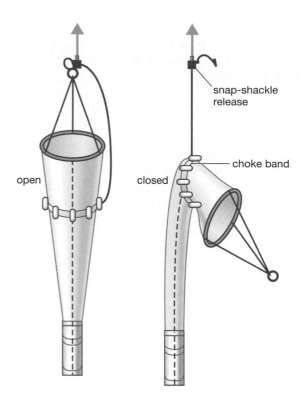

open

closed

snap-shackle release

choke band

of line or cable paid out, as well as the towing speed. A marked line, the use of an inclinometer (to measure the line angle), and a little trigonometry are all you'll need to estimate the depth of the net during the horizontal tow (**Figure 7.15**).

TECHNICAL BOX 7.1

How to Troubleshoot a Deep Horizontal Tow with a Little Help from Pythagoras

Let's say you were interested in conducting a deep horizontal plankton tow at the discrete depth of 15 m (d = 15 m). Before setting out from port, you were clever enough to measure that the height of the research vessel's A frame, relative to the water line of the vessel, was 4.5 m (h = 4.5 m).

$$d = adj - h \quad \therefore adj = d + h$$

$$adj = 15 + 4.5 = 19.5 \text{ m}$$

You toss your closed plankton net overboard, paying close attention to the total length of cable paid out from the A frame (hyp). After paying out 100 m of cable (hyp = 100.0 m), you use an inclinometer and find that the angle of the steel cable coming off the A frame is 76° from vertical (θ = 76°).

$$adj = hyp \cdot cos\ \theta$$
$$adj = 100 \cdot cos\ 76° = 100 \cdot 0.24 = 24 \text{ m}$$

But 24 m is a little too deep. Luckily, you were using a closed net and had not opened it yet. You reel in 25 m of cable (hyp = 75.0 m) and note that the angle has changed, ever so slightly, to 75°. A quick recalculation is performed, and you find:

$$adj = 75 \cdot cos75° = 75 \cdot 0.26 = 19.5 \text{ m}$$

As long as the captain maintains current speed, your horizontal tow will be conducted at your preferred depth of 15 m.

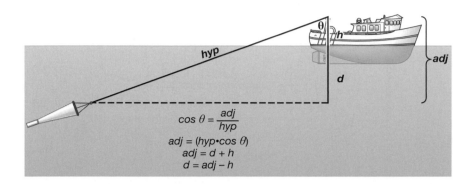

$$\cos \theta = \frac{adj}{hyp}$$
$$adj = (hyp \cdot \cos \theta)$$
$$adj = d + h$$
$$d = adj - h$$

Figure 7.15 In order to estimate the depth of a deep horizontal tow, the total length of cable paid out represents the hypotenuse (*hyp*) of the resultant right triangle. If the line angle (*θ*) can be determined using a simple inclinometer, the cosine of *θ* can then be used to determine the length of the adjacent (*adj*) arm of the triangle. Wherever the line is deployed from, its height above sea level (*h*), when subtracted from *adj* will yield the depth *d* at which the tow occurred.

Benthic tows (or "sleds" as they are sometimes called) are a special variety of deep horizontal tow, and require the use of a conical plankton net suspended in a wireframe cage, fitted with struts or "skis" so the entire device can be literally dragged along the bottom (**Figure 7.16**). Benthic sleds work well over sandy or shelly substrates, and the more robust designs can even be deployed over cobble or rocky substrates. Occasionally, benthic sleds can be used over soft mud substrates, but such soft sediments are easily disturbed by the dragging action of the sled, so there is significant risk that the collected plankton samples will be clogged with suspended sediment as well.

Another clever alternative to the deep horizontal tow is to use a pump to haul the water up to the net rather than to haul the net down into the water. This is usually accomplished by rigging a powerful water pump to rigid PVC plumbing on the deck of the ship (**Figure 7.17**). The intake vent on a specific length of rigid tubing is lowered to the desired depth, and the pump is engaged. The seawater (and suspended plankton) are then pumped up to the deck of the ship, where the water samples can easily be captured in sample bottles, or the water can be pumped through a screen mesh of the desired size, where the plankton are collected for further processing and study. This method becomes technically challenging at greater depths, as longer lengths of rigid tubing become susceptible to current shear, and the weight of the water column being pumped to the deck of the ship can easily overwhelm the capacity of the pump.

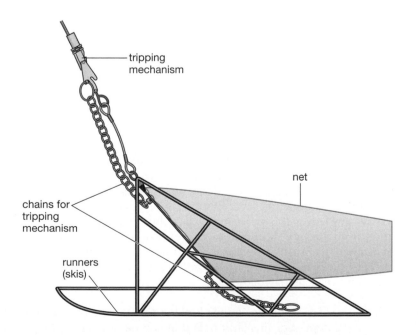

tripping mechanism

chains for tripping mechanism

runners (skis)

net

Figure 7.16 A typical benthic sled, used for horizontal tows just above the substrate. The style of runners (skis) is usually engineered specific to the substrate over which the sled will be pulled. Some sleds are equipped with a net-opening mechanism (sliding net bars), which can be triggered with a weighted messenger to open or close the net at a specific time or location during the tow.

Figure 7.17 A deep horizontal "tow" can also be accomplished by lowering a length of rigid pipe to the desired depth and pumping the water up to the deck while the ship maintains constant forward speed.

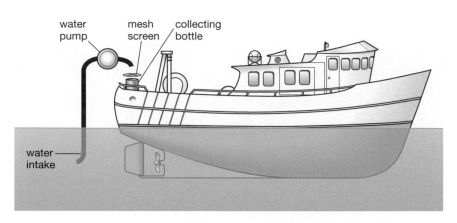

The Vertical Tow Is Best for Integrated Sampling of the Entire Water Column

Vertical tows (see Figure 7.13) are also quite simple to execute in the field, as the net is simply lowered to the desired depth (usually just above the bottom) and hauled vertically to the surface. For best results, the net should be lowered in the "closed" position, opened at depth, and hauled vertically at a consistent speed of approximately 1.0 m s^{-1}. If the plankton net in use does not feature a mechanism for opening and closing the net, excellent results can still be obtained with a "fixed-mouth" net if significant weight (1–2 kg) is added to the cod-end bucket. The weighted net, when lowered to depth at a very brisk pace, will essentially be "flushed out" on its way down.

Vertical tows are an excellent way to collect the broad community of plankton that occupy the entire water column, regardless of depth. This is particularly important with regard to species that exhibit daily vertical migration behaviors, because it is often very difficult to know exactly where in the water column they may be located at any given time of day. By using a vertical haul, the plankton will be captured in the net regardless of what depth they happen to be occupying when the plankton tow is executed.

One significant drawback to the vertical tow is the relatively meager amount of water that is actually being filtered for plankton collection. Whereas horizontal tows are typically conducted over distances of several hundred meters to a kilometer or more, the vertical tow distance is of course limited by the depth of the water column; in coastal waters and estuaries, these distances can be 10 m or even less! To compensate for such short tow distances, it is usually advisable to use a plankton net with the largest mouth diameter that can be reasonably handled. It is also useful to perform several replicate tows at each location; for example, three 20-m tows will theoretically filter the same amount of seawater as a single 60-m tow.

The Double-Oblique Tow Offers the Best of Both Worlds, Combining Features of the Horizontal and Vertical Tows

Horizontal tows, although very effective at screening plankton over very long towing distances, suffer from the fact that they can only sample at a single, discrete depth for each deployment (potentially "missing" a wide variety of plankton that might occupy several different layers of the water column). The vertical tow solves this problem by sampling the entire water column, but suffers from the very short distances over which the vertical tow is being hauled (limited of course by the local bathymetry). The double-oblique tow is a very effective hybrid strategy that combines the best features of both the horizontal and vertical tow methods.

Through the use of specially designed plankton nets that possess a combination of weights, depressor fins, and nets that can be opened and closed in order to collect plankton and modify the hydrodynamics of the net assembly, the double-oblique tow can perform a single (or a series) of undulating tow paths (see Figure 7.13). In this manner, the net can be towed over significant horizontal distances (to maximize the amount of seawater filtered for plankton), and its undulating path can collect plankton from all depths (except very near to the bottom). Although double-oblique tows require a more sophisticated net assembly than the standard conical plankton net, for extensive plankton surveys, the double-oblique method truly is the "gold standard" among plankton census methods.

Plankton Quantitation Requires Knowledge of the Filtered Volume of Seawater

As with virtually all biological census methods, it is absolutely critical that the species within the collection can be competently identified and quantified. Quantification is usually accomplished by taking a numerical count of individuals that belong to a particular species. Sometimes a different biometric (like biomass) is used instead of an actual count of individuals, but the concept is always the same: the quantitation of each individual in the collection, identified and grouped by species, is the keystone of all biological research.

The quantitation of plankton is most commonly expressed in terms of natural population density; that is, as an expression of the total number of individuals (n) belonging to a particular species (i) and present in a fixed volume of seawater (V). Thus, the population density for each plankton species (N_i) is calculated using Equation 7.1 as

$$N_i = \frac{n_i}{V} \qquad (7.1)$$

If the total biomass for each species (m_i) is used instead of an actual count, Equation 7.1 is expressed instead as a species-specific biomass concentration M_i, as defined by Equation 7.2:

$$M_i = \frac{m_i}{V} \qquad (7.2)$$

Performing plankton counts, or determining biomass, is a relatively straightforward process (at least, in concept). However, determining the true volume of water that was filtered through the plankton net during each deployment requires a bit more effort to calculate. If a submersible pump was used, the volume of water that was pumped and subsequently filtered for plankton can be easily collected in a large container below the filter screen and its volume measured directly. Otherwise, if the flow rate of the pump is known (Q, in units of $cm^3\,s^{-1}$ or $m^3\,s^{-1}$), a stopwatch can be used to measure the total time elapsed (t) during each pumping cycle. The volume V of water pumped (and filtered) is then calculated using Equation 7.3:

$$V = Q \cdot t \qquad (7.3)$$

If the water was filtered through a plankton net deployed in the water, the process involved in calculating V requires a bit more effort. Regardless of the shape of the plankton net used, if the cross-sectional area A of the net mouth can be calculated, the distance over which that net was towed is all that is required to calculate the volume of seawater filtered by that net. If a standard conical plankton net is used (**Figure 7.18**), it is a simple task to use

Figure 7.18 A plankton net whose mouth area A is towed a specific distance d will filter a discrete volume of seawater, represented by the geometry of a column (where $V = d\pi r^2$).

Figure 7.19 A digital flowmeter, used to estimate the distance traveled by a plankton net during deployment, based on the number of revolutions counted by the flowmeter's propeller.

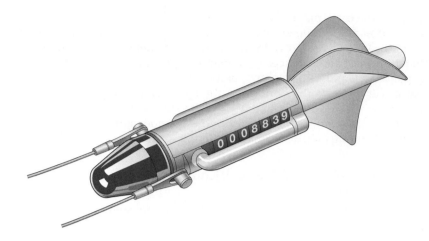

the mouth radius r to calculate the cross-sectional area of the net mouth (A), just as we would the area of a circle, using Equation 7.4:

$$A = \pi r^2 \tag{7.4}$$

Although it may be a simple concept, accurately measuring the distance of a plankton tow is quite challenging without the right equipment. The preferred method is to use a digital flowmeter, a device that resembles a tiny torpedo and is designed to count the total number of revolutions of its tiny propeller (**Figure 7.19**). The flowmeter should be mounted on the net bridle, ideally in such a way that it is completed suspended within the mouth of the plankton net.

Before each deployment, the digital counter on the flowmeter is set to zero. When the net is deployed, it is important to use a stopwatch to start the timer the instant the net is in the water. While towing the net, the water flowing past the flowmeter will turn its propeller in direct proportion to the velocity of the water. When the plankton net is recovered, the timer is stopped and the number of propeller revolutions on the flowmeter is counted. Each flowmeter is specifically calibrated so that the number of revolutions displayed on the flowmeter and the time elapsed (t) can be used to estimate the average velocity $\bar{v}$ of the net tow. Distance traveled (d) is then calculated using Equation 7.5:

$$d = \bar{v} \cdot t \tag{7.5}$$

Keep in mind that most plankton nets cannot filter the same volume of water through the net as quickly as it enters the mouth of the net; in fact, the **filtration efficiency** of towed nets will always decrease with mesh size. In order to account for this phenomenon, it is usually necessary to deploy the digital flowmeter under the exact same towing conditions, but without any of the attached net gear. The calibration distance $d°$ is then calculated using Equation 7.5, but is based on the number of revolutions displayed by the flowmeter when completely unimpeded by the plankton gear. In nearly all cases, the flow rate past the "naked" flowmeter is greater, so $d° > d$. Thus, the filtration efficiency F of a particular rig can be calculated as

$$F = \frac{d}{d°} \tag{7.6}$$

where F represents a dimensionless scaling factor, ranging between 0.0 and 1.0, that is used to correct for a particular net gear's impedance. For any tow (horizontal, vertical, or double oblique), Equation 7.7 can then be used to calculate the total volume of seawater (V) actually filtered by the net tow:

$$V = A \cdot d \cdot F \qquad (7.7)$$

If a flowmeter is not available, a much more crude method to estimate the tow distance is to simply instruct the captain of the research vessel to maintain constant speed while performing the plankton tow. A stopwatch is used to measure the time elapsed, and Equation 7.5 is used to estimate the distance traveled based on the ship's estimated velocity ($\bar{v}$).

In either case, once the distance d of the plankton tow is known, the total volume V of seawater filtered to collect the plankton sample is calculated using Equation 7.7. Then, it is a relatively straightforward task of counting the total number of individuals grouped by species (n_i) in the sample to determine the population density of each species (N_i) using Equation 7.1.

The Sample Volume from a Plankton Tow Can Be Used to Derive the Population Density of Plankton in Natural Waters

When a plankton sample is collected in the field, it will of course be concentrated in a very small volume of seawater trapped in the cod-end bucket of the plankton net. It is critical that the volume collected from the cod-end bucket is measured very accurately, as it will be necessary to know the sample volume V_S in order to later determine the natural population density of each plankton species (N_i).

Consider that the plankton in the sample were originally collected from a very large volume of seawater (V) filtered by the plankton net (calculated using Equation 7.7), but these plankton are now artificially concentrated in a very small volume of seawater in the cod-end bucket, which comprises the sample volume V_S. If both the volume filtered (V) and the volume collected (V_S) are known, you can use Equation 7.8 to calculate the volume scaling factor (f_V):

$$f_V = \frac{V}{V_S} \qquad (7.8)$$

When the number of plankton are ultimately counted in the collected sample (n_S), it is a relatively easy task to convert those counts into an estimate of the true population density (N, as the number of individuals per unit volume) using Equation 7.9:

$$N = \frac{\left(\dfrac{n_S}{V_S}\right)}{f_V} \qquad (7.9)$$

Plankton Samples Must Be Properly Preserved for Later Analyses

In most circumstances, it will be impossible to process the plankton sample(s) immediately after they are collected. In order to prevent damage and degradation to the plankton samples, it is often necessary to preserve them in some fashion, so they can be stored and analyzed later. Of course, if live plankton are needed (for photosynthetic, respiration, or some other metabolic study), they cannot be preserved at all; they must be utilized, in their natural state, as quickly as possible.

If any sort of biochemical analyses will be conducted on the plankton, you should avoid the introduction of any chemical preservative. In such cases, it is often best to freeze the sample for later processing. Unfortunately, the mere act of freezing many plankton will cause damage to their bodies, as the

Figure 7.20 An example of the clear acrylic Folsom plankton splitter. A plankton sample is poured into the central drum of the splitter and gently agitated to thoroughly mix the sample. When the drum is rotated, a single blade divides the sample exactly in half, and each half is then poured into matching collecting trays for further processing or preservation.

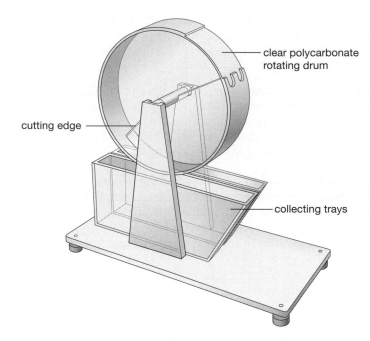

clear polycarbonate rotating drum

cutting edge

collecting trays

fluid within their cells will expand when frozen, causing cell or tissue rupture. If damage to the organism is significant enough to alter their anatomy, it may make species identification more difficult.

In some cases, the best way to preserve plankton for microscopic analysis and species identification is by using a chemical fixative (such as 2–4% buffered formalin) or a preservative (like 70% ethanol). Preserved samples can be kept at room temperature without fear of cell degradation for many, many years—and the microscopic anatomy of each organism shall remain largely unaffected. Of course, the use of any artificial preservative will also change the molecular chemistry of the plankton (and can even alter their wet, dry, and organic biomass).

Since there are significant advantages to using both freezing and chemical preservation methods, the use of a "plankton splitter" device delivers several different options for sample processing. The plankton splitter is a very simple device that uses a single watertight compartment that can be subdivided into two, identical subcompartments (thereby "splitting" the plankton sample). Although there are several designs, the Folsom splitter (**Figure 7.20**) performs well and is quite easy to use. As long as the plankton sample is well mixed, the "split" samples are assumed to be identical to each other. Using the splitter, one half of the plankton sample can be frozen (for biochemical analyses) and the other can be chemically preserved for later species identification and long-term archiving.

The Methods for Processing Plankton Rely On Representative Subsamples

Because most plankton are microscopic, a very small volume of seawater can contain thousands (if not millions) of individuals. This is more the case when plankton are collected by filtering seawater through a mesh screen or net, concentrating them in an even smaller volume than they naturally occupied in the ocean. As a practical matter, very seldom is it possible to count and identify every single organism in a concentrated plankton sample. Usually, 3–5 aliquots are taken from each plankton sample for analysis, and the resultant counts (and species identifications) from the aliquots are "scaled up" to represent the whole sample from which they were taken.

To maximize consistency, it is important that the volume of each aliquot is exactly the same. It is also important that the device chosen to recover such small volume aliquots does not damage the delicate microscopic plankton during collection. This is most often accomplished using a Hensen–Stempel pipette, an inexpensive device specifically engineered to collect a consistent volume every time, without damage to the planktonic organisms (**Figure 7.21**).

For plankton samples collected in estuaries or in shallow coastal waters, plankton densities are usually high enough to not require any further concentration of the sample prior to analysis. However, for samples that have very low plankton densities (or for those samples dominated by the smallest plankton size classes), it may be necessary to let the plankton "settle out" of the sample, so they become even more concentrated by the very act of sinking to the bottom of the counting chamber.

Utermöhl settling chambers are simple but ingenious devices that consist of a series of sample tubes, each of which are seated atop a special recessed microscope slide (**Figure 7.22**). When the plankton sample is delivered to the sample tube, the plankton will naturally settle to the bottom over time. A glass plate is then slid between the sample tube and the microscope slide underneath it, trapping the settled plankton in a tiny recess in the microscope slide. The slide, which contains all of the plankton that settled out of the sample, can now be removed from the Utermöhl device and placed on a microscope for plankton enumeration and identification.

Counting and Identifying Plankton Is a Visual Enterprise

Small aliquots of the plankton sample, dispensed from a Hensen–Stempel pipette or settled from a Utermöhl settling chamber, will eventually be deposited into plankton-counting chambers, which resemble thick-bodied microscope slides that have been constructed to hold a very precise volume of water. There are several different varieties of plankton-counting chambers, but the most commonly used are the 1000-μL Sedgwick–Rafter cell, the 100-μL Palmer cell, and the 50-μL haemocytometer cell. Samples with plankton that are few in number, or that have very large body sizes, are best accommodated in the larger counting cells. Samples that are very concentrated will almost certainly require the use of a haemocytometer, which is designed to hold a very small volume and has a microscopic, multidimensional grid when viewed under a microscope (to aid in counting cells).

Because plankton-counting cells are quite thick, it may not be possible to use the higher magnification lenses on standard microscopes, which are

Figure 7.21 An example of the Hensen–Stempel pipette, specifically engineered to provide consistent, high-precision aliquots of small volumes (1.0, 5.0, 10.0, or 20.0 mL). The unique "sealed plunger" mechanism ensures that microscopic cells remain undamaged when the aliquot is taken from the larger plankton sample.

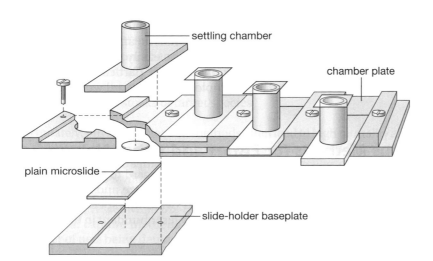

Figure 7.22 An example of the Utermöhl settling chamber, a device used to concentrate plankton on microscope slides as they sink to the bottom of the settling chambers. (Adapted, courtesy of Aquatic Research Instruments.)

Virtual collection methods eliminate the possibility of species mortality associated with the sampling method, as such methods typically rely on live observations, photographs, or video footage while performing the census. More sophisticated imaging systems can use water-penetrating lasers or high-frequency sound waves (sonar) to collect "images" of the nekton using the light or sound waves being reflected from the organisms (see Figure 7.6). Because virtual methods are noninvasive (and are presumed to be nondestructive), they are also useful in capturing the natural behavior of nekton.

Although virtual methods can be used to preserve images and data of the "captured" specimens, without physical access to the organism it is impossible to accurately measure the full panoply of biometric measurements that could be performed on a specimen that was truly "in hand." In some cases, the positive identification of a species can only be done from a physical collection of specimens. The technical aspects of virtual collection methods can also lead to significant (and expensive) technology issues, and most imaging systems (or observers) have a relatively limited field of view within the region where the collection is taking place. Such collections are also difficult to calibrate against effort, so CPUE statistics may not be possible.

For these reasons, it is far more common that nekton surveys will rely on the physical collection of organisms; thus, the census methods discussed for fishes and other nekton shall focus on the various types of net-collection strategies hereafter. Nonetheless, the reader is encouraged to research the various virtual methods of collection, as there are distinct and obvious benefits when a less destructive sampling method can be identified. Keep in mind that the "mark-and-recapture" methods previously discussed in Chapter 6 are easily adaptable (and are quite commonly used) for nekton surveys. That being said, CPUE measures are arguably the most prevalent (and most important) metric used to survey the nekton.

Biometry Statistics Are Critical to Nekton Surveys

Not surprisingly, nekton surveys require the identification (to species whenever possible) and the enumeration of all collected specimens. However, because of the large size of most nekton, species identification does not require the aid of a microscope and can typically be performed by trained personnel at the time of collection. Beyond a simple count of individuals within each species, the most important biometrics to be measured at the time of collection for most nekton (and especially for fishes) include

- Live (wet) biomass

and

- Standard fork length for all other fishes

or

- Lower jaw fork length (LJFL) for billfishes only

While the live (wet) biomass of each specimen can be easily measured in the field using a balance or scale, length measurements are most often conducted using a nylon measuring tape for large specimens, stainless-steel calipers for small specimens, or a flat measuring board for everything in between (Figure 7.24). Length measurements that utilize either calipers or a flat measuring board will always provide a "straight body length" value. Length measurements that rely on the use of a fiberglass tape represent a "curved body length," since the tape must follow the body contour of the

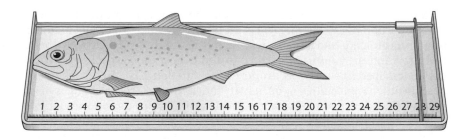

Figure 7.24 A flat measuring board (sometimes called a "fish board") is typically fashioned from noncorrosive materials and is built with an integrated ruler to allow swift length measurements of nekton specimens in the field.

specimen. Since there are obvious differences between straight and curved body length measures, it is critical to record the type of body length measure (and equipment) being used.

Choosing the appropriate **fork length (FL)** measurement is quite simple. For nearly all fishes, the regular fork length (FL) simply measures the distance from the tip of the upper jaw to the tail fork (Figure 7.25). For billfishes, the lower jaw fork length (LJFL) is used instead of the regular fork length, and is measured as the projected straight distance from the tip of the lower jaw to the posterior tip of the shortest caudal fin ray (that is, the "fork" of the tail). On occasion, collected specimens may have a damaged (or missing) caudal fin, making it impossible to perform any fork length measurements. In these rare cases, it may be necessary to measure the standard length (SL), predorsal length (LD_1), or the pectoral anal length (PAL) and to consult the literature for the appropriate conversions to the fork length for that particular species.

The Natural Abundance of Nekton Can Be Estimated Using "Catch Per Unit Effort" (CPUE) Data

Using live (wet) weight and/or fork length biometrics, it is possible to assess the relative abundance (and age classes) of particular nekton species based on an analysis of how much effort is ultimately expended in the capture of those specimens—this is what is known as a measure of "catch per unit effort" (CPUE). In its simplest form, CPUE can be defined using Equation 7.10:

$$CPUE = q_i \cdot N_i \qquad (7.10)$$

where the natural abundance of a particular species (N_i, as the total number of individuals available for capture among species i) is related to CPUE according to an empirical proportion of catchability (q_i) for that same species. This same relation can be used to estimate the total number of individuals captured within a specific size class, using fork length or some other body length metric. Equation 7.10 can also be adjusted to calculate the CPUE in terms of biomass M_i instead of numerical abundance N_i.

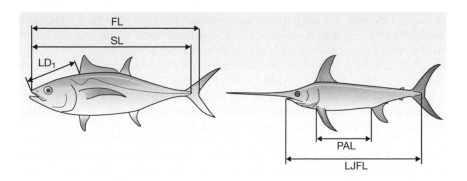

Figure 7.25 Typical fork length measurements for fishes, including alternate measurements for billfishes.

Of course, the apparent simplicity of Equation 7.10 is misleading. First of all, the proportion of catchability (q_i) is an unknown value that must be determined empirically, and its value will change depending on the species being collected and the gear used in the effort to collect those specimens. Using the generalized linear model proposed in Equation 7.10, the "catchability" of a specimen is presumed to be density independent; that is, changes in the natural abundance N_i of the target species will induce a proportional (linear) change in CPUE, but catchability remains constant. Although several fisheries researchers have suggested that alternate, nonlinear relationships between CPUE and the natural abundance N_i are more realistic, the linear model used in Equation 7.10 is by far the most heavily utilized mathematical model in fisheries research.

Since CPUE measures are so prevalent in nekton stock assessments, it is critical that all of the factors that could potentially affect the proportion of catchability (q_i) are normalized to remove as much bias as possible. This typically requires that the exact same gear, and the exact same methodologies, be used throughout all collections of a particular species. For this reason, it is absolutely essential that investigators report the specific gear and deployment strategies used at the time of collection, and that the collection methods never veer from these particulars. It is for this reason that comparisons of CPUE data (performed at different times, or in different regions, or with slightly different gear, etc.) are very difficult to make. In order to make sense of long-term or repeat-sampled CPUE data, the collection gear and methods used must be exactly the same, in every detail.

Nets or Hook-and-Line Capture Methods for Nekton

Like the plankton methods discussed earlier, nekton collections also rely heavily on the filtration of seawater (and the efficient retention of specimens) using mesh screens or nets, but deployed on a much larger scale. Unlike the plankton, the sometimes large and fast-swimming nekton offer unique challenges when attempting their capture. This introduces the possibility that nekton might exhibit different avoidance behaviors, depending on the species, the type of gear, and the collection strategy used. All of these factors make nekton surveys far more challenging to execute in the field than plankton collections; at least, the catchability of nekton is far more complex than plankton. Therefore, it is often necessary to use more specialized gear, and more creative collection methods, to adequately survey the nekton.

Seine Nets Can Be Easily Deployed in Shallow Water by Boat or by Hand

Seine nets are long rectangular nets that are fitted with weights along the bottom edge of the net (so that the lower edge of the net drags along the bottom) and a series of floats attached to the top edge, deployed in open water. The mesh size of the seine is carefully selected so as to prevent the escape of the target species; it is not to physically ensnare the nekton (which should be avoided, as ill-sized nets can easily damage organisms that become lodged in the net). Seine nets are typically deployed from a ship or small boat, but smaller (2.5–5 m) "beach seines" can be deployed in shallow water by two people dragging the net towards shore (**Figure 7.26**). Standard seine nets are particularly effective in capturing relatively small to moderately sized nekton that are also somewhat sluggish swimmers (and are therefore easily ensnared by a static or slowly deployed seine net).

To capture all nekton (including large schools of swift swimmers), a purse seine is an excellent alternative. Purse seines are typically tall enough such

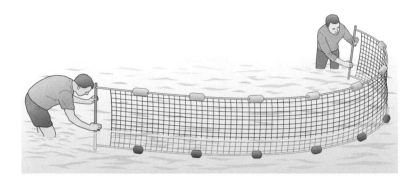

Figure 7.26 A typical seine net, with characteristic "wall-like" construction. The weighted line is dragged along the bottom (to prevent escape underneath the net), while the float line at the top keeps the net from collapsing during the tow.

that the float line remains at the surface and the weighted line reaches all the way to the bottom. One end of the purse seine is deployed, and while the seine is continually paid out from the deck of the ship, the vessel makes a quick circle in open water, creating a top-to-bottom wall of net that curls back on itself (**Figure 7.27**). The weighted line at the bottom is then hauled aboard much more quickly than the float line, causing the bottom of the net to close off and form a purse, in which all nekton are trapped with no possibility of escape (unless they leap over the float line prior to capture).

Trawl Nets Are Used in Deeper Waters and Require the Use of a Vessel

Trawl nets are mid-water (or near-bottom) nets that are usually so large that they must be towed by a powered vessel. Designed very similarly to the seine, trawls possess the same weight- and float-line construction of a seine, but they typically possess a sacklike extension that extends far behind the main body of the net and acts like a funnel to concentrate all of the collected nekton at the cod end (**Figure 7.28**).

Among the different types of trawl nets, the most common is the otter trawl, which possesses large wooden "otter boards" on either side of the mouth of the net. As the vessel moves forward and the doors first enter the water, the water resistance causes the boards to swing open, keeping the mouth of the net open as long as the vessel maintains forward velocity. Beam trawls are similar in design, except they lack the otter boards and instead possess a single, rigid beam across the top of the net, which forces the net to remain open regardless of the towing vessel's forward speed.

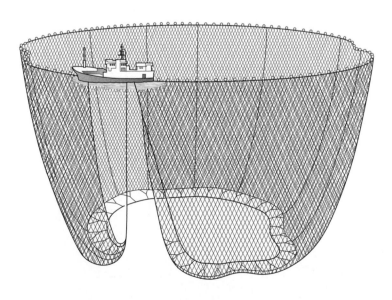

Figure 7.27 A large purse seine, gathered from the bottom, is a particularly effective way to trap large, swift-swimming nekton.

Figure 7.28 A typical otter trawl, showing the characteristic "otter boards" located on either side of the net.

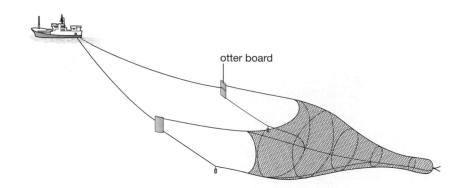

otter board

Because of their weight (and relatively slow tow speed), trawls are best suited for collections of **demersal** nekton. Because they are dragged along the bottom, trawls will also collect a substantial number of benthic species as well. Because of their relatively fragile construction, otter and beam trawls should only be performed in regions where the substrate is known to be mud, sand, or gravel (as hard bottoms such as exposed rock or oyster or coral reefs will undoubtedly destroy the trawl net). For hard-bottom trawling, armored cage nets or dredges are required.

Dredges Must Be Pulled Along the Bottom to Collect Benthic Organisms

Dredges are a special type of rigid trawling "net," usually constructed from heavy-gauge steel resembling a cage (**Figure 7.29**). Owing to their heavy construction, they are best utilized over and around hard substrates where traditional nylon seines or trawl nets would be ripped to shreds. Their heavy weight is also ideal for digging into the soft benthos, so they are also used for the collection of large infaunal species (mostly bivalve molluscs, like clams and scallops). Oysters are also commonly collected using dredges, although the construction of an oyster dredge is designed to specifically snag on the oyster reef. When increased tension is applied to the dredge cable, entire sections are pried off the living reef and hauled aboard.

Cast and Push Netting are Best Suited for Use in Bays, Tidal Creeks, and Other Small Coastal Areas

For nekton collection in small, enclosed spaces, the use of hand-held nets may be sufficient. These can come in a variety of forms, from long-handled scoop nets to banner-sized push nets. Of course, hand-held nets are only useful in shallow bays, estuaries, and coastal waters, and are best used for the targeted capture of single individuals.

In slightly deeper waters, small cast nets are more effective at capturing small nekton. The cast net is made from a circular net with weights along

Figure 7.29 Towed dredges are often used to collect shellfish from the benthos. They can also be used over hard substrates without fear of damaging the sampling gear, due to their sturdy metal construction.

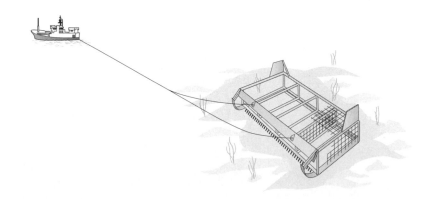

the circumference of the net. A choke collar is woven into the circumference of the net as well, so that the net will tighten at the bottom when tension is applied to the cast line (**Figure 7.30**).

Gill Netting is Used to Collect Specimens from Vast Open-water Areas, Over Extended Periods of Time

Gill nets differ from seines and trawls in that the gill net is specifically designed to ensnare the nekton rather than to merely impede movement. Gill nets are designed as walls of low-visibility filaments arranged to produce a mesh with diamond-shaped apertures. When deployed in open water, anchored in place, and buoyed with floats to keep the net fully extended and erect at specific depths, the net itself is almost invisible. The apertures of the net are designed so that a fish swimming into the net will pass partially through the aperture until its body is ensnared by the net filaments and are prevented from "backing out" of the net, as the fish's gill covers become lodged as well (**Figure 7.31**).

Since gill nets are quite effective at trapping a wide variety of nekton, they are particularly useful as a census tool because their catch will better represent the entirety of the nekton community. However, they are a particularly lethal method of sampling the nekton, and there are typically a tremendous amount of collateral losses (including marine birds, mammals, and reptiles).

Trapping Allows the Collection of Specimens that are Unharmed during their Capture

Similar to the mesh traps that are so effective at trapping benthic species (as discussed in Chapter 6), when traps are positioned at various depths within the water column (using anchor lines and/or floats), such traps can also be used to collect pelagic nekton. It is also possible to anchor a series of submerged seine nets in such a way as to create "fish fences," which are not designed to ensnare fishes but are instead used to corral them into dead-end traps (**Figure 7.32**). Although fish-fence traps are most effectively used for the collection of demersal species, they can also be employed in mid-water depths using a complicated array of anchor lines. Passive, nonensnaring nets such as these do not induce panic among the nekton, so they tend to follow the fence line into the trap rather than to simply swim over or under the blocking seine.

Hook and Line Methods are Typically More Selective than Gill Netting

Hook-and-line capture is generally a less destructive collection method than using gill nets, simply because only one (or a few) organisms are being captured at any given time. However, such methods are only useful in the collection of certain nekton (primarily large game fishes), as the method itself requires that the collection device be baited. Since there are many intangible variables that will influence the effectiveness of the hook-and-line method (not the least of which is bait selection), it has its limitations as a community assessment tool. However, it is one of the few effective methods for capturing large game fishes (including sharks).

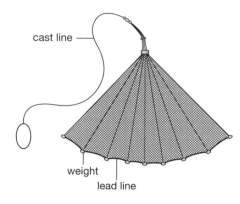

Figure 7.30 A standard design for a cast net, showing the unique closable lead line when tension is applied to the cast line.

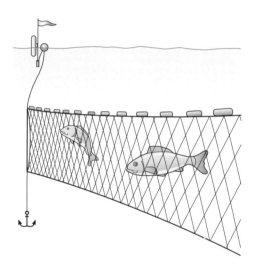

Figure 7.31 Gill nets are a particularly effective method of collecting the broadest community of pelagic nekton.

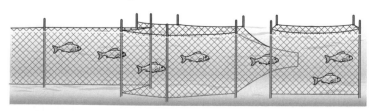

Figure 7.32 Traps, if well designed and well positioned, can be extremely effective at collecting nekton (especially demersal species).

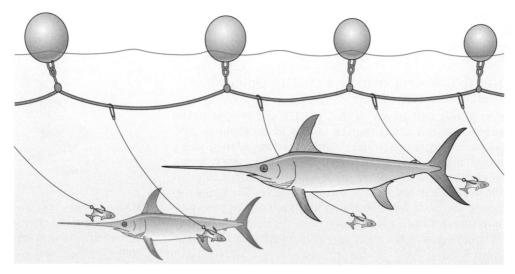

Figure 7.33 Long-lining is an extreme example of the "hook-and-line" collection strategy. Because of the explicit method of capture, it is only useful on predatory fishes (collections of which can be heavily biased by the researcher's choice of bait).

In order to increase catch rates, it is sometimes useful to employ a "long-line" method. Long-lining is essentially an extremely long rope with several baited hooks extending below the main line at different depths (**Figure 7.33**). Although long-lining is an effective method for collecting certain species, the line must be attended carefully and quickly so that captured specimens are recovered in a timely fashion. If the long lines are too long, the effective capture rate can be significantly reduced because hooked fishes, when left too long on the line, are themselves eaten by opportunistic predators.

How to Define "Effort" Within CPUE Analyses, Based on the Capture Method

Regardless of the gear ultimately selected for collecting nekton, the catch is relatively easy to define, either as (1) a count of all specimens collected (n); (2) a count of specimens belonging to some predetermined size class (x), usually based on fork length ($n_{FL} = x$); or (3) the total biomass or live weight of specimens captured (m). When these measures are normalized against the effort (E) associated with their catchment, Equation 7.11 can be used to calculate the *CPUE*:

$$\frac{n}{E} \quad or \quad \frac{n_{FL} = x}{E} \quad or \quad \frac{m}{E} = \text{CPUE} \tag{7.11}$$

Since "effort" is an inherently subjective measure, it is important to standardize how it is defined. As you can well imagine, different gear require different effort in its proper deployment. Below are some guidelines to help you define a standard measure of effort relative to the type of gear you have chosen (please note that the recommended units of effort are listed in descending order of preference for each gear type).

Seine, trawl, dredge, or net (including gill nets)
- Total "soak time" of net deployment(s)
- Total distance seined, trawled, or dredged
- Total number of net deployments
- Total number of trips made

Traps
- Total "soak time" of traps(s)
- Total number of trap deployments

Hook and line
- Total "soak time" of hook-and-line deployment(s)
- Total number of hooks used
- Total length of hook-and-line gear
- Total number of trips made

Analysis of Pelagic Collections

Similar to the benthic census methods discussed in Chapter 6, the types of ecological analyses that are possible from collections of pelagic organisms (be they viruses, bacteria, phytoplankton, zooplankton, or nekton) are virtually limitless, so long as you are able to (1) identify, (2) enumerate, and (3) perform targeted biometric measurements of all the species within your collection. This chapter is meant to provide the reader with a thorough introduction to the variety of methods for collecting pelagic organisms. The types of measurements (and subsequent analyses) you might choose to perform on the specimens in your collection cannot be anticipated, nor should they be prescribed. Instead, the reader is encouraged to refer to Appendix F of this text, which serves as a detailed compendium of the fundamental biometric and analytic methods to utilize, once your specimens have been collected.

References

Bakus GJ (2007) Quantitative Analysis of Marine Biological Communities: Field Biology and Environment. John Wiley and Sons.

Blackwell SM, Moline MA, Schaffner A, Garrison T, & Chang G (2008) Sub-kilometer length scales in coastal waters. *Continental Shelf Research* 28:215–226.

Cochran WG (1977) Sampling Techniques, 3rd ed. John Wiley and Sons.

Coyer J & Witman J (1990) The Underwater Catalog: A Guide to Methods in Underwater Research. Cornell University, Shoals Marine Laboratory.

Harris R, Wiebe P, Lenz J, Skjoldal H-R, & Huntley M (eds) (2000) ICES Zooplankton Methodology Manual. Academic Press.

Margalef R (1978) Sampling Design: Some Examples. In Phytoplankton Manual (A Sournia ed), pp. 17–31. *Monographs on Oceanographic Methodology* 6, UNESCO.

Omori M & Ikeda T (1992) Methods in Marine Zooplankton Ecology. Krieger Publishing.

Southwood TRE (1978) Ecological Methods. Chapman and Hall.

Sutherland WJ (ed) (1996) Ecological Census Techniques: A Handbook, 2nd ed. Cambridge University Press.

Further Reading

Boistel R, Swoger J, Krzic U, Fernandez V, Gillet B, & Reynaud EG (2011) The future of three-dimensional microscopic imaging in marine biology. *Marine Ecology* 32(4):438–452.

Cotter J (2009) Statistical estimation of mean values of fish stock indicators from trawl surveys. *Aquatic Living Resources* 22(2):127–133.

Crippen RW & Perrier JL (1974) The use of neutral red and evans blue for live-dead determinations of marine plankton. *Stain Technology* 49:97–104.

Kemp PF, Cole JJ, Sherr BF, & Sherr EB (eds) (1993) Handbook of Methods in Aquatic Microbial Ecology. CRC Press.

Fleming JM & Coughlan J (1978) Preservation of vitally stained zooplankton for live/dead sorting. *Estuaries* 1:135–137.

Godoy EAS, Gerhardinger LC, Daros F, & Hostim-Silva M (2006) Utilization of bottom trawling and underwater visual census methodologies on the assessment of the fish communities from Arvoredo Biological Marine Reserve, SC, Brazil. *Journal of Coastal Research* 39:1205–1209.

Haddon M (2011) Modelling and Quantitative Methods in Fisheries, 2nd ed. CRC Press.

Horinouchi M, Nakamura Y, & Sano M (2005) Comparative analysis of visual censuses using different width strip-transects for a fish assemblage in a seagrass bed. *Estuarine Coastal and Shelf Science* 65(1–2):53–60.

John SG, Mendez CB, Deng L, Poulos B, Kauffman AKM, Kern S, Brum J, Polz MF, Boyle EA, & Sullivan MB (2011) A simple and efficient method for concentration of ocean viruses by chemical flocculation. *Environmental Microbiology Reports* 3(2):195–202.

Mullin MM & Brooks ER (1976) Some consequences of distributional heterogeneity of phytoplankton and zooplankton. *Limnology & Oceanography* 15:748–755.

Quinn TJ & Deriso RB (1999) Quantitative Fish Dynamics. Oxford University Press.

Wassenberg TJ, Blaber SJM, Burridge CY, Brewer DT, Salini JP, & Gribble N (1997) The effectiveness of fish and shrimp trawls for sampling fish communities in tropical Australia. *Fisheries Research (Amsterdam)* 30(3):241–251.

Zale AV, Parrish DL, & Sutton TM (eds) (2013) Fisheries Techniques, 3rd ed. American Fisheries Society.

Unit 3
Methods of Data Analysis

Contents

Chapter 8

Introduction to Univariate Analysis

"To be is to be the value of a variable." – Willard van Orman Quine

Remember the good ol' days when you were first learning algebra? The equations with only a single variable were always the easiest to solve. Turns out the same is true with statistics: **univariate** (single-variable) analyses are the easiest to set up for hypothesis testing, and they also result in very obvious, straightforward conclusions.

Univariate comparisons are at the very heart of scientific inquiry. After all, the cardinal rule of science is to limit the number of variables in an experiment so as to limit any sources of bias, while at the same time limit the confounding effects of multiple influences on the outcome. So the ultimate experiment should have only one variable. Many, many experimental groups, but only one variable.

Unfortunately for us, Mother Nature can rarely be convinced to keep things that simple; at least, not in a natural field setting. But even if your field program is designed to measure lots and lots of different variables, we can still choose to analyze each variable, one by one, when we make our comparisons between our sampled populations. Sticking to one variable makes things easy, but "easy" can still yield very important results.

Key Concepts

- Investigations of single-variant dynamics require that only a single variable be analyzed, separate from the rest.
- The *t* test is the most common statistical test used to make single-variable comparisons between different populations of data.
- The *t* test can be used to compare measures of central tendency (that is, the mean, median, or mode) between two populations.
- The *t* test can also be used to compare measures of the standard deviation or variance between two populations.
- Comparisons between multiple (three or more) populations can be done simultaneously using the one-way analysis of variance (ANOVA).

The Foundations of Univariate Analysis

As we have seen in the previous chapters, great attention has been paid to the many different kinds of variables that are deemed important in the aquatic sciences. And while we have also explored the practical methods by which those variables can (and should) be measured, it is critical that we look beyond the mere numbers on the page and engage ourselves in the effort to analyze those measurements. In most circumstances, the data we have gathered throughout the course of our laboratory and/or field efforts are most useful to us in the context of "comparative analysis." Univariate analyses represent the simplest, most straightforward method for making comparisons between populations.

If we wished to use inferential statistics (Chapter 2) to make those comparisons, there are a variety of statistical tests designed to determine equality (or inequality) between two or more sets of measurements of the same variable. For example, if we were to focus our attention on a single variable, we could use the unique distribution of our data (using central tendency, standard deviation, or variance) to compare our measurements against some other theoretical value or distribution.

As we learned in Chapter 2, the basic test of equality is accomplished using a two-tailed test, where the null hypothesis H_o assumes the compared measures are not significantly different from each other (in other words, they are equal to each other). In the course of our statistical tests, if we are able to confirm the alternative hypothesis H_a that the compared measures are in fact not equal, then we can then follow up with a right- or left-tailed test to determine which population is significantly greater than the other.

As elementary as those concepts seem, don't let their simplicity fool you—statistics need not be complicated in order to be powerful or insightful. There is sublime elegance in the ability to determine equality (or the degree of inequality) based on the comparative measures using a single variable. Well, that may be romanticizing things a bit, but the point should be well taken: always opt for simple when simple will do.

That being said, there are definite limits to what information we can glean from single-variable analyses. One of the most notable weaknesses of univariate analysis is the fact that without the use of more than one variable, we are also limited in our ability to determine correlations. By definition, any associative relationship between one variable and another usually requires a minimum of two variables. Likewise, if we are interested in establishing a directional association that implies causation, we typically have to use two or more variables in that analysis.

For example, let us assume that we are using biometric data from juvenile fishes collected from a seine net survey (**Figure 8.1**). From our collection, we have measured the standard length SL and standard depth SD among all Atlantic menhaden (*Brevoortia tyrannus*) and Atlantic croaker (*Micropogonias undulatus*) specimens collected (**Table 8.1**).

If we were interested to explore the growth dynamics of Atlantic menhaden, it would be reasonable for us to expect that as the fish grows, that growth would affect each fish's length (SL) and its depth (SD), and that the two measures are in some way related to (correlated with) each other. Thus, univariate analysis would not be possible, because we would need to compare two variables in order to establish that relationship. In most circumstances,

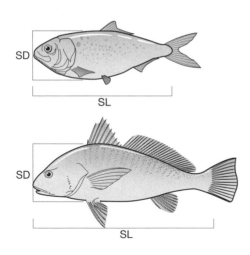

Figure 8.1 Biometric dimensions of Atlantic menhaden (*Brevoortia tyrannus*, top) and Atlantic croaker (*Micropogonias undulatus*, bottom), measured as standard length (SL) and standard depth (SD) for each fish collected in the sample. (Courtesy of the Maryland Department of Natural Resources.)

Table 8.1 *Standard Length (SL) and Standard Depth (SD) Measurements Taken from Juvenile Atlantic Menhaden (B. tyrannus) and Atlantic Croaker (M. undulatus)*

	B. tyrannus		M. undulatus	
	SL (mm) N = 14	SD (mm) N = 14	SL (mm) N = 14	SD (mm) N = 14
	102	44	89	36
	97	41	62	29
	78	36	90	36
	52	28	68	31
	112	46	54	27
	91	39	88	35
	36	18	65	26
	44	20	78	33
	27	17	65	25
	60	31	58	28
	89	38	47	24
	118	47	102	45
	98	41	86	36
	<u>107</u>	<u>45</u>	<u>74</u>	<u>32</u>
Mean ($\bar{X}$) =	79	35	73	32
Std Dev (s) =	30	11	16	5.7

correlations and causative relationships within our data can only be investigated within the context of multivariate analysis (which we shall explore in the next chapter).

Instead, let's say you wanted to compare the standard lengths SL between the two species in your collection. Then it would be a relatively simple matter to determine the central tendency, standard deviation, and variance of SL among your menhaden and croaker specimens and make your comparisons. In this case, you would be using only one variable to make your comparisons (that is, the standard length of each species), so univariate analysis would be appropriate.

You may be asking yourself—wouldn't "species type" be considered a second variable in this example? The answer is: it depends on how the comparisons are structured, and whether that second variable is intended to be used as a grouping variable or as a scaled variable. Let's consider this more closely.

Univariate Comparisons Must Be Structured to Possess Only One Scaled Variable

Recall from our initial discussion in Chapter 2 that there were two fundamental types of measurements: scaled measures and nominal measures. Scaled measures are those that can be ordered according to a continuous scale (which would also include ordinal measures). In our earlier example using menhaden and croaker biometrics (see Table 8.1), it is easy to see that the standard length and/or depth of a particular specimen will vary

Figure 8.3 Site-to-site comparisons can be performed using univariate analysis as long as the stations are discriminated from each other using a nominal (categorical) measure, like Transect # - Station #. In that case, pairwise comparisons can be made between individual stations (for example, Station 2-1 versus Station 2-7, or Station 2-7 versus Station 5-7). Comparisons can also be made between entire transects (for example, Transect 2 versus Transect 5).

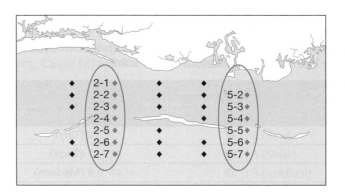

you have established a number of field stations along a number of cross-shelf transects (Figure 8.3). If you wanted to use Transect 2 and compare the SL of menhaden captured at the nearshore station (Station 2-1) against those from the offshore station (Station 2-7), the use of a nominal variable like "Station #" would permit univariate analysis because "Station #" is being used as a category, not as a scale measurement. However, if you decided to discriminate your stations by using something like "distance from the coastline," your comparisons between Station 2-1 and Station 2-7 would instead be based on a continuous scale measurement (that is, distance from the shore). In that case, you would have to use multivariate analyses (see Chapter 9) to look at the correlation between menhaden SL (scaled variable #1) and distance from shore (scaled variable #2).

Common Statistical Methods for Univariate Analysis

Just because we are focusing our attention on a single variable does not mean that the many, many measurements of that variable will be uniform. Quite the contrary, we expect there to be differences in those measurements. As we learned in Chapter 2, one of the most fundamental ways to describe a nonuniform dataset is by its central tendency; that is, by its mode, median, or mean.

When performing comparisons between two datasets, it is a common practice to compare the central tendencies of those datasets; chief among these are comparisons of the mean ($\bar{X}$). Conceptually, this makes the most sense to investigators, as the mean represents the "average condition" of the measured variable. The central tendency is usually the best way to quantify the variable in question, so it naturally follows that it would also be the best way to make comparisons between populations.

However, it is also possible (and in some cases more desirable) to compare the distribution of data rather than the central tendency of that data. Recall that the standard deviation s represents the basic tendency of the measured data to depart from the mean. Thus, a small standard deviation is indicative of minimal variability within the dataset. Another important descriptor of data distribution is the variance s^2, which is simply defined as the square of the standard deviation (so it is an even more sensitive method to describe departure from the mean).

Typically, comparisons of central tendency are used when investigators wish to test how accurately the datasets represent the populations from which the measurements were taken. Thus, comparisons of central tendency are thought to best represent the population from which the data were taken. For example, the mean standard lengths of our juvenile menhaden and croaker from Table 8.1 are 79 mm and 73 mm, respectively. These values represent the central tendency (in this case, the mean SL) for juvenile menhaden and croaker, so any comparisons made between the two are comparisons made between the "average condition" of each sampled population. Comparisons between the standard deviations (or variances) can also be made, but they are only concerned with how the data are distributed about the central tendency, not what the data actually represent.

Although it would at first seem preferable to compare central tendency rather than the data distribution, it is important to consider both. After all, it is conceivable that two populations might share the same mean as a coincidence, but have very different distributions. Likewise, two populations might have identical distributions, but very different central tendencies (**Figure 8.4**). Each will have its own meaning relative to the other, so it is important that you carefully consider which strategy offers the most value to your analyses. For example, you may not care about the average standard length of menhaden or croaker juveniles; it may be more important to study the overall variability within your measures of SL. If that were the case, you might be more interested to compare the standard deviations (or the variances) between the two populations.

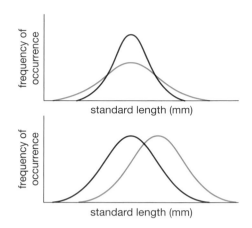

Figure 8.4 It is possible for two populations to share the same central tendency despite having very different distributions (top), just as it is possible for two populations to exhibit identical distributions but with different central tendencies (bottom).

The Most Common Statistical Tool for Univariate Analysis Is the *t* Test

As we have discussed in Chapter 2, there are a wide variety of statistical tests that are available to the researcher, depending on whether the data are normally distributed or not (see Figures 2.10–2.11). Although it is impossible to offer a "one size fits all" statistical test for the reader, what can be offered is a clear, concise review of one of the most common statistical methods used in the analysis of single-variant datasets. This of course would be the ***t*** **test**.

In the parlance of statistics, the *t* test is warranted for the analysis of single-variant data from one or more sampled populations, provided that the measures being compared are (1) normally distributed, (2) homoscedastic (exhibit equal variances), and (3) sampled independently from the two populations being compared. Hence, the *t* test is a parametric test, and a very powerful one at that. The power of the *t* test lies not only in its mathematical robustness, but more importantly, in its simplicity (thus making it rather easy to interpret the results). Essentially, when the *t* test is performed, a test statistic *t* is calculated from the sample mean ($\overline{X}$, using Equation 2.2) and analyzed for the population's deviation *d*, relative to the estimated standard error *SE*, as defined in Equation 8.1:

$$t = \frac{\overline{X} - d}{SE}, \quad \text{and} \quad SE = \frac{s}{\sqrt{N}} \quad (8.1)$$

where *N* represents the total number of observations in the dataset and assumes that the standard deviation *s* can be estimated from Equation 2.3. However, it is unnecessary for the reader to perform these calculations longhand using Equation 8.1, as all modern statistical software programs are capable of performing the *t* test statistic calculations as a routine operation.

In order for the *t* test to be of any value to the researcher, it is critical that an appropriate null and alternative hypothesis (H_o and H_a, respectively) are defined prior to performing the statistical test. In Chapter 2, we learned that these hypotheses are always predefined for us, so we don't need to worry

will not influence (or be influenced by) any other sample. In the context of population independence, you must carefully consider whether your populations might have a significant influence on each other. For example, if we were measuring wind speed at a number of ocean buoys within a few kilometers from each other, there might be some dependence there (such that the wind at one buoy might be blowing toward another buoy, thereby influencing the wind speed measured there). If we wanted to investigate winds at one buoy compared to another, we might have to assume those "populations" are dependent on one another. However, if we wanted to compare the winds on April 1st to those on May 1st, it's probably a safe bet to consider those populations as independent.

Standard Deviation Equality: If two different populations share the same standard deviation, they must also share the same variance, and therefore the exact same probability distribution. But just because two populations have the same distribution does not mean that they also share the same central tendency. That means we can still test for differences in the mean, median, and mode of our two populations.

The Single-Sample *t* Test Is Used to Compare Measurements Against a Historical or Theoretical Value

If the *t* test is used to make statistical comparisons between populations, it might seem a bit odd to consider using a *t* test on a single population. The single-sample *t* test is quite useful when you have data from only a single population but you wish to compare that population to some idealized example or to some theoretical value for which you do not have data.

For example, if you had collected daily precipitation data from a weather station in your hometown over an entire year, you would likely have enough data to ensure that they are normally distributed, thanks to the central limit theorem. Based on your sampling strategy, it is clear that we only have one population in our dataset: the daily precipitation data from a single location. From our rainfall data, we might conclude that the mean daily precipitation for the entire year was 0.76 cm day^{-1}. Of course, with 365 measurements of daily rainfall, we could also easily calculate the standard deviation and variance if we were so inclined.

If we look in the record books, we might be able to find a historic value that indicates that the mean daily precipitation, for our particular location, is 0.62 cm day^{-1}. This is a value for which we do not have a matching dataset; however, it is a historical value against which we can compare our data using the single-sample *t* test to determine whether our mean daily precipitation value (0.76 cm day^{-1}) is significantly different from the historical value of 0.62 cm day^{-1}. Thus, we define our $\bar{X}_o$ using the historical value and our $\bar{X}_a$ using the value derived from our own data:

$$\bar{X}_o: \ 0.62 \text{ cm day}^{-1}$$
$$\bar{X}_a: \ 0.76 \text{ cm day}^{-1}$$

Then it is a relatively simple task to define H_o and H_a for hypothesis testing.

The Paired-Sample *t* Test Is Used to Test the Dependence Between Two Populations

As we have already discussed, populations that are dependent on each other will have that dependence reflected in the data. Because the data are in some ways "related" to each other, we must first attempt to separate the effects of one population on the other prior to our analysis. This is most effectively

SINGLE-SAMPLE *t* TEST ASSUMPTIONS

1. Independence: The result of one observation must not influence the result of another observation in the dataset.

2. Single Population: Every observation within the dataset was collected from the same population.

3. Normal Distribution: The distribution of values within the dataset follows a normal (Gaussian) curve.

done by determining the magnitude of the differences between the two samples and conducting our test on those differences (rather than on the original measurements themselves).

Earlier in our discussion of population independence, we used the example of wind speed data collected at a variety of ocean buoys and theorized that the wind field at one buoy might influence the wind field at a neighboring buoy, thereby making our populations dependent. To investigate that potential dependence, we might calculate the difference between the average hourly wind speed ($\bar{v}$) measured at buoy 1 versus buoy 2 (**Figure 8.6**).

If we define $\Delta = \bar{v}_2 - \bar{v}_1$, we can perform our t test on the calculated variable Δ to determine whether the difference between $\bar{v}_1$ and $\bar{v}_2$ is significant or not. What we're really doing here is a single-sample t test on the difference between $\bar{v}_1$ and $\bar{v}_2$ (as $\bar{X}_\Delta$) and determining whether $\bar{X}_\Delta = 0$. Our hypotheses would then be stated as:

$$H_o: \bar{X}_\Delta = 0$$
$$H_a: \bar{X}_\Delta \neq 0$$

Of course, the example given in Figure 8.6 is meant to illustrate the general application of the paired t test. In reality, a sample size of $N = 7$ at each of our buoys is far too few for statistical analysis, and we would most likely be in violation of the assumption of normality.

Variance Is Used to Determine Whether to Use the Pooled-Sample or Welch's t Test

The classic pooled-sample t test and Welch's variation are each different from the previous t tests in that they both use the sample variance s^2, rather

PAIRED-SAMPLE t TEST ASSUMPTIONS

1. Independence: The data pairs used to calculate Δ must not influence (or be influenced by) any other data pairs in the dataset.

2. Single Population: Each pair of observations within the dataset was collected from the same population.

3. Normal Distribution: The distribution of values for Δ follows a normal (Gaussian) curve.

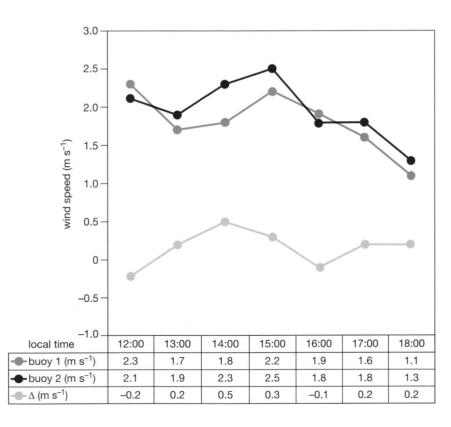

local time	12:00	13:00	14:00	15:00	16:00	17:00	18:00
buoy 1 (m s⁻¹)	2.3	1.7	1.8	2.2	1.9	1.6	1.1
buoy 2 (m s⁻¹)	2.1	1.9	2.3	2.5	1.8	1.8	1.3
Δ (m s⁻¹)	−0.2	0.2	0.5	0.3	−0.1	0.2	0.2

Figure 8.6 The wind fields at buoy 1 and buoy 2 can be tested for dependence by taking the difference (Δ) between the average hourly wind speeds recorded at the two sites and performing a paired t test. If the winds recorded at buoy 1 have no influence on the winds recorded at buoy 2 (and vice versa), the paired t test should return a result indicating that Δ = 0 (no dependence).

than central tendency, to test hypotheses of population equality. However, before you can decide which of these two tests is most applicable, you must first determine whether the standard deviations s in your populations are equal ($s_1 = s_2$). Fortunately, we already have all the tools we need to proceed.

Recall that the single-sample t test can be used to test the central tendency, standard deviation, or variance of any population against a theoretical value. If we have two populations, all that is necessary is to gather the descriptive statistics of population 1 in order to calculate its standard deviation (s_1) using Equation 2.3. This value can then be used as your theoretical value, which you can use to determine whether $s_2 = s_1$ using a single-sample t test to test the hypotheses:

$$H_0:\ s_2 = s_1\ (s_2^2 = s_1^2)$$
$$H_a:\ s_2 \neq s_1\ (s_2^2 \neq s_1^2)$$

If the resultant p-value $\geq \alpha$, you must accept H_0 (that is, the standard deviations are indeed equal) and therefore proceed with the pooled-sample t test. Of course, if the p-value $< \alpha$, you must reject the null hypothesis H_0 and accept H_a instead, which states that $s_2 \neq s_1$. If this is the case, you must proceed using Welch's t test.

The Pooled-Sample t Test Assumes Equal Variances Between the Populations Being Compared

The pooled-sample t test, sometimes called the "independent-samples t test," simply pools the standard deviation from both populations in order to calculate the test statistic t using Equation 8.2:

$$t = \frac{\overline{X}_1 - \overline{X}_2}{s_P \sqrt{\dfrac{1}{N_1} + \dfrac{1}{N_2}}} \tag{8.2}$$

Note that the means $\overline{X}_i$ and the total number of observations (N_i) of each population are used in conjunction with the pooled standard deviation s_P, calculated in Equation 8.3 as:

$$s_P = \sqrt{\frac{(N_1 - 1)\, s_1^2 + (N_2 - 1)\, s_2^2}{(N_1 - N_2) - 2}} \tag{8.3}$$

Fortunately, the solutions to these equations are typically calculated by statistical software, which will automatically provide the user with a value for the test statistic t and the associated p-value; Equations 8.2–8.3 are provided merely for the reader's interest and information.

Remember that the only reason we can pool the standard deviations of both populations is because we just confirmed that $s_1 = s_2$; that is, the standard deviations are mathematically indistinguishable from each other. This means that the variances s^2 of each population are also equal. Hence, the only real functionality of the pooled-sample t test is to test the equality of the central tendency (usually $\overline{X}$) of population 1 versus population 2.

POOLED-SAMPLE t TEST ASSUMPTIONS

1. Independence: Every observation recorded in population 1 is independent from (that is, not influenced by) the observations recorded in population 2.

2. Equal Variance: The variances s^2 of both sampled populations are equal to each other.

3. Normal Distribution: Both populations exhibit normal (Gaussian) curves.

Welch's *t* Test Is Used When the Variances of Two Populations Are Not Equal

Welch's *t* test is very similar to the pooled-sample *t* test, except for the very important distinction that Welch's *t* test should be used for independent samples that do not share equal standard deviations or variances. For Welch's *t* test, the test statistic *t* is calculated using Equation 8.4:

$$ t = \frac{\overline{X}_1 - \overline{X}_2}{\sqrt{\dfrac{s_1^2}{N_1} + \dfrac{s_2^2}{N_2}}} \tag{8.4} $$

Remember that the applicability of Welch's *t* test presupposes that neither the standard deviations *s* nor the variances s^2 of our populations are equal. So, if we already know that $s_1 \neq s_2$ (and therefore $s_1^2 \neq s_2^2$), we can use Welch's *t* test to take these conclusions one step further and perform one-tailed tests on the standard deviations and variances to investigate:

Right tailed	H_o: $\overline{X}_1 \leq \overline{X}_2$, or $s_1 \leq s^2$, or $s_1^2 \leq s_2^2$	
	H_a: $\overline{X}_1 > \overline{X}_2$, or $s_1 > s_2$, or $s_1^2 > s_2^2$	
Left tailed	H_o: $\overline{X}_1 \geq \overline{X}_2$, or $s_1 \geq s_2$, or $s_1^2 \geq s_2^2$	
	H_a: $\overline{X}_1 < \overline{X}_2$, or $s_1 < s_2$, or $s_1^2 < s_2^2$	
Two tailed	H_o: $\overline{X}_n = \overline{X}_o$	
	H_a: $\overline{X}_n \neq \overline{X}_o$	

The One-Way Analysis of Variance (ANOVA) Is Used to Compare Multiple Populations Simultaneously

Regardless of the actual name of the test, the one-way analysis of variance has absolutely nothing to do with the variance. Nor does it have anything to do with the standard deviation. In fact, the one-way analysis of variance (or one-way ANOVA) should actually be called the analysis of means (ANOME?), since that's exactly what it's used for.

The good news is that as far as statistical tests go, the ANOVA is remarkably easy to set up and perform. Essentially, the one-way ANOVA is used to test whether multiple sample means are in fact equal to each other. Thus, the ANOVA tests the hypotheses:

H_o: $\overline{X}_1 = \overline{X}_2 = \overline{X}_3 = \cdots = \overline{X}_n$
H_a: Not all $\overline{X}$ are equal

Essentially, the ANOVA is used to assess the differences among several means to investigate whether those differences are due to random variations in the populations (H_o) or whether those differences are significant enough to be attributed to something other than natural variation (H_a).

The one-way ANOVA can be a real time-saver when it comes to hypothesis testing multiple populations. If an ANOVA is performed and the *p*-value $> \alpha$, the result would indicate that the researcher must accept H_o and thereby proclaim that all of the means, from all sampled populations, are mathematically indistinguishable from each other.

WELCH'S *t* TEST ASSUMPTIONS

1. Independence: All observations in population 1 are independent of all observations in population 2.

2. Single Population: All observations taken from population 1 belong to a single population, which may or may not be the same as population 2. But all observations taken from population 2 must also belong to a single population.

3. Normal Distribution: The distribution of values in both populations follow normal (Gaussian) curves.

However, if the p-value $< \alpha$, the result of the one-way ANOVA would indicate that at least one of the means is different from all the rest. It is also entirely possible that all of the means are different from each other. Unfortunately, the ANOVA cannot be used to determine which of the means are different from the rest, so any confirmation of H_a would dictate that the researcher would then have to compare one population to the next using the appropriate t test, in a stepwise fashion, until all possible comparisons have been made.

As an example, let us assume that we were comparing the means of six different populations. If this were the case, our hypotheses would be:

H_o: $\bar{X}_1 = \bar{X}_2 = \bar{X}_3 = \bar{X}_4 = \bar{X}_5 = \bar{X}_6$
H_a: Not all $\bar{X}$ are equal

If we had adopted $\alpha = 0.05$ and our one-way ANOVA returned a p-value $< \alpha$, we would be forced to accept H_a and would then have to compare all six populations, using an independent-samples t test (pooled or Welch's), for all possible combinations:

$\bar{X}_1 : \bar{X}_2$	$\bar{X}_2 : \bar{X}_3$	$\bar{X}_3 : \bar{X}_5$
$\bar{X}_1 : \bar{X}_3$	$\bar{X}_2 : \bar{X}_4$	$\bar{X}_3 : \bar{X}_6$
$\bar{X}_1 : \bar{X}_4$	$\bar{X}_2 : \bar{X}_5$	$\bar{X}_4 : \bar{X}_5$
$\bar{X}_1 : \bar{X}_5$	$\bar{X}_2 : \bar{X}_6$	$\bar{X}_4 : \bar{X}_6$
$\bar{X}_1 : \bar{X}_6$	$\bar{X}_3 : \bar{X}_4$	$\bar{X}_5 : \bar{X}_6$

That's 15 separate t test comparisons! What's worse, we are accepting a 5% chance of error ($\alpha = 0.05$) for each of the 15 comparisons we are making. If we use Equation 8.5 to consider the accumulation of all of those chances of error together (as an "experimentwise error rate," ε), our error becomes:

$$\varepsilon = 1 - (1 - \alpha)^n \tag{8.5}$$

where the total number of separate comparisons made (n) has a profound effect on ε. For our particular example, our experimentwise error rate would be:

$$\varepsilon = 1 - (1 - 0.05)^{15} \approx 1 - (0.46) \approx 54\% \tag{8.6}$$

A 54% error rate seems positively lousy.

Actually, ε refers to the probability that at least one of the comparisons would lead the researcher to mistakenly claim a significant difference between the means when in fact there was no significant difference (a Type I error). In that context, $\varepsilon = 54\%$ is not as disastrous as it sounds. Keep in mind that $\varepsilon = \alpha$ for each separate comparison, so on a case-by-case basis, our error rate is constrained to α.

Of course, if you wanted to define a smaller acceptable experimentwise error rate, it would require that you dramatically reduce your chosen value for α. In our example, if we wanted to limit ε to 10%, we simply rearrange Equation 8.5 and solve for α:

$$\alpha = 1 - (1 - \varepsilon)^{1/n} = 1 - (1 - 0.10)^{1/15} \approx 0.007 \tag{8.7}$$

ONE-WAY ANOVA ASSUMPTIONS

1. Independence: Every observation, regardless of the population from which it came, is independent from every other observation in the dataset.

2. Equal Variance: The variances s^2 of all sampled populations are equal to each other.

3. Normal Distribution: All sampled populations exhibit normal (Gaussian) curves.

Community Comparisons Using Univariate Analysis

As we have discussed in earlier chapters, there are a great variety of methods that can be used to collect physicochemical data from the substrate and/or aquatic medium. As long as we focus on a single variable at a time, these data can be analyzed using univariate analysis. Likewise, if we wished to focus on a single biometric variable (such as standard length), we could use univariate analysis for those comparisons as well. However, when we start looking at entire communities of organisms, those assemblages become quite complex, and the organisms themselves may have very different biometric variables. For example, the standard length SL may be a perfectly appropriate biometric for a variety of fish species, but our seagrass species would require a different biometric (such as leaf-blade length). If we wanted to investigate the entire community using univariate analysis, we would have to pick a single variable that would have universal applicability.

Generally, this is why ecologists use measures that include either counts (the number of individuals per species) or biomass (the amount of organic mass per species) to assess the biological community under investigation. It's a simple but very important concept: no matter what kind of critter we encounter, their "presence" can be measured as a count, and the relative "quantity" of that critter can be measured as biomass. These measures can then be easily related in terms of the two-dimensional (areal) or three-dimensional (volumetric) space they occupy, which gives us some idea of their species density within the community:

$$\text{Density based on counts:} \quad \frac{\#\ individuals\ (unitless)}{area\ or\ volume\ occupied} \implies \frac{\#}{m^2} \quad \text{or} \quad \frac{\#}{m^3}$$

$$\text{Density based on mass:} \quad \frac{biomass\ (g)}{area\ or\ volume\ occupied} \implies \frac{g}{m^2} \quad \text{or} \quad \frac{g}{m^3}$$

Species density (our scaled variable) can then be calculated for each species (our nominal variable). Since we're only analyzing for one scaled variable in this scenario, we can do so using univariate analysis.

Measures of Species Richness Are the Easiest Way to Compare Communities

Depending on the size of your study site, it may be possible to conduct a complete count, or complete collection, of all the organisms in the area. This method is particularly effective for areas with obvious boundaries or well-contained areas. However, we are seldom fortunate enough to be able to conduct direct and complete counts, as most study areas are far too expansive for us to attempt a complete count. Occasionally, the number of organisms counted or collected is far too large to justify the time commitment to count and/or measure the biometrics of each and every specimen. In such cases, we are often compelled to perform some kind of subsampling regime, such as the more typical plot and plotless methods described in Chapter 3.

Although the simplest census method is to survey a variety of locations and simply determine whether a particular species is present or absent, such qualitative studies are rarely useful because they lack ecological context and provide no data regarding the relative abundance or biomass of a particular species in relation to any other species in the survey. Still, presence/absence (P/A) data are useful in providing a count of the total number of different species found in the area or system under investigation: this is what's known as **species richness** (S). Of course, richness can be assessed at any taxonomic level, which can be particularly useful if the investigator has difficulty in identifying organisms to the species level.

Simple Community Comparisons Can Be Performed Using Ranked Abundance Data and the DAFOR/SACFOR Methods

Beyond using simple P/A data to calculate species richness, the collection of frequency data (that is, how often a particular species is encountered) can be used to establish the relative abundance of each species within the community. This is typically performed using a ranking scheme that seeks to define the whether each recorded species is dominant, abundant, frequent, occasional, or rare (DAFOR) within the community. Although DAFOR ranks are established using a scaled variable to define the difference(s) between each rank, DAFOR frequencies are still considered to be semiquantitative assessments because it is entirely up to the observer to set the frequency conditions that define DAFOR.

For example, an investigator may choose to define DAFOR based on the frequency or biomass of each species in relation to the total amount assessed (that is, as a proportion of the total):

% of survey	Rank
80+	Dominant
40–79	Abundant
15–39	Frequent
5–14	Occasional
1–5	Rare

What this method lacks in objective quantitation, it more than makes up for in simplicity and ease of application in the field (**Figure 8.7**). Comparisons can still be made across study sites, or over time, so long as the observer maintains a consistent survey methodology and does not redefine the DAFOR conditions. This method can be used in either plot or plotless surveys and is well suited for simple, qualitative monitoring projects.

Timed DAFOR surveys can also be used to gather information on species richness and relative abundance, but they cannot be used to determine population densities or most other biometrics. To conduct a timed DAFOR survey, the observer defines a fixed time period (one hour, for example), which is then divided into equal time intervals that are given numerical ranks of decreasing magnitude:

Rank 6 (dominant) = 0–10 minutes

Rank 5 (abundant) = 11–20 minutes

Rank 4 (frequent) = 21–30 minutes

Rank 3 (occasional) = 31–40 minutes

Rank 2 (rare) = 41–50 minutes

Rank 1 (very rare) = 51–60 minutes

Figure 8.7 A photo-quadrat showing two different species of benthic organisms. We can see that there are 3 *Patella* limpet snails and 160 *Chthalamus* barnacles in the photo-quadrat ($N = 163$ total individuals, $S = 2$ species richness). If we were using simple counts for our DAFOR analysis, we would classify *Chthalamus* as Dominant (160 of 163, or 98.2%) and *Patella* as Rare (3 of 163, or 1.8%) in this particular example. (Courtesy of Mark A. Wilson.)

Once the observer begins the survey, the entire period is spent searching for previously unrecorded species that are recorded as belonging to the time interval in which they were first encountered. The fundamental assumption of this method is that the more abundant species are likely to be recorded in the earlier time intervals, whereas rare species will take much longer to find. Of course, this method does not account for organism behaviors, so timid or reclusive species may be recorded as "rare" regardless of whether they are truly less abundant. Taking several replicate surveys in the same area can improve the accuracy of results and is always recommended for DAFOR frequency assessments.

In some cases, it may be preferable to add additional ranks to the DAFOR scale. Another scale commonly used is the SACFOR scheme, where the ranks indicate whether a particular species is superabundant, abundant, common, frequent, occasional, or rare (SACFOR). Which of these is used, and how the ranks are defined, is completely at the discretion of the investigator. However, if previous DAFOR (or SACFOR) assessments have been made for your particular area of interest, you would be well advised to choose the identical scheme—that way, you would at least have the flexibility to compare your own survey results with previous investigations.

Biodiversity and Its Related Measures Are the Cornerstones of Community Comparison Methods

For decades, ecologists have sought to develop quantitative methods for assessing the "ecological health" of a particular area or habitat—an inherently qualitative judgment. In order to study community dynamics, it swiftly became evident that anecdotal observations of changes to the ecosystem were too subjective for scientific applicability. In an effort to objectively define the community structure within a particular ecosystem (and compare those results to a variety of investigations performed at different times or locations), ecologists devised several methods to assess community structure according to strict but simple mathematical protocols.

The most widely used indices of community structure are those that determine **species diversity** as a quantitative measure of the variety of species and their numerical contribution to the community as a whole. Diversity analyses can take many mathematical forms, and as we have demonstrated, species richness S is the simplest index of diversity for us to measure. Although richness has its uses, it contains no information about the total number of individuals belonging to each of the species encountered, nor does it contain any information about the total number of individuals counted in the entire community. For example, if we can describe a community of fishes collected from trawl nets as having a species richness S of 8, all we would know is that there are 8 different species identified from that survey (**Figure 8.8**); we would know nothing about the number of individuals belonging to each species, nor would we know how many total specimens were collected from the trawls.

So it is clear that a better measure of species diversity (also known as **biodiversity**) should take into account both the number of species in the collection as well as the number of individuals belonging to each species. Just as we saw in the example of DAFOR surveys, some species in a collection will be abundant while others will be less common. If the species were all represented evenly in the collection, we would arguably have a perfectly diverse population. Conversely, when we see that a certain species is

common name	species	
Atlantic croaker (croaker, hardhead)	*Micropogonias undulatus*	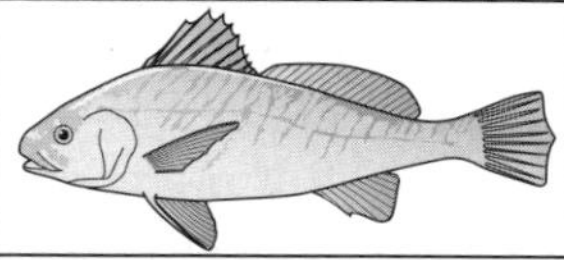
Atlantic menhaden (alewife, bunker, pogy, bugmouth, fat-back)	*Brevoortia tyrannus*	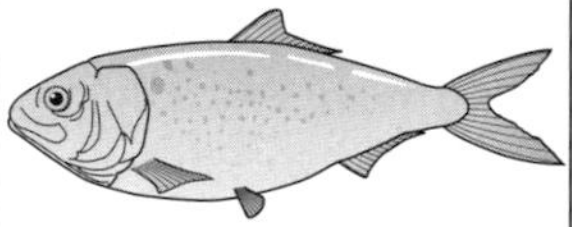
butterfish	*Peprilus triacanthus*	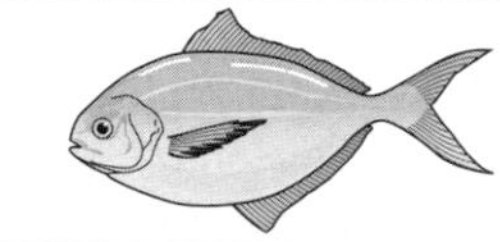
northern sea robin (sea robin)	*Prionotus carolinus*	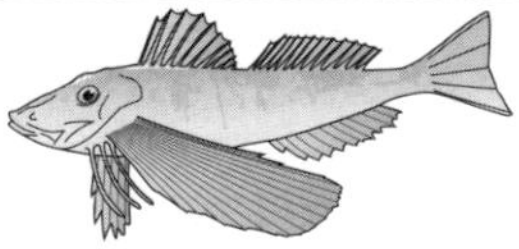
silver perch (perch, sand perch)	*Bairdiella chrysoura*	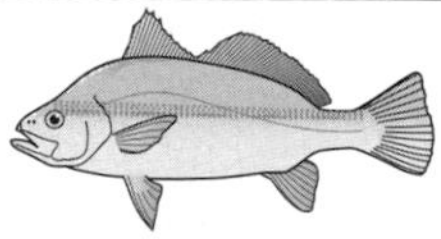
spot (norfolk spot, yellowbelly)	*Leiostomus xanthurus*	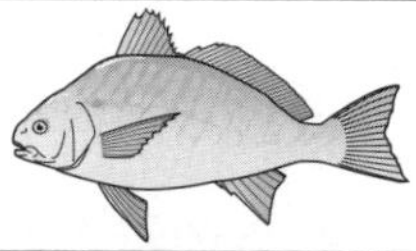
spotted seatrout (speckled trout, spotted trout, speckle)	*Cynoscion nebulosus*	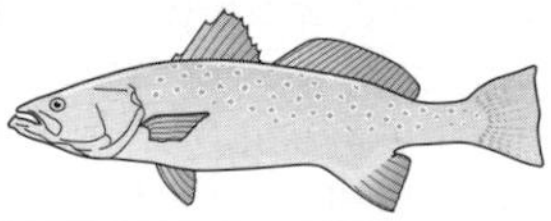
striped mullet (mullet, jumping mullet)	*Mugil cephalus*	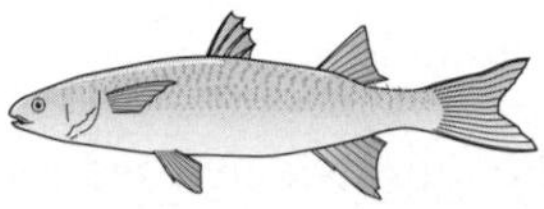

Figure 8.8 The number of different species identified within a particular community can provide a very simple index of biodiversity, known as the species richness S. This particular collection of fishes from the Atlantic Ocean includes eight ($S = 8$) different species. (Courtesy of the Maryland Department of Natural Resources.)

dominant while others are quite rare (a condition of unequal representation), the population is said to be less diverse.

Fortunately, there are several methods for calculating the unequal representation of species within a collection, as species diversity. One of the most widely used indices of diversity is the Shannon–Weaver index (see Equation 8.8), which provides a quantification of species diversity (H') as:

$$H' = -\sum \left(\frac{N_i}{N}\right) \cdot log_{10}\left(\frac{N_i}{N}\right) \tag{8.8}$$

where N_i is the number of individuals belonging to the ith species and N is the total number of individuals in the collection. As an alternative to H', **species dominance** calculates the probability that two individuals randomly selected from the community will actually belong to the same species. This probability is typically calculated using Equation 8.9, which represents the Simpson index of species dominance (ℓ):

$$\ell = \sum \left(\frac{N_i}{N}\right)^2 \tag{8.9}$$

Simpson diversity (D_S), which is analogous to H', can be easily calculated using Equation 8.10, once ℓ is known:

$$D_S = 1 - \ell \tag{8.10}$$

Since both values of H' and D_S are sensitive to the overall species richness S of the collection, maximum species diversity is reached as $N_i \rightarrow (N/S)$. This is an important point, because the values for H' and D_S will be meaningless unless you calculate just how diverse the population is (H' and D_S) when compared to the theoretical maximum diversity (H'_{max} and D_{max}), defined in Equations 8.11–8.12 as:

$$H'_{max} = log_{10} S \tag{8.11}$$

$$D_{max} = 1 - \left(\frac{1}{S}\right) \tag{8.12}$$

Once the diversity measures (H' and D_S) and their theoretical maxima (H'_{max} and D_{max}) are calculated, it is a relatively simple task to determine **species evenness** using Equations 8.13–8.14, which define species evenness as the mathematical "nearness" of the observed diversity to the theoretical maximum diversity:

$$J' = \frac{H'}{H'_{max}} \tag{8.13}$$

$$E_S = \frac{D_S}{D_{max}} \tag{8.14}$$

where J' and E_S are the calculated evenness values for the Shannon–Weaver (H') and Simpson (D_S) indices of diversity, respectively.

EXAMPLE BOX 8.1

Using Specimen Counts to Compute Species Richness, Diversity, Dominance, and Evenness

Let's assume we had conducted a number of trawls at three different coastal sites and carefully identified and enumerated each collected specimen to yield the data below. If we were interested in characterizing the diversity and evenness of each site, we would then be able to make some univariate comparisons between the sites. Not only would we be able to compare the sites immediately, if we returned to the same locations at a later date and conducted the same trawl surveys, we would then be able to make some "before-and-after" comparisons as well. Consider our data in Table 8.4.

In the context of univariate analysis, diversity and evenness calculations are extremely valuable because they represent an entire community of organisms that have been mathematically reduced to a single scaled variable. As long as the method of data collection is consistent, diversity and evenness measures can be compared between different sites, or between different time periods, using univariate statistical methods like the t test or ANOVA (assuming of course that our data do not violate any of the critical assumptions of those parametric tests). Even if statistical comparisons are not possible, diversity and evenness measures still represent summary data of complex biological communities, ripe for comparison.

Table 8.4 Identification and Enumeration of Nekton Specimens Collected from Coastal Research Stations A–C

Common name	Species name	N_i at site A	N_i at site B	N_i at site C
Atlantic croaker	*Micropogonias undulatus*	51	27	11
Atlantic menhaden	*Brevoortia tyrannus*	8	32	17
butterfish	*Peprilus triacanthus*	9	0	0
northern sea robin	*Prionotus carolinus*	0	2	5
silver perch	*Bairdiella chrysoura*	0	14	18
spot croaker	*Leiostomus xanthurus*	44	21	28
spotted seatrout	*Cynoscion nebulosus*	9	2	0
striped mullet	*Mugil cephalus*	3	1	2
N		124	99	81
S		6	7	6
H'		0.600	0.664	0.679
H'_{max}		0.778	0.845	0.778
J'		0.771	0.786	0.873
ℓ		0.310	0.245	0.236
D_S		0.690	0.755	0.764
D_{max}		0.833	0.857	0.833
E_S		0.828	0.881	0.917

Although site A had the greatest abundance of fish (N), it was more heavily dominated by a few particular species, as evidenced by the largest value for Simpson dominance ($\ell = 0.320$). Indeed, site A has the lowest values for Shannon–Weaver ($H' = 0.600$) and Simpson ($D_S = 0.690$) diversity, compared to the other sites.

Although site B had a slightly higher species richness ($S = 7$) than the other locations, site C actually exhibited the highest species diversity, regardless of whether the Shannon–Weaver ($H' = 0.679$) or Simpson ($D_S = 0.764$) diversity indices were used. This is particularly evident if we consider that the evenness calculation at site C estimates that the biodiversity measured there was within 87.3% ($J' = 0.873$) to 91.7% ($E_S = 0.917$) of the theoretical maximum, the using Shannon–Weaver and Simpson diversity measures, respectively.

Using the Before-After-Control-Impact (BACI) Approach

One of the simplest and most widely used methods for performing univariate comparisons is the before-after-Control-Impact (BACI) design. This method involves the collection of data before some event (to serve as the baseline), followed by subsequent measures of the same variable immediately after the event occurs. In this way, the measures of a single variable can be assessed in pairwise fashion to detect the "before-and-after" differences using a two-sample. However, one of the major difficulties in this type of experimental setup is that it is unlikely you will be able to anticipate exactly when (or where) the event will occur. In order to compensate for this, it is often necessary to establish at least two study sites: a Control site and an Impact site. Of course, the impact site serves as the location where the event is anticipated to occur. Your Control site should be as similar as possible to the Impact site, but located where it is highly unlikely, in any circumstance, to be affected by the event you wish to study. This way, you have all your bases covered: time-sensitive changes will be detected in the before versus after data, whereas spatially sensitive changes will be detected in the Control versus Impact data.

As an example, let's assume there are plans to build a titanium dioxide (TiO_2) plant near the coast, and there are some concerns that TiO_2 pollution may adversely affect the nearshore environment. If you suspect there may be an Impact to the environment after the factory is built and becomes operational, it would be wise to analyze the coastal environment for various pollutants before the plant is built, to serve as your baseline data. To make sure there are no spatial differences in the baseline pollutant data, you decide to locate your Control site 10 km upcoast from the anticipated Impact site. Once the baseline data at both sites are measured, any event-related changes (for example, one year after the TiO_2 plant became operational) should be relatively easy to quantify and test for significance against the baseline (**Figure 8.9**).

Event-related changes (sometimes called perturbations) to the system are essentially the same as performing an experiment in the laboratory, or delivering a treatment to a test subject—in either case, you are essentially looking to quantify the "before-and-after" changes. Although our pollution example is most certainly a negative effect on the ecosystem, keep in mind that not all perturbations are negative. It is possible to perform the same kind of perturbation analysis on positive effects, like increased survivability against infection post-treatment with antibiotics, or increased water clarity after the creation of managed oyster reefs.

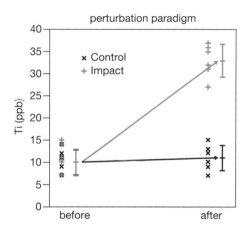

Figure 8.9 The classic BACI perturbation paradigm, where the Control and Impact sites exhibit very similar baseline measures, before the system is disturbed by some event. Time-related changes, unrelated to the disturbance event, can be measured by making comparisons between the "before-and-after" conditions within the Control site. Spatially related changes (presumably from the perturbation event) can be measured by comparing differences between the Control and Impact sites. In this example, the concentration of dissolved titanium (Ti, in parts per billion) measured in the coastal waters near a TiO_2 plant strongly indicates an Impact to the system when compared to the Control.

Figure 8.10 The classic BACI monitoring paradigm, where the Control and Impact sites may exhibit very different baseline measures, but the time-related changes measured at each site can provide context to the "natural" state of change at each site. Monitoring efforts are also valuable because they essentially establish the new baseline for each site under investigation. In this example, the concentration of dissolved titanium (Ti, in parts per billion) measured at the Impact site exceeds the contamination measured at the Control site, but the time-related changes at each site are relatively stable. If a new perturbation event occurred, these data would serve as the "new" baseline for all future analyses. Without the benefit of baseline data acquired from monitoring programs, many temporal and spatial comparisons would not be possible.

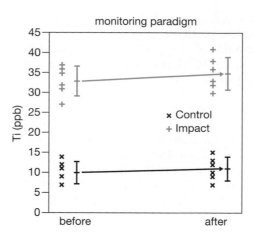

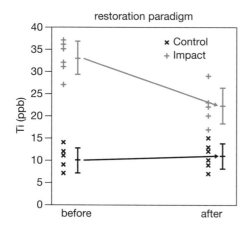

Figure 8.11 The classic BACI restoration paradigm, where the Control and Impact sites may exhibit very different baseline measures, but the effectiveness of restoration efforts (or natural recovery) can be quantified as the Impact site returns to its original, pristine state (represented by the Control site). In this example, coastal clean-up efforts (and reductions in waste outfalls) have significantly reduced the titanium contamination (Ti, in parts per billion) at the Impact site. Although the Impact site has not yet fully recovered (using the Control site as our standard), this data could be used to test the effectiveness of the clean-up efforts, or provide an estimate of how much longer it will take before the Impact site is fully recovered.

The BACI design is also used quite heavily in monitoring programs, where the intent is not to describe event-related changes between the Control and Impact site, but to simply monitor the sites over time. Such studies may be appropriate for locations where a perturbation event has already occurred, but it is still necessary to monitor how the system is responding over time. Monitoring programs are often underappreciated, because there is the misconception that there is no research going on; that data are merely being collected for the sake of collection. However, monitoring studies do provide a means to demonstrate how similarly (or dissimilarly) the Control and Impact sites will change over time, under their respective "natural" conditions (**Figure 8.10**). Monitoring programs are also beneficial because we can rarely anticipate when (or where) perturbation events will occur next, so data collected from monitoring programs can be used as the critical baseline when new perturbations occur.

The BACI design is also quite effective in measuring the effectiveness of restoration efforts and/or ecosystem recovery after a perturbation event. Whether they benefit from natural or anthropogenic intervention, natural systems are capable of recovering from all but the most grievous perturbation events. Measuring the capacity for ecosystem recovery, either as a function of time or location, can be accomplished using the BACI design by simply analyzing the differences between the control and impact sites before recovery, and then again at some future time (**Figure 8.11**). Such analyses are used to estimate the amount of time necessary for the impact site to return to its natural state. If specific restoration efforts were employed in an effort to return the site to its original, pristine condition, the effectiveness of those methods can also be analyzed in this fashion.

What is critical to the BACI design, whether it is applied as a perturbation, monitoring, or restoration paradigm, is that all comparisons must be conducted in pairwise fashion (control vs. impact, or before vs. after) to provide the necessary univariate evidence of correlation and/or causation. The inclusion of multiple sites, multiple treatments, or multiple time series does not make such analyses impossible. Quite the contrary: multivariate analyses often provide much stronger (and more useful) evidence of correlation and causation. But those are methods that involve more than one scaled variable and, as such, require a completely different class of statistical tests for hypothesis testing.

Moving Beyond the Simplicity of Univariate Analysis

As we have seen, univariate analyses can be quite useful, especially if we are interested in making simple comparisons between one population and another. These comparison strategies can be applied to any field within the natural sciences, whether you're comparing grain size between two different sediment cores, the concentration of mercury within two different estuaries, the significant wave heights measured at two different ocean buoys, or the number of fish species collected in April compared to November. Such pairwise comparisons can be used to establish the statistical foundation on which more thoughtful and more rigorous comparisons can be made.

As we have already discussed, the greatest weaknesses of univariate comparisons are twofold: (1) we cannot compare more than one variable at a time; and (2) it is more difficult to establish association or causation without the use of multivariate comparisons. Think about that second point for a moment. How can we possibly hope to form even the simplest association (if A then B) if we cannot analyze A and B in tandem as scaled variables?

For example, with univariate analyses, we might be able to demonstrate mathematically that the concentration of mercury in estuary 1 is greater than estuary 2. We might follow up our analyses and demonstrate that disease rates among fishes in estuary 1 are also greater than in estuary 2. We might be tempted to infer that those increased disease rates are coincident with increased mercury levels, but without the capacity to analyze both variables simultaneously, our conclusions would be little more than innuendo. It would be a reasonable assumption, to be sure; but there would be no direct statistical evidence of that correlation. For that, we will need to expand our horizons to include multivariate analyses and the power to resolve such associations.

As we will discover in the next chapter, multivariate analyses ultimately allow us to establish both association and causation. And by the very act of demonstrating cause and effect in the language of mathematics (that is, statistics), we shall discover that we are at the same time embarking on the process of modeling our natural world.

References

Jones ER (1996) Statistical Methods in Research. Edward R. Jones.

Kanji GK (1999) 100 Statistical Tests. SAGE Publications.

Keeping ES (1995) Introduction to Statistical Inference. Dover Publications.

Kenny DA (1979) Correlation and Causality. John Wiley and Sons.

Linton M, Gallo Jr PS, & Logan CA (1975) The Practical Statistician: Simplified Handbook of Statistics. Wadsworth Publishing Company.

Mandel J (1964) The Statistical Analysis of Experimental Data. Dover Publications.

Salkind NJ (2007) Statistics for People Who (Think They) Hate Statistics: The Excel Edition. SAGE Publications.

Steiner F (ed) (1997) Optimum Methods in Statistics. Akadémiai Kiadó.

Thompson SK (1992) Sampling. John Wiley and Sons.

Further Reading

Jaisingh LR (2006) Statistics for the Utterly Confused, 2nd ed. McGraw-Hill.

Kachigan SK (1986) Statistical Analysis: An Interdisciplinary Introduction to Univariate and Multivariate Methods. Radius Books.

Keller DK (2006) The Tao of Statistics. SAGE Publications.

Newman I & Newman C (1977) Conceptual Statistics for Beginners. University Press of America.

Timm NH & Mieczkowski TA (1997) Univariate and Multivariate General Linear Models. SAS Institute.

Turner JR & Thayer JF (2001) Introduction to Analysis of Variance: Design, Analysis, and Interpretation. Sage Publications.

Wei WWS (2005) Time Series Analysis: Univariate and Multivariate Methods, 2nd ed. Pearson.

Chapter 9

Introduction to Multivariate Analysis

"The ocean: multiple, to a blinding oneness."– Archie Randolph Ammons

As we learned in the previous chapter, the ability to mathematically test the statistical significance of values within a single variable certainly has its uses, and forms the foundation for hypothesis testing in the sciences. But the natural world is a very complex place, and rarely will a single variable be sufficient for our investigative needs. In truth, it is far more realistic to expect that the phenomena we seek to describe shall require mathematical models that imply some kind of interrelationship between two or more variables.

Multivariate analyses allow us to perform more complicated (and more meaningful) investigations. Multivariate methods typically try to establish either correlation or causation using a variety of statistical tests. Correlation simply attempts to establish that two (or more) variables are somehow related to each other, and that the value of one variable somehow exerts an influence on the value of a different variable. Correlation analysis doesn't explore how or why the variables are associated—it simply seeks to establish that an association exists. Causation analysis is a different and more powerful tool, because it seeks to define how the variables are related to each other, so we can use that relationship in a predictive manner (so the values of one variable can be used to predict the values of an associated variable). Not only does this process provide insight as to how the natural world functions, but it also allows us to develop mathematical models of complex phenomena we observe in the field or in the laboratory. If you can model it, you can simulate it.

Key Concepts

- Investigations of multivariate dynamics require that two or more variables be analyzed simultaneously.

- Correlation analyses are required to establish whether the variables under consideration are interrelated and covary in some way.

- Linear regression analyses can be used to develop a mathematical formula to predict the value of a dependent variable based on the measurement of a different, independent variable.

- Multiple linear regression analyses employ several different variables simultaneously to predict the value of a dependent variable.

- Curve estimation is used when the mathematical model that describes interrelated variables is nonlinear.

Introduction to Correlation Analysis

As we learned in the previous chapter, there are a wide variety of statistical methods (particularly the *t* tests) that we can use if all we want to do is focus on a single variable among different samples and simply compare one central tendency against another. In the context of statistics, this is necessary because we must be able to distinguish between two measurements that are "close enough" to be equal, and those that are truly different from each other. Where (and how) you draw the line between "statistically similar" and "statistically different" is the reason why *t* tests are so important, because they establish the statistical foundation that allows us to compare different values within a single variable. Then it is a relatively simple matter to expand our comparisons to include measurements of more than just one variable.

As a practical matter, the natural world is a complicated place so most scientific investigations require some kind of **multivariate analysis**, when we must compare two or more variables at the same time if we want to make any sense about the way the natural world functions. But it's also important that we limit our analyses to include only those variables that are germane to the hypotheses we are trying to test. In other words, don't waste your time measuring every variable under the sun—try instead to focus only on those variables that have some relevance to the question at hand.

That means the first order of business in any multivariate analysis is to establish whether the variables under investigation are (1) associated with each other and (2) associated with the phenomenon you are attempting to explain. Ideally, both conditions should be satisfied, since it is of little use to discover two variables that are associated with each other but have nothing to do with your hypothesis.

So how do we establish **correlation**? Fortunately, there are a host of statistical methods that we can use to mathematically define whether the variables in our dataset are indeed associated with each other. These same statistical tests can also be used to determine the relative "strength" of each association. This is especially useful in cases where there are several variables associated with each other, because it allows the researcher to immediately determine which associations are most significant and focus on those.

Correlation Analysis Can Only Be Performed Among Continuous Variables

Demonstrating correlation between variables is a very powerful tool in scientific research, because correlation allows us to explore the connection between those variables, based on the results of our statistical tests. Not only can correlation analyses provide us with information as to how strong or weak the correlation is, they can also help us understand whether the association is positive or negative.

A **positive correlation** implies direct association between two or more variables. If we're dealing with continuous data, a positive correlation means that the measurements of the variables will "follow" each other (**Figure 9.1**). In other words, if the measurements in one variable get larger, the measurements of the other variable(s) will follow suit and also increase. Likewise, if the numerical values of one variable decline, the others would also decline.

A **negative correlation** implies an inverse relationship; that is, as the measurements of one variable increase, the measurements of the other variable(s) will decrease. Of course, if there is zero correlation (that is, neither positive nor negative), it simply means that the variables are completely independent of each other and are not associated at all. This condition would

BE CAUTIOUS WITH CORRELATIONS

Correlations only imply that the variables are somehow associated with each other. Correlations can provide insight as to the strength of those associations, but they do not imply causation.

For example, an analysis of the chemical constituents of seawater would indicate an extremely strong correlation between dissolved Cl^- and dissolved SO_4^{2-}, simply because they are the two most common anions in the world ocean. Although we might be able to use that association to predict the relative concentrations of Cl^- and SO_4^{2-} in other seawater samples, it would be incorrect for us to claim that the Cl^- concentration is somehow caused by SO_4^{2-}.

essentially look like the data for each variable are distributed randomly with respect to each other.

Correlation Analysis Can Be Used on No More Than Two Dependent Variables at a Time

At the heart of every correlation analysis is the intent to establish some kind of association between variables. If we take a moment to think about the logic of association, we have two fundamental possibilities. On one hand, it is conceivable that we might have a particular variable whose measured values are completely unaffected by any of the other variables we are measuring. We would call this an **independent variable** (IV). In contrast, there is also the distinct possibility that the value of some variable is being influenced by another variable. Not surprisingly, the variable exposed to this influence is called the **dependent variable** (DV). These distinctions are critically important, because correlation analysis requires that we first define what we think our DVs and IVs really are, so we can use correlation analysis to test whether our DVs are really determined by (dependent on) the IVs we've selected.

As an example, let's consider the density of seawater (ρ) as measured in kilograms of seawater per cubic meter. Based on our previous education and knowledge of ocean properties, we might be tempted to assert that ρ will vary as a function of many different measureable parameters, such as

- Water temperature (because of thermal expansion or contraction of the water volume)
- Salinity (due to increased or decreased mass of dissolved solutes)
- Hydrostatic pressure (due to the compressibility of the water volume)
- Evaporation (due to the loss of fresh water from the water volume)
- Precipitation (due to the gain of fresh water to the water volume)

Since we are already making the claim that ρ is influenced by these variables, we have already defined ρ as our DV. For the other variables listed above, they can only be considered to be IVs if they are not influenced by each other in any way. We may be getting ourselves into trouble by using evaporation and precipitation here, simply because they could each affect our measures of water temperature and/or salinity. However, if we remove them from our list of IVs, we can still assume their effects would be included in measures of temperature and salinity. And if we make things even simpler and take our measurements from relatively shallow water, we can likely ignore the influence of hydrostatic pressure.

So that leaves us with only two IVs: water temperature and salinity. If we can convince ourselves that the salt content of a seawater sample would not influence its temperature, and that the temperature of a seawater sample would in no way determine its salt content, we can confidently claim that they are indeed independent. In this rather simple example, we count one DV (ρ) and two IVs (temperature and salinity). In the context of correlation, we would first test whether the two IVs are even associated with our DV, ρ. If an association can be demonstrated as either a positive or negative correlation (**Figure 9.2**), we would then use correlation analysis to explore whether either (or both) of the IVs can actually predict the value of our DV.

It is important to note that although we used only two independent variables in our simple example, there is no theoretical limit to the number of independent variables you can use in a correlation analysis (although practicality

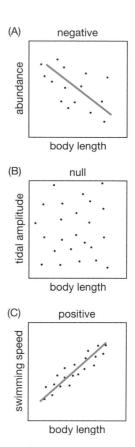

Figure 9.1 Correlation between variables is most easily visualized when the association is linear. Note that when the data are plotted on a Cartesian graph, negative correlations exhibit a negative slope (A) whereas positive correlations exhibit a positive slope (C). Technically speaking, it is not possible to describe a line for a null correlation (B) because x and y vary independently of each other.

Figure 9.2 The measure of seawater density, as a function of temperature and salinity, is a well-established correlation in marine science. Note how the depth profile of seawater density (the pycnocline, A) shows an inverse relationship with temperature (the thermocline, B). This is indicative of a negative correlation between density and temperature. In contrast, the depth profile of salinity (the halocline, C) closely follows the same pattern of the pycnocline and is thus indicative of a positive correlation between density and salinity. Although these relationships may be easy to demonstrate by simply plotting the measurements, correlation analysis can be used to test the statistical significance of these relationships and may even be used to predict density values based on the strength of its positive correlation with salinity or its negative correlation with temperature.

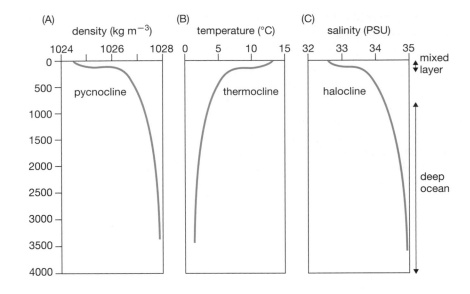

will likely limit you to only a handful). And although you may use any number of independent variables, the statistical methods most commonly used for correlation analyses will limit you to no more than two dependent variables at a time. So take care to define the number of dependent variables you wish to test, as well as the number of different independent variables you will use as predictors of each dependent variable. When it comes time to choose the most appropriate type of correlation analysis, it is easy to provide guidance as long as you know how many dependent and independent variables you wish to include (**Figure 9.3**). There are several different correlation methods to explore, but the basic structure of the linear regression provides the foundation for all of the other correlation analyses.

Linear Regression Analysis

Statistically, the simplest model of association is the **linear regression**. For two variables (X and Y), we can use Equation 9.1 to express their association simply as

$$Y = mX + b \tag{9.1}$$

where X and Y represent the measurements for each of the two variables we want to compare, m represents the slope of the resultant line, and b represents the Y-intercept (the special condition that defines the magnitude of Y when X is set to zero). In reality, our measurements will always exhibit some degree of random error (ε), so the linear correlation function is more correctly defined in Equation 9.2 as

$$Y = mX + b + \varepsilon \tag{9.2}$$

It is important to note that not all associations are linear, so the linear regression model described by Equation 9.2 cannot be applied in all cases. However, it does represent the simplest example of a multivariate association between a single independent variable and a single dependent variable. Using the linear regression model (see Equation 9.2), we might even be able to use the independent variable X) to predict the value of the dependent variable (Y, presuming a strong correlation exists between X and Y. So before we get ahead of ourselves, it would seem that the first task of our correlation analysis is to determine whether an association between X and Y even exists.

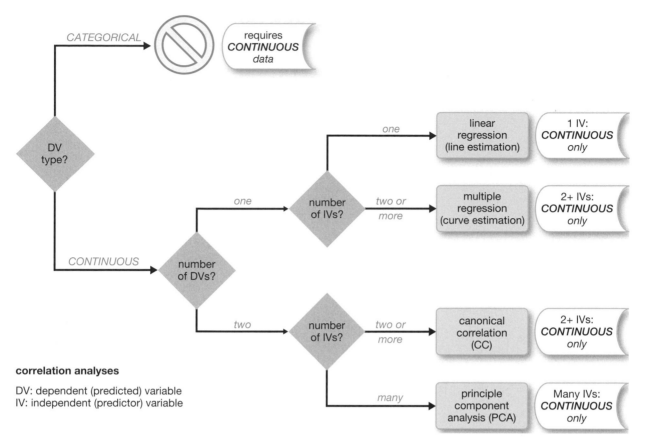

Figure 9.3 A simplified flowchart of the multivariate correlation analyses most commonly used in the natural sciences. Although all correlation analyses require continuous data, the most appropriate method is determined by the number of dependent variables (DVs) being predicted by correlation, as well as the number of independent variables (IVs) being used as predictors.

Correlations Among Normal Data Can Be Tested Using Pearson's Correlation Coefficient

The most common statistical method used to test the strength of a linear correlation between two variables is Pearson's correlation coefficient r, as defined in Equation 9.3:

$$r = \frac{\sum_{i=1}^{n}\left(Y_i - \overline{Y}\right)\left(X_i - \overline{X}\right)}{\sqrt{\sum_{i=1}^{n}\left(Y_i - \overline{Y}\right)^2} \cdot \sqrt{\sum_{i=1}^{n}\left(X_i - \overline{X}\right)^2}} \tag{9.3}$$

where X_i and Y_i represent the ith measurements of the X and Y variables, respectively, and $\overline{X}$ and $\overline{Y}$ are the means. Don't let the complexity of Equation 9.3 spook you: Pearson's coefficient is actually quite easy to calculate from spreadsheet data using modern statistical software.

As an example of correlation analysis, let's use a simple hydrography dataset that contains four variables: water temperature T, salinity S, density ρ, and wind speed U. In reality, we would want to use a much more robust dataset, but for the sake of this example, we can get away with using such a small dataset (**Table 9.1**).

If we assume that our data are normal, and that our pairwise correlations are indeed linear, we can use Pearson's r to investigate how our variables are

Table 9.1 Physical Properties Measured at Select Oceanographic Stations, Including Water Temperature (T, °C), Salinity (S, PSU), In-Situ Density (ρ, kg m^{-3}), and Average Wind Speed (U, m s^{-1})

Station	T (°C)	S (PSU)	ρ (kg m^{-3})	U (m s^{-1})
1	28.1761	19.4654	1010.87	3.3975
2	24.3896	15.9064	1010.90	3.4422
3	29.4930	24.8525	1013.28	2.8610
4	20.8722	29.1413	1021.07	2.0564
5	22.3083	33.0060	1022.42	2.9505
6	19.8067	31.7843	1023.15	3.1740

associated. Based on our example hydrography data in Table 9.1, the results of Pearson's r (using Equation 9.3) would be

$$T\,|\,\rho \qquad r = -0.829$$
$$p = 0.041$$
$$S\,|\,\rho \qquad r = 0.949$$
$$p = 0.004$$
$$U\,|\,\rho \qquad r = -0.513$$
$$p = 0.298$$

For all datasets, the value and sign of Pearson's coefficient r will always fall in the range of $-1 < r < +1$. A positive correlation is indicated when $r > 0$ and a negative correlation when $r < 0$. The strength of the correlation can be determined by the magnitude of r. Very weak associations are indicated as $r \rightarrow 0$; thus, very strong associations can be deduced as $r \rightarrow \pm 1$.

When performing correlation analyses, our null (H_o) and alternate (H_a) hypotheses are always stated as

$$H_o: r = 0 \qquad (X \text{ and } Y \text{ are not correlated})$$
$$H_a: r \neq 0 \qquad (X \text{ and } Y \text{ are correlated})$$

From the results of Pearson's analysis, we can see that water temperature T is strongly associated with water density ρ, as a negative correlation ($r = -0.829$). In other words, the inverse association indicates that as water temperatures increase, water density will decrease. The fact that $p < 0.05$ indicates that the result is statistically significant, so we must accept H_a and reject H_o with respect to the association between T and ρ.

The correlation between salinity S and water density ρ is even stronger ($r = +0.949$). Note that for this particular correlation, the sign of r indicates a positive correlation between salinity and density; therefore, as salinity values increase, we should expect density values to increase as well. The highly significant p-value ($p = 0.004$) indicates an even stronger statistical significance, so we must accept H_a and reject H_o with regard to the association between S and ρ as well.

At first glance, it would seem as though wind speed U and water density are negatively correlated ($r = -0.513$). However, since our p-value is much larger than 0.05, we cannot confirm that the result is statistically significant. In other words, we cannot confirm that the correlation does indeed exist, so we must reject H_a and accept H_o and therefore treat wind speed and water density as truly independent variables that are not at all associated with each other.

ASSUMPTIONS OF PEARSON'S LINEAR CORRELATION

1. Linearity: Variables X and Y must be associated according to the mathematical function $Y = mX + b + \varepsilon$.

2. Normality: The distribution of values for X and Y must follow a normal (Gaussian) curve, with a clearly discernible central tendency and standard deviation.

Table 9.2 Physical Properties Measured at Select Oceanographic Stations and Ranked According to Ascending Magnitudes of Water Temperature (T, °C), Salinity (S, PSU), and In-Situ Density (ρ, kg m^{-3})

Station	T (°C, rank)	S (PSU, rank)	ρ (kg m^{-3})	ρ (rank)
1	28.1761, 5	19.4654, 2	1010.87	1
2	24.3896, 4	15.9064, 1	1010.90	2
3	29.4930, 6	24.8525, 3	1013.28	3
4	20.8722, 2	29.1413, 4	1021.07	4
5	22.3083, 3	33.0060, 6	1022.42	5
6	19.8067, 1	31.7843, 5	1023.15	6

Correlations Among Nonnormal Data Can Be Tested Using Spearman's Ranked Correlation Coefficient

If our data are presumed to exhibit a linear correlation but are not normally distributed (or their distribution is unknown), we must rank our data from smallest to largest and compute Spearman's correlation coefficient based on the ranks rather than the actual measurements. For the sake of simplicity, let's use our earlier hydrography dataset as an example, but rank only the temperature T, salinity S, and density ρ measurements (Table 9.2).

The results of Spearman's correlation would then be

T (ranked) | ρ (ranked) $r = -0.771$

$\qquad\qquad\qquad\qquad\qquad\quad p = 0.072$

S (ranked) | ρ (ranked) $r = 0.886$

$\qquad\qquad\qquad\qquad\qquad\quad p = 0.019$

From the results of Spearman's correlation analysis, we can see that the ranked values for the water temperature T are still negatively associated with water density (ρ), but those results are no longer statistically significant ($p > 0.05$). This is most likely a consequence of using so few data for our example of ranked correlation analysis; if we included more data (that is, more ranks), Spearman's results would likely comport well with Pearson's. Of course, if we had more data, we might even satisfy the central limit theorem and therefore be safe in assuming our data were normally distributed (allowing us to use Pearson's correlation in the first place).

With regard to our ranked salinity (S) data, salinity is still positively correlated with water density ($r = 0.886$), and the result remains significant ($p = 0.019$) even with such a small dataset. Hence, salinity and density are clearly the strongest association, based on our available data and the diversity of correlation analyses used.

Correlations Can Also Be Used to Make Predictions Using Regression Analysis

These sorts of correlation analyses are extremely critical to field researchers because they (ideally) allow us to eliminate superfluous variables and focus only on those that have the strongest associations. If we want to use those associations to make predictions about how the natural world functions, it is not enough to simply say they are correlated—what is really important is to define a mathematical model where the measured values of certain variables can be used to predict the values of some other variable. That process is commonly called **regression analysis**. In the simplest case, that mathematical model would be a familiar one: the equation for a line, where $Y = mX + b + \varepsilon$.

Recall that for linear regression analyses, we must assume that our data are continuous, and that we are investigating the correlation between two variables (X and Y) that exhibit a linear relationship according to the mathematical model defined earlier in Equation 9.2. Also keep in mind that the variables X and Y are true unknowns—these are the measurements we must take in the field or laboratory. The way Equation 9.2 is written, we can see the value of Y can only be determined if we know the value of X, as well as the slope m, the intercept b, and the random error ε associated with the $X|Y$ relationship. In this context, Y is considered to be the dependent variable, because its value is dependent on the function involving m, X, b, and ε. That means that X is considered to be the independent variable, because its value is completely unaffected by Y, m, b, and ε.

If this is the first time we're taking a crack at solving the linear regression equation (as defined in Equation 9.2), we will first have to take measurements of variables X and Y, but we won't know the values for m, b, or ε. Although the values of m, b, and ε are currently unknown, they are not variable—they actually represent different mathematical constants. A **constant** is a mathematical value that does not vary and is critical to the solution of a mathematical equation. And the nice thing about constants is that once you figure out what their value should be, that value will never change.

So let's take another look at Equation 9.2 and think about this for a bit. We know that both X and Y are variable, but we should be able to predict the value for Y (the dependent variable) if we can measure the value of X (the independent variable) and somehow know the constant values for m and b. If the correlation between X and Y is significant and ordered, we can use that association to estimate the values of m and b by simply gathering enough measurements of X and Y and letting that relationship "reveal itself" to us (**Figure 9.4**).

Note that we have completely ignored any attempts to mathematically define ε. This is because ε represents the random errors in all our measurements of X and Y and in the relationship defined in Equation 9.2. Since ε represents random error, its randomness also makes it unknowable—so there's no point in trying to quantify it. In mathematics, we can cheat a little bit and say that our expectation for error, $E(\varepsilon)$, is zero. So that also means it is impossible to determine the "true" value of Y; what we're really trying to do is estimate the value of Y, based on all our measurements of X and Y (error and all). So we should modify Equation 9.2 and assert that our predicted value of Y (as Y') is more correctly stated in Equation 9.4, now as

$$Y' = \hat{m}X + \hat{b} \tag{9.4}$$

where $\hat{m}$ and $\hat{b}$ are the statistical estimates for the slope and intercept, respectively.

The method for estimating $\hat{m}$ and $\hat{b}$ is a statistical one that uses our measurements of X and Y to estimate the slope $\hat{m}$ and intercept $\hat{b}$, and then tests how well those estimates predict the value of Y (as Y') compared to the actual measurements of Y. Mathematically, $\hat{m}$ and $\hat{b}$ can be determined using Equations 9.5–9.6:

$$\hat{m} = \frac{n \sum XY - \sum X \sum Y}{n \sum X^2 - \left(\sum X\right)^2} \tag{9.5}$$

$$\hat{b} = \overline{Y} - \hat{m}\overline{X} \tag{9.6}$$

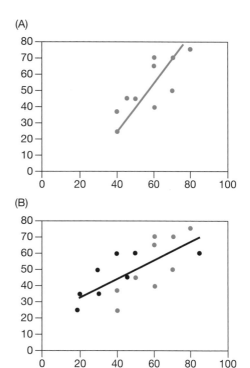

(A)

(B)

Figure 9.4 As more data are gathered, the true relationship between X and Y is revealed in the data themselves. Although there will always be error (ε) associated with our measurements, our original estimates of the slope m and intercept b of a linear regression (A) are based on few data points (blue). The original regression will improve (B) with the incorporation of more data points (black) as these new data will "correct" the original estimates of the slope m and intercept b of the regression. When the addition of new data does not significantly change the slope or intercept, it means we have finally determined them as constants.

As you can see, the number of measurements (n) of X and Y affects the estimate of $\hat{m}$, which in turn is used to estimate $\hat{b}$ (using the mean values of X and Y). Fortunately, these estimates are routinely calculated by statistical software as part of the linear regression analysis, so one would rarely need to calculate these values by hand. But it is nonetheless instructive to take a look at the mathematical model used in Equation 9.5 to estimate $\hat{m}$, because it assumes that any random errors in our measures of X and Y will eventually cancel themselves out ($\varepsilon \to 0$) if enough observations (n) are made. Since $\hat{m}$ is then used to estimate $\hat{b}$ (in Equation 9.6), this same assumption applies to our estimate of $\hat{b}$.

Regression Analysis Is Used to Test Hypotheses and Define Confidence Intervals Within Predictive Mathematical Models

The most common reason for testing a hypothesis using linear regression is to test whether X and Y are truly associated with each other. As discussed earlier, the null hypothesis H_o in any correlation analysis assumes that X and Y are not at all associated with each other, in which case the estimated slope $\hat{b}$ would be zero. The alternate hypothesis H_a assumes that X and Y are indeed associated, and that Y can be predicted by using some function of X and the constants $\hat{m}$ and $\hat{b}$ (as defined in Equation 9.4). For brevity, this can be written as

H_o: $\hat{m} = 0$ (no association between X and Y')

H_a: $\hat{m} \neq 0$ (association exists; X predicts Y')

As standard output for any linear regression, the statistical software will provide the regression estimates of both the slope $\hat{m}$ and intercept $\hat{b}$, as well as the results of the two-tailed t test for $\hat{m}$. As usual, any $p < \alpha$ would indicate a statistically significant result, and the investigator would be compelled to accept H_a and reject H_o. This would also mean that the estimates of $\hat{m}$ and $\hat{b}$ could then be used with confidence to predict Y, according to Equation 9.4.

Depending on the statistical software used to conduct the linear regression analysis, the output might include the correlation coefficient r or the r-squared (r^2) value. The r^2 is simply that—the square of the correlation coefficient r, which represents the total percentage of Y values that are accurately predicted from X, using the regression estimates $\hat{m}$ and $\hat{b}$ (**Figure 9.5**).

Let us take another look at the hydrography data we used in Table 9.1 when discussing correlation analyses. If these same data are used in a linear regression analysis, the results are shown in Figure 9.5.

In the first case, our values for salinity S are predictive of the density ρ according to the model described in Equation 9.7, based on the fundamental linear regression model (see Equation 9.4) fitted with the statistical estimates of $\hat{m}$ and $\hat{b}$ given in Figure 9.5:

$$\rho = 0.808\,S + 996.185 \tag{9.7}$$

Since the p-value for the slope is 0.004, we must accept H_a and conclude that the mathematical relationship expressed in Equation 9.7 is statistically significant. In fact, the r-squared (r^2) value of 0.901 indicates that 90.1% of all our measurements for ρ can be accurately predicted from our measurements of S. Generally speaking, all values of r^2 will range between 0 and 1, so the higher the r^2 value, the better the predictive power of the regression equation.

This is easily demonstrated by the results of the regression analysis for the temperature T and density ρ. When Equation 9.4 is refitted with our estimates

Figure 9.5 Typical output from a statistical software program, indicating the estimated slope $\hat{m}$ and intercept $\hat{b}$ for each regression analysis, as well as the tests of statistical significance (p-values), r-squared (r^2) values, and 95% confidence intervals for the predictive functions. In this example, our measurements of salinity S and water temperature T were analyzed for their ability to predict water density ρ, the dependent variable, using a linear regression.

model summary

	r	r^2	statistics		
model			F	df	p-value
LIN	0.949	0.901	36.321	4	0.004

predictors: (constant), S

coefficients

Mdl		coefficients					95% confidence interval	
		value	std. error	r	t	p-value	lower bound	upper bound
LIN	intercept ($\hat{b}$)	996.185	3.547		280.840	0.000	986.336	1006.033
	slope ($\hat{m}$)	0.808	0.134	0.949	6.027	0.004	0.436	1.180

dependent variable: ρ

model summary

	r	r^2	statistics		
model			F	df	p-value
LIN	0.829	0.688	8.810	4	0.041

predictors: (constant), T

coefficients

Mdl		coefficients					95% confidence interval	
		value	std. error	r	t	p-value	lower bound	upper bound
LIN	intercept ($\hat{b}$)	1046.787	10.164		102.991	0.000	1018.568	1075.006
	slope ($\hat{m}$)	−1.234	0.416	−0.829	−2.968	0.041	−2.389	−0.080

dependent variable: ρ

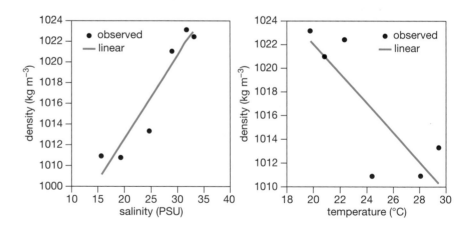

of $\hat{m}$ and $\hat{b}$, based now on the $\rho \mid T$ association (see Figure 9.5), an alternate regression equation for ρ can be defined in Equation 9.8, now as

$$\rho = -1.234\,T + 1046.787 \qquad 9.8$$

In this case, the result of our regression analysis is still significant ($p = 0.041$), so we must accept H_a and assume that Equation 9.8 is also a significant predictor of ρ. That being said, the weaker value for r^2 indicates that only 68.8% of our measurements of ρ can be accurately predicted from T.

Perhaps the most powerful aspect of the regression analysis is its predictive power. Although it is always preferable to collect real-world measurements of critical variables, our regression analyses indicate that we might not need to measure density at all. If we measure salinity or temperature instead, we can use the regression equations (see Equations 9.7 and 9.8) to predict the density values without actually having to measure them! Since Equation 9.7 exhibits a smaller p-value (stronger significance) and a higher r^2 value (better predictive capability), our regression analyses indicate that we would be well advised to use the salinity S as a more accurate predictor of the density ρ, rather than the water temperature T.

As standard output in most statistical programs, the linear regression will also provide 95% confidence intervals for the estimates of the slope $\hat{m}$ and intercept $\hat{b}$. By examining the upper and lower bounds of $\hat{m}$ and $\hat{b}$, the investigator can demonstrate the **confidence belt** that describes 95% of the data, where the main regression line serves as the central tendency (mean) of the regression analysis (**Figure 9.6**).

Multiple Regression and Curve Estimation

Linear regression analysis is most useful when the dependent variable in the correlation can reliably be estimated with a single independent variable. However, natural systems are usually quite complex and are rarely estimable using a single variable. More often than not, we are faced with the extremely daunting task of predicting some natural behavior that is forced by several different variables, all operating at the same time and in very different ways.

Fortunately for us, the simple linear regression can be expanded to include more than one independent variable. Theoretically, it is possible to have an infinite number of independent variables that could act on and therefore determine the behavior of the dependent variable. A slight modification to Equation 9.4 is needed to account for this possibility, resulting in the multiple regression equation defined in Equation 9.9:

$$Y' = \hat{m}_1 X_1 + \hat{m}_2 X_2 + \hat{m}_3 X_3 + \cdots \hat{m}_i X_i + \hat{b} \qquad (9.9)$$

ASSUMPTIONS OF THE LINEAR REGRESSION METHOD

1. Linearity: Y must be a linear function of X and can be estimated as $Y' = \hat{m}X + \hat{b}$.

2. Independence: Each measurement of Y must be independent of all other measurements of Y.

3. Normality of Error: The random errors ε associated with the relationship between X and Y must be normally distributed with the same standard deviation.

4. Equal Variance: The variance s^2 must be equal for all ε.

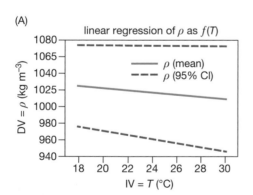

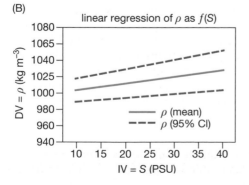

Figure 9.6 The regression equation that predicts the water density ρ as a function of the salinity S had a stronger statistical significance (lower p-value) and better correlation coefficient (higher r^2) than the function using the water temperature T (see Figure 9.5). We can plot the central regression line (mean) and the upper/lower bounds of the 95% confidence intervals of the y-intercept ($\hat{b}$) for each of the T (A) and S (B) regressions and analyze the variance predicted by each model. In this case, we can plainly see that the S regression (B) has a much lower predicted variance, as illustrated by the narrow confidence belt constrained between the upper/lower bounds of the 95% confidence intervals for $\hat{b}$. In contrast, the T regression (A) has a very broad confidence belt and would therefore be a less reliable predictor of ρ.

This is fundamentally the same as the simple linear regression in Equation 9.4, except that we now have to analyze several different slopes (and several different variables) instead of just one. That also means each slope must be tested (using a t test) to determine whether it is statistically different than zero. Just as before,

H_o: $\hat{m}_i = 0$ (no association between X_i and Y')

H_a: $\hat{m}_i \neq 0$ (association exists; X_i predicts Y')

As an example of a multiple regression, let us return to the simple hydrographic dataset in Table 9.1 that includes the unranked measurements of temperature T, salinity S, density ρ, and wind speed U. Earlier in this chapter, we were able to demonstrate through Pearson's correlation analysis that within our example dataset, wind speed is not correlated with density, but both salinity and temperature are. We then performed two separate linear regressions: the first to determine the linear equation where density is predicted using salinity, and the second where density is predicted instead by water temperature. Both salinity and temperature were found to be correlated with, and predictive of, density. However, they were each analyzed separately. Perhaps we could improve our ability to predict density if salinity and temperature were considered together.

Now, instead of analyzing all of the independent variables piecemeal, we might instead opt to perform a **multiple regression** analysis to determine if (and how) our measurements of density can be predicted using measurements of salinity, temperature, and wind speed (that is, multiple independent variables) simultaneously. Assuming they are all relevant to density, the relationship could be expressed as the multiple regression equation defined in Equation 9.10:

$$\rho' = \hat{m}_S S + \hat{m}_T T + \hat{m}_U U + \hat{b} \tag{9.10}$$

When all three independent variables (S, T, U) are analyzed together in a statistical software package, the output will look something like **Figure 9.7**.

It would at first seem like ρ can be predicted perfectly ($r^2 = 1.00$) using measurements of water temperature, salinity, and wind speed together (see Figure 9.7A). However, since the slope for wind speed ($\hat{m}_U$) has $p > 0.05$, we cannot confidently claim that this particular correlation is significant and we must accept the null hypothesis H_o that $\hat{m}_U = 0$ and can therefore be ignored with regard to the multiple regression equation for ρ (see Equation 9.10). That would give us a new generalized equation for ρ, which is now more accurately described as Equation 9.11:

$$\rho' = \hat{m}_S S + \hat{m}_T T + \hat{b} \tag{9.11}$$

If the multiple regression analysis is run again, this time by eliminating the wind speed U as one of the independent variables, the standard error of the estimate for ρ does unfortunately increase. The good news is that the increase in the standard error of the estimate for ρ is very slight, our r^2 is still 1.00, and all the p-values have remained very strongly significant now that the prediction equation has been streamlined (see Figure 9.7B), such that only the relevant variables (namely, salinity, and water temperature) are included. The result is a prediction equation for ρ, where Equation 9.11 is now fitted with our estimates of $\hat{b}$, as well as our simultaneous estimates of both $\hat{m}_S$ and $\hat{m}_T$, to become Equation 9.12:

$$\rho' = 0.601S - 0.592T + 1015.817 \tag{9.12}$$

This is a vast improvement over the simple linear regression analysis, because we have been able to establish that ρ is actually correlated with

(A)

model summary

model	r	r^2	std. error of the estimate
LIN	1.000	1.000	4.837E-02

predictors: intercept ($\hat{b}$), windspd, T, S
dependent variable: ρ

coefficients

Mdl		coefficients				95% confidence interval	
		value	std. error	t	p-value	lower bound	upper bound
LIN	intercept ($\hat{b}$)	1016.215	0.312	3260.418	0.000	1014.187	1017.556
	S	0.597	0.004	136.047	0.000	0.578	0.616
	T	−0.592	0.007	−85.385	0.000	−0.622	−0.562
	U	−0.103	0.050	−2.056	**0.176**	−0.318	0.112

dependent variable: ρ

(B)

model summary

model	r	r^2	std. error of the estimate
LIN	1.000	1.000	6.969E-02

predictors: (constant), T, S
dependent variable: ρ

coefficients

Mdl		coefficients				95% confidence interval	
		value	std. error	t	p-value	lower bound	upper bound
LIN	intercept ($\hat{b}$)	1015.817	0.352	2886.337	0.000	1014.697	1016.937
	S	0.601	0.006	105.257	0.000	0.583	0.619
	T	−0.592	0.010	−59.311	0.000	−0.624	−0.560

dependent variable: ρ

Figure 9.7 Typical output from a statistical software program, indicating the estimated slope and intercept for every independent variable analyzed in the multiple regression analysis (A). By eliminating variables that are not significantly correlated with the dependent variable (B), the multiple regression can be run again and improved.

water temperature and salinity together, as variables that are complementary to each other in their power to predict the dependent variable ρ. In fact, if we use the upper and lower bounds of the confidence interval of the y-intercept ($\hat{b}$) of the multiple regression compared to Figure 9.6, we can visually demonstrate the superiority of our new multiple regression model that now boasts a very narrow confidence belt (**Figure 9.8**). Had we not performed the multiple regression analysis, we would have erroneously used either salinity or water temperature to estimate density, when we should have been using both.

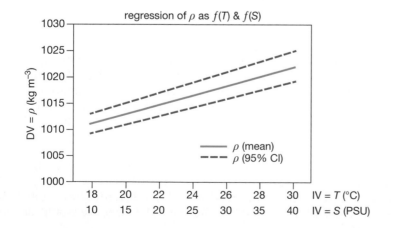

Figure 9.8 A plot of the central tendency and the upper/lower bounds of the 95% confidence interval of the y-intercept ($\hat{b}$) for the dependent variable ρ, as predicted by the multiple regression of both independent variables T and S, considered simultaneously. Indeed, the multiple regression of ρ boasts a much lower predicted variance, as illustrated by the narrow confidence belt constrained between the upper/lower bounds of the 95% confidence interval of $\hat{b}$. Note the difference in scale of the y-axis for the multiple regression, where the variance in the predicted ρ values ranges from 1009 to 1025 kg m⁻³, while the ρ values from the linear regressions in Figure 9.6 range anywhere from 940 to 1080 kg m⁻³.

Multiple Regressions That Are Nonlinear Must Be Determined Using Curve Estimation

Most statistical applications also possess the capability to perform more complicated, nonlinear regressions as a function of a single independent variable. Although nonlinear multiple regression methods do exist, they require a significant level of mastery to perform and are admittedly beyond the introductory scope of this text. The reader is invited to consult more advanced statistical methods for nonlinear regression analyses involving two or more independent variables.

Within the variety of nonlinear mathematical relationships that require a single independent variable for predictive capability, the most common are defined as follows:

$$Y' = \hat{b} + \left(\hat{m} \cdot \ln X\right) \qquad \text{Log} \qquad (9.13)$$

$$Y' = \hat{b} + \frac{\hat{m}}{X} \qquad \text{Inverse} \qquad (9.14)$$

$$Y' = \hat{b} + \hat{m}_1 X + \hat{m}_2 X^2 \qquad \text{Quadratic} \qquad (9.15)$$

$$Y' = \hat{b} + \hat{m}_1 X + \hat{m}_2 X^2 + \hat{m}_3 X^3 \qquad \text{Cubic} \qquad (9.16)$$

$$Y' = \hat{b} X^{\hat{m}} \qquad \text{Power} \qquad (9.17)$$

$$Y' = \hat{b} \hat{m}^X \qquad \text{Compound} \qquad (9.18)$$

$$Y' = e^{\hat{b} + \left(\frac{\hat{m}}{X}\right)} \qquad \text{S} \qquad (9.19)$$

$$Y' = \frac{1}{\left(1/u\right) + \left(\hat{b}\hat{m}^X\right)}; \quad u > |Y_{max}| \qquad \text{Logistic} \qquad (9.20)$$

$$Y' = e^{\hat{b} + \hat{m}X} \qquad \text{Growth} \qquad (9.21)$$

$$Y' = \hat{b} \cdot e^{\hat{m}X} \qquad \text{Exponential} \qquad (9.22)$$

Curve estimation simply involves entering the independent and dependent variables into the statistical curve-fitting routine and selecting from the available models (see Equations 9.13–9.22). Depending on the statistical software being used, the results of curve estimation will be produced in an output file similar to **Figure 9.9**.

In this example, the salinity S was once again analyzed for its capacity to accurately predict the density ρ. According to this curve-fit analysis, all of the nonlinear models exhibit $p < 0.05$; therefore, they are all suitable as a predictive model for ρ. However, the model with the highest r^2 value should be chosen, as it represents the model that most accurately estimates the dependent variable ρ.

Recall from our earlier example that the simple linear regression between salinity and density had $r^2 = 0.901$ (see Figure 9.5). Our curve-fit analysis would allow us to improve the early model, particularly if we choose the quadratic model with $r^2 = 0.941$. If this were the case, we would use the

independent: S										
dependent	Mdl	r^2	df	F	Sig	upper bound	b	m1	m2	m3
ρ	**LIN**	**0.901**	4	36.32	**0.004**		996.185	0.8082		
ρ	LOG	0.856	4	23.68	0.008		957.118	18.6232		
ρ	INV	0.794	4	15.46	0.017		1033.88	−405.44		
ρ	**QUA**	**0.941**	3	23.71	**0.015**		1018.21	−1.1120	0.0390	
ρ	CUB	0.936	3	22.00	0.016		1010.29	−0.1285		0.0005
ρ	**COM**	**0.901**	4	36.44	**0.004**		996.377	1.0008		
ρ	POW	0.856	4	23.76	0.008		958.814	0.0183		
ρ	S	0.795	4	15.51	0.017		6.9412	−0.3988		
ρ	**GRO**	**0.901**	4	36.44	**0.004**		6.9041	0.0008		
ρ	**EXP**	**0.901**	4	36.44	**0.004**		996.377	0.0008		
ρ	LGS	0.878	4	28.67	0.006	1030.0	6.2E-05	0.930		

Figure 9.9 Typical output from a statistical software program, indicating the results from a variety of curve-fitting models. The model with a significant p-value and the highest r^2 represents the best possible curve fit for the data. In this example, the LIN, COM, GRO, and EXP models all share the lowest p-value (0.004), but the QUA model has the best r-squared (r^2) value (0.941). Since the QUA model has the best r^2 value and still has a strong p-value (0.015), the QUA model is most likely to produce the best results as a predictive model of ρ when using S as the independent variable.

fundamental quadratic model (as defined in Equation 9.15) and refit the model with our own statistical estimates of $\hat{b}$ and $\hat{m}_i$ from Figure 9.9 to yield Equation 9.23:

$$\rho' = 1018.21 - 1.112S + 0.039S^2 \qquad (9.23)$$

Although the quadratic model in Equation 9.23 is certainly an improvement from the simple linear model we derived as Equation 9.7, we have already demonstrated that the multiple regression equation that describes ρ as a function of both salinity and water temperature (see Equation 9.12) is an improvement beyond all of the curve-fit models. Thus, we should stick with Equation 9.12 as our preferred model for estimating ρ.

Correlation Between Two Dependent Variables

In our previous examples, we have focused our attention on the correlation methods used to investigate the correlation between one or more independent variables (a set of predictors) and a single dependent variable. But what do we do if we have two dependent variables we'd like to examine for correlation? That's when we must invoke more complicated **multidimensional scaling** methods in order to analyze our data. And although that sounds impressive (or scary), we already have practice in multidimensional scaling. Recall the different slopes we derived for S and T in our multiple regression of ρ in Equation 9.12. Each of those slopes was essentially a scaling factor, applied to each of the independent variables in our equation in order to provide the best possible prediction of ρ. The more variables we have to work with, the more dimensions we have to scale.

Canonical Correlation (CC) Is Used to Explore Correlations Between Two Sets of Dependent Variables

Consider the situation we have just examined using multiple regression analysis, where multiple independent variables like salinity S and temperature T are used to predict a single dependent variable, such as ρ. If we wished to expand our oceanographic analyses to include a new multiple regression, we might be also be interested to predict phytoplankton biomass (as chl, a single dependent variable) as a function of several independent variables represented as dissolved nutrients (like NH_4^+, NO_3^-, and PO_4^{3-}). In much the same fashion, we would simply use the general multiple regression model (see Equation 9.9) and adapt it for our specific interests, becoming Equation 9.24:

$$\rho' = \hat{m}_S S + \hat{m}_T T + \hat{b} \quad \& \quad chl' = \hat{m}_{NH_4} NH_4^+ + \hat{m}_{NO_3} NO_3^- + \hat{m}_{PO_4} PO_4^{3-} + \hat{b}$$

$$(9.24)$$

Figure 9.11 A simplified flowchart of the multivariate analyses most commonly used in the natural sciences to test group differences. These methods can be used to analyze either continuous data or categorical data, but the most appropriate method is determined by the number of dependent variables (DVs) being predicted by correlation, as well as the number of independent variables (IVs) being used as predictors.

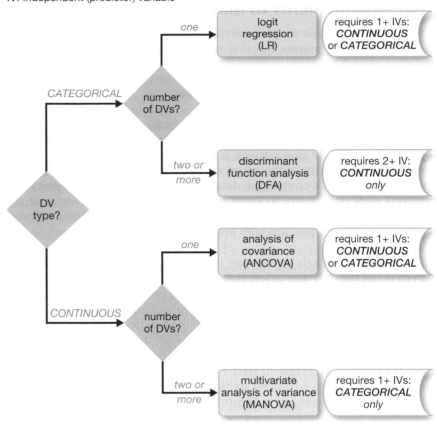

testing group differences

DV: dependent (predicted) variable
IV: independent (predictor) variable

References

Jones ER (1996) Statistical Methods in Research. Edward R. Jones.

Kanji GK (1999) 100 Statistical Tests. SAGE Publications.

Keeping ES (1995) Introduction to Statistical Inference. Dover Publications.

Keller DK (2006) The Tao of Statistics. SAGE Publications.

Kenny DA (1979) Correlation and Causality. John Wiley and Sons.

Linton M, Gallo Jr PS, & Logan CA (1975) The Practical Statistician: Simplified Handbook of Statistics. Wadsworth Publishing Company.

Mandel J (1964) The Statistical Analysis of Experimental Data. Dover Publications.

Newman I, & Newman C (1977) Conceptual Statistics for Beginners. University Press of America.

Rencher AC & Christensen WF (2012) Methods of Multivariate Analysis, 3rd ed. John Wiley & Sons.

Salkind NJ (2007) Statistics for People Who (Think They) Hate Statistics: The Excel Edition. SAGE Publications.

Steiner F (ed) (1997) Optimum Methods in Statistics. Akadémiai Kiadó.

Thompson SK (1992) Sampling. John Wiley and Sons.

Further Reading

Afifi A, May S, & Clark VA (2011) Practical Multivariate Analysis, 5th ed. CRC Press.

Everitt B & Hothorn T (2011) An Introduction to Applied Multivariate Analysis with R. Springer.

Gittins R (1985) Canonical Analysis: A Review with Applications in Ecology. Springer-Verlag.

Jackson JE (2003) A User's Guide to Principal Components. John Wiley & Sons.

Jolliffe IT (2002) Principal Component Analysis. Springer-Verlag.

Kachigan SK (1986) Statistical Analysis: An Interdisciplinary Introduction to Univariate and Multivariate Methods. Radius Books.

Mari DD & Kotz S (2001) Correlation and Dependence. Imperial College Press.

Parkhomenko E (2009) Sparse Canonical Correlation Analysis: Data Integration for Regular and High Dimensional Studies. VDM Verlag.

Timm NH & Mieczkowski TA (1997) Univariate and Multivariate General Linear Models. SAS Institute.

Wei WWS (2005) Time Series Analysis: Univariate and Multivariate Methods, 2nd ed. Pearson.

Unit 4
Methods of Data Assimilation (Modeling)

Contents

Chapter 10

Fundamental Concepts in Modeling

"Mathematical reasoning may be regarded rather schematically as the exercise of a combination of two facilities, which we may call intuition and ingenuity."– Alan Mathison Turing

From the earliest days of human existence, inquisitive minds have always sought to relate their observations of the world around them in the context of causality. Were the universe not so unerringly committed to "cause and effect" over chaos, the very practice of the scientific method would be an impossible dream, as would be any attempt to mathematically describe all those ordered system(s) that define the machinations of our natural world. We've come a long way from estimating the arrival of the summer solstice using the megaliths of Stonehenge; now, there are a multitude of apps for that on your smartphone. The Druids never had it so good.

Without a doubt, there are processes at work within the natural world that we have scarcely begun to discover and subsequently describe. The entire scientific enterprise compels us not merely to make observations or gather data, but to explain the functioning of the universe in the past, present, and future. As a scientist, that should be a daunting (and humbling) realization. Take heart, for there is good news: any system that exhibits order can be studied, and described, using mathematical models—that's where your genius is so desperately needed.

Key Concepts

- A model is an object, method, or mechanism used to simulate some aspect of reality; numerical models accomplish this using mathematical expressions.

- The variables and processes that define complex systems are all sensitively interconnected (and must be modeled as such).

- Mathematical models can be used as scientific tools to test a variety of alternate hypotheses in simulated systems.

- The creation of a conceptual diagram is critical to the construction of the numerical model it represents.

- All mathematical models consist of five elements: forcing functions, external variables, state variables, fixed parameters, and universal constants.

So What Is a Model?

A model is simply an object, method, or mechanism used to simulate some aspect of reality. You may have some first-hand experience with model making: whether building (and populating) a dollhouse in your youth or assembling an airplane or automobile from an assortment of plastic parts, you essentially constructed a physical model of a real object and undoubtedly used it to simulate some aspect of reality that interested you at the time. In the context of science, physical models are used all the time (primarily in engineering applications) to investigate interaction phenomena without going to the trouble and expense to perform the experiment at full scale. For example, naval engineers might wish to examine the hydrodynamics of a new hydrofoil design. Instead of building a full-scale prototype, engineers might opt to build a small-scale physical model of the hydrofoil in order to test the stability of the new design (**Figure 10.1**).

A **mathematical model** is a representation of a physical object, method, or mechanism but in "computational space." Mathematical models attempt to reduce reality (or our simulation of it) to an assemblage of numbers, each of which is representative of some variable we are interested in. For instance, if naval engineers create a mathematical model of streamflow over the chassis of their new hydrofoil, they might be able to predict design instabilities simply by analyzing the numerical results within their mathematical model.

Although most of us probably feel more comfortable creating physical models rather than mathematical ones, there are many applications in which the mathematical model is a much simpler option.

Figure 10.1 Physical (*left*) and mathematical (*right*) model representations of the streamflow associated with a new propeller design. (Left, courtesy of Alan C. McClure Associates.)

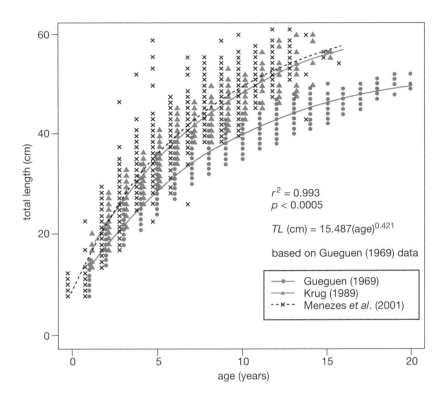

Figure 10.2 A mathematical model of total body length (cm) among blackspot seabream (*Pagellus bogaraveo*) in the Bay of Biscay, based on age. Using these data, a researcher would be able to establish a mathematical model that describes the age-to-body-length correlation among this particular species of seabream. Once the mathematical model is defined, we can use that equation to either estimate seabream length based on measured age, or we could rearrange the equation and estimate seabream age based on measured lengths. (From Lorance P, *ICES J. of Marine Science* doi: 10.1093/icesjms/fsq072. With permission from Oxford Journals.)

As an example, let us assume that a fish biologist might be interested in tracking the total length of a particular species of fish as a function of age. Throughout the years, several researchers gather the critical body length and age structure data (**Figure 10.2**) that can be used quite easily to construct a mathematical model (that is, an equation) that defines the relationship between these two variables. In this case, it would be much simpler to describe the growth dynamics of this fish species using a mathematical model than it would be to construct an actual physical model of each size class.

Of course, the model that defines the growth parameters of our example fish species would only be possible if there were a significant body of statistical data that specifically relate age to body length, so that an ichthyologist could determine whether any given measurement is "usual." Although the more traditional methods of statistical analysis are indeed important and can define interrelationships within certain data, a well-designed model can reveal a deeper understanding of the entire system and how it functions overall.

Although the simplest mathematical models can be solved using a hand-held calculator, most require the use of a computer. In years past, the complexity of mathematical models was limited by the availability of significant computing power. However, with the tremendous advances in high-performance computing technology in recent decades, mathematical models are now much more likely to be limited by a modeler's lack of insight or modeling ability than by processor speed. But model complexity is not the grail we seek; the true value of a mathematical model lies in its ability to provide context and insight to the complexity of the natural world.

A well-designed model is one that faithfully represents the dynamics observed in the natural world, and requires that we, as modelers, are capable of assimilating the existing body of knowledge into our particular model. At its core, a model (and its creator) must possess some knowledge of (1)

The power of models to test alternate hypotheses stems from a model's inherent mathematical precision: that it behaves according to the very strict rules of arithmetic.

Consider a very simple mathematical model of the form $2x = y$. Don't let the simplicity of this model fool you—there are actually an infinite number of solutions to it. But if our original premise is that $x = 2$, the solution to our model (that is, the value of y) must be 4. We could test our model, over and over again, and consistently arrive at the same solution: $2x = 4$ when $x = 2$.

But what happens if we loosen up the model a bit and allow x to vary? If our alternate hypothesis is that $x = 3$, the solution to our model becomes $2x = 2(3) = 6$. As you can see, the fundamental structure of our mathematical model ($2x = y$) did not change at all, but the solution to our model was sensitive to the new hypothesis. This basic methodology allows modelers to test alternative hypotheses using the model to calculate different numerical outcomes to compare with data gathered in the field.

Using Conceptual Diagrams

The initial focus of scientific research should always be on defining the nature and scope of the problem you wish to investigate. Since the resources necessary to perform research are always limited, it is critical that you carefully consider the boundaries of your research and/or model. Similar to the steps taken to define the limits of your research when designing a field experiment, a well-constructed model also demands that you constrain your spatial and temporal scales to fit the problem. Beyond placing boundaries in space and time, it is critical that you include only those elements that are relevant to the dynamics you wish to model.

Modelers usually begin this task by constructing a conceptual diagram of the dynamics they wish to simulate. More than a brainstorming exercise, a conceptual diagram is a graphic representation of the elements that are important to the problem at hand and how these elements are connected by processes. Such diagrams are often used to represent complex dynamics in a very simplified, graphical form; almost like an outline or blueprint from which a modeler may begin the construction of a model.

Consider the conceptual diagram for the nitrogen cycle (**Figure 10.5**), in very simple terms:

Figure 10.5 A simple conceptual diagram of the nitrogen cycle. Note that in this conceptual model, there are many different processes (boxes) that dictate how nitrogen is recycled in natural systems (blue arrows). Each of those processes will require specific input/output variables (circles) which can either be calculated by a numerical model or measured in the field.

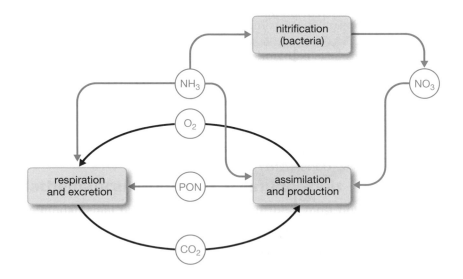

Conceptual diagrams are extremely valuable in that they provide a visual summary of the important variables and processes that define the dynamics you may wish to model. A **variable** is a model element that can be described as a scalar quantity, the magnitude of which may change over time. For instance, NH_3 concentrations might change dramatically over time, but regardless of that variability, we can always assign some quantitative value to a momentary measurement of NH_3. Quantification of variables can be done either by taking direct measurements in the field and inserting that data into a numerical model, or having the model itself calculate (predict) those values.

If we choose to have our model calculate the value of a particular variable, we must first know something about the processes that affect that variable. Processes are much more difficult to understand, as they represent any number of factors that combine to induce change in the system. Although processes can also be described mathematically, these are usually defined by equations rather than by simple quantities.

Challenging as it may be to mathematically define all of the variables and processes within the nitrogen cycle, at least we can use the conceptual diagram as a checklist of the features our model must possess. This diagram can help us to decide which variables we might wish to measure ourselves (as part of our larger field research) and which variables can be estimated by the model. The diagram can also be used to inform the modeler as to which processes must be included in the model, provided that the equations that define those processes are known.

If the equation that defines a particular process is not known, additional field or laboratory research will be required. For example, if nitrogen excretion rates are not well known, it will be necessary to design an experiment to specifically quantify the rate(s) of metabolic conversion of particulate organic nitrogen (PON) to NH_3. This process, by itself, is undoubtedly complex and will be affected by its own set of variables and processes (like a model within a model). When all of the variables and processes within the conceptual diagram have been defined, the modeler may finally begin to tie those elements together into a unified numerical model, which is essentially a mathematical expression of the conceptual diagram.

The Fundamental Elements of Modeling

Regardless of scientific discipline, so long as the variables and processes contained within your conceptual diagram can be translated into the language of mathematics, there are no limits on what dynamics you may hope to model. But just as there are universal limits to the logic and structure of mathematics, so too are there limits to the logic and structure of numerical models. These are the fundamental elements of modeling that are common to all numerical models, regardless of the model's designated purpose.

Forcing Functions Are the Fundamental Equations That Define the Mathematical Structure of the Model

Forcing functions are, as the name suggests, mathematical functions that influence the state of the system you wish to model. In the context of the conceptual diagram, forcing functions are the processes that ultimately influence the variables (and perhaps other processes) within the model. Defining a forcing function as a mathematical equation is perhaps the most difficult task of modeling, as each forcing function is, in itself, a mini-model nested within the larger model.

In a conceptual diagram, forcing functions are represented by the boxes and arrows—not only do they provide connectivity between the variables

SCIENCE IS PROCESS ORIENTED

Although it is a relatively simple task to measure the concentration of ammonia (NH_3) in water, what can we glean from these seemingly chaotic results?

Time (UTC)	NH_3 (μM)
06:17:37	2.17
07:13:44	2.12
09:38:11	0.56
11:56:31	1.19
13:05:16	0.14
15:28:52	3.82
18:46:29	0.87

According to the conceptual diagram of the nitrogen cycle (see Figure 10.5), NH_3 concentrations are affected by assimilation and production, respiration and excretion, and nitrification. Unless we can define how each and every one of those processes is affecting NH_3 concentration over time, our measurements will do very little to reveal the dynamics of this system.

in your model, but they also define the rules by which those variables will change over time. For example, in our conceptual diagram of the nitrogen cycle (see Figure 10.5), the conversion of ammonia (NH_3) to nitrate (NO_3) is mediated by bacteria in a process called "nitrification." We know from our conceptual diagram that the process of nitrification ties NH_3 to NO_3. Thus, if nitrification is taking place, our model should reflect a reduction in NH_3 concentrations while NO_3 concentrations will rise (as bacteria convert NH_3 to NO_3). In this way, the process of nitrification will force our model to recalculate NH_3 and NO_3 concentrations over time. Note that since NH_3 and NO_3 concentrations are tied to other variables in the nitrogen cycle, the forcings caused by nitrification will also affect other forcing factors in the model.

Because of the complexity inherent in most forcing functions, modelers usually must develop separate submodels to mathematically define each of these processes. In the nitrification example, it may not be enough to simply consider nitrification as a simple function of NH_3 and NO_3 concentrations. Since bacteria are the mediators of this process, we may be required to know something more about the biology of these organisms and what determines the rate at which these bacteria will convert NH_3 to NO_3. Those may be details that are extremely important to the function and fidelity of our model.

Variables Are Those Elements Within the Model That Can Be Quantified, and Whose Values Are Likely to Change

External (explicit) variables are those model elements that can be described as quantities that vary over time but are determined outside and independent of the model. These quantities are usually measured in the field and provide the critical link between the natural world in reality (most faithfully represented by field data) and the natural world in simulation space (your model).

External variables are incredibly useful in testing our understanding of the forcing functions that influence those same variables in the natural world. For instance, if we wished to determine the forcing function that defines nitrification, we might measure NH_3 and NO_3 concentrations over time, taking great care to consider the relationship between these two external variables in an effort to mathematically define the process by which bacteria convert NH_3 to NO_3. Ultimately, that relationship will be reflected in our measurements of NH_3 and NO_3 concentrations.

External variables can also be used to make predictions about other variables within our model. If, through previous research, we were able to mathematically define the forcing function that describes nitrification, we could simply measure NH_3 concentrations in the field and use our forcing function of nitrification to estimate what the concentration of NO_3 should be. In that case, only NH_3 would be considered an external variable (since it was the only variable determined outside and independent of the model).

State (implicit) variables are those model elements that can be described as quantities that vary over time but are calculated by the model. Selecting the proper balance between external and state variables is critical to model performance. Choosing to measure too many external variables from the field can be time consuming and expensive, and essentially negates the need for a numerical model. By the same token, calculating too many state variables has the effect of disconnecting your model from reality. Unfortunately, there is no simple guide to follow when choosing the appropriate balance between which variables should be measured externally and which should be calculated in your model as state variables—that analysis must come later, when your simulation is complete and you begin the process of testing model fidelity (discussed later in Chapter 14).

Parameters and Constants Differ from Variables in That They Represent Fixed Quantities That Do Not Change

Parameters are those model elements that can be described as quantities but do not vary over time. These are quantities that may be considered to be constant in the specific context of your model. Occasionally, parameters are used to simplify the external variables within a model in order to prevent the model from becoming too complex (that is, a model containing too many external and/or state variables). This may be an attractive option to simplify a model, especially when a particular variable is determined to have a small variance and/or standard deviation.

For example, if our original intent was to include NH_3 concentration as an external variable within our model of the nitrogen cycle, we would obviously have to externally measure NH_3 concentrations in the system of interest. If, through the course of our research, we determined that NH_3 concentrations were always measured within a strict range of 0.73–0.79 μM, it would be far simpler to include NH_3 in our model as a fixed, "once and done" parameter rather than as a constantly changing variable that must be calculated, time and again, within our model. In this example, a good choice might be to treat NH_3 as a model parameter, with a fixed concentration of 0.76 μM (corresponding to the median of our measured NH_3 concentration range).

Although parameters can be used to simplify variables within a model, in rare cases, forcing functions can also be simplified in this fashion. If we also measured nitrification rates under a wide variety of environmental conditions, we might find that the bacterial conversion of NH_3 to NO_3 is sensitive only to ambient NH_3 concentrations. Since we found that the range of NH_3 concentrations was very small, we might also find that the range of nitrification rates is also very narrow (0.021–0.023 $\mu M\ NO_3\ \mu M^{-1}\ NH_3\ day^{-1}$). If this were the case, we may wish to simply use the median nitrification rate (0.022 $\mu M\ NO_3\ \mu M^{-1}\ NH_3\ day^{-1}$) as a fixed parameter rather than go to all the trouble to develop a very complicated forcing function to calculate the instantaneous nitrification rate as a variable in our model.

As a general rule, any variable (or forcing function) that exhibits very low variance and/or standard deviation can be "parameterized" without fear of oversimplifying the dynamics we might wish to simulate. Parameters may be measured in the field or gleaned from the literature, but are typically found as a range of values exhibiting a Gaussian, Poisson, or χ^2 distribution (discussed earlier in Chapter 2). Depending on the nature of the model element being parameterized, modelers can use the central tendency of the distribution.

Universal constants are similar to parameters in that they too can be described as simple quantities that do not vary over time, but these are global constants regardless of the specific context of the model. For instance, the molar weight of the NH_3 molecule (17.0307 g mol^{-1}) does not vary, nor is it specific to our particular model of the nitrogen cycle—it is applicable to and valid for all possible models. Hence, it is considered to be a universal constant.

From Concept to Creation: What Now?

The key to any successful modeling effort is to first develop a detailed conceptual diagram of the model you wish to create. Toward this end, it is absolutely critical that you perform an exhaustive literature search of every conceivable variable and forcing function outlined in the conceptual diagram. Frequently, this will lead you to amend your conceptual diagram, adding new variables and forcing functions for consideration. For each element

 FORCING A PARAMETER

Just because a particular forcing function may exhibit a low variance and standard deviation does not necessarily mean it is a good candidate for parameterization. Forcing functions are inherently complex and should represent a cause for concern if they are "too easy" to quantify or exhibit very little sensitivity to a wide variety of environmental conditions.

The measured results of complex processes (such as the nitrification rates of bacteria) can easily become biased by a critical variable, important to the process but unknown to (and unmeasured by) the researcher. Unless the forcing function in question has been exhaustively investigated and consistently exhibits low variance, external variables are a safer choice for parameterization than forcing functions.

within your conceptual diagram, you must determine whether the literature, your field data, or the model itself can be used to mathematically describe each forcing function and variable within—only then can you proceed in the actual creation of your model.

Ultimately, all of your model elements must be woven together into a cohesive numerical model. As the model itself is a simplification of natural dynamics, certain errors and omissions are inherent to every model—the trick is to limit their impact on model fidelity. Luckily, several computational strategies exist that can greatly assist in the process of translating the conceptual diagram into a functioning numerical model. It is precisely these "rules of engagement" that we shall discuss in the next chapter.

References

Fennell W & Neumann T (2014) Introduction to the Modelling of Marine Ecosystems, 2nd ed. Elsevier.

Jorgensen SE & Bendoricchio G (2001) Fundamentals of Ecological Modelling, 3rd ed. Elsevier Science.

Lorance P (2011) History and dynamics of the overexploitation of the blackspot sea bream (*Pagellus bogaraveo*) in the Bay of Biscay. *ICES Journal of Marine Science* 68(2):290–301.

Royle JA & Dorazio RM (2008) Hierarchical Modeling and Inference in Ecology. Elsevier.

Schmitt RJ & Osenberg CW (eds) (1996) Detecting Ecological Impacts: Concepts and Applications in Coastal Habitats. Academic Press.

Further Reading

Bender EA (2000) An Introduction to Mathematical Modeling. Dover Publications.

Dale VH (ed) (2002) Ecological Modeling for Resource Management. Springer-Verlag.

Grant WE & Swannack TM (2007) Ecological Modeling: A Common-Sense Approach to Theory and Practice. Blackwell Publishing.

Haefner JW (1996) Modeling Biological Systems: Principles and Applications. Springer-Science.

Jopp F, Reuter H, & Breckling B (eds) (2011) Modelling Complex Ecological Dynamics: An Introduction into Ecological Modelling for Students, Teachers, and Scientists. Springer-Verlag.

Jorgensen SE, Chon T-S, & Recknagel FA (eds) (2009) Handbook of Ecological Modelling and Informatics. WIT Press.

Murray JD (2002) Mathematical Biology I: An Introduction, 3rd ed. Springer-Verlag.

Murray JD (2003) Mathematical Biology II: Spatial Models and Biomedical Applications, 3rd ed. Springer-Verlag.

Shiflet AB & Shiflet GW (2014) Introduction to Computational Science: Modeling and Simulation for the Sciences, 2nd ed. Princeton University Press.

Smith J & Smith P (2007) Environmental Modelling: An Introduction. Oxford University Press.

Chapter 11

Model Structure

"You have to learn the rules of the game. And then you have to play better than anyone else." – Albert Einstein

Modeling ocean dynamics is a tricky business. After all, marine science occupies the nexus between all of the natural science disciplines—biology, chemistry, geology, and physics—and we might as well throw mathematics into the mix. That means an accomplished marine scientist has to be proficient at all of these disciplines, simultaneously. And if you want to be an ocean modeler? Add computer programmer to the list.

Given these facts, it should come as no surprise that the fundamental rules of ocean modeling are, in a word, arcane. Before we can embark on our lofty goal of simulating complex ocean processes, we must first consider the spatial and temporal dimensions of the problem at hand, and how we might construct a model that is representative of reality. This will require a bit of preparation before we can get started. As we will soon discuss, our model's ability to ingest what data we do possess, and propagate that information throughout the simulation space, is what will ultimately allow it to make estimates of unmeasured quantities in the far reaches of our domain and to make bold predictions about what shall come to pass.

Key Concepts

- The spatial dimensions of aquatic models are typically defined in three dimensions: two dimensions in the horizontal (x,y) and one dimension in the vertical (z).
- The stability of any numerical model is determined by its spatial (grid) resolution, relative to its temporal resolution.
- The model domain must be contained by explicit boundary conditions that dictate the propagation of information into (and out of) the simulation space.
- Prior to model initialization, the grid must be populated either with true measurements or with data interpolated from those measurements.
- The numerical method of finite differencing will enable the model to predict future behavior based on current values.

Defining Space and Time

The first order of business when developing a numerical model is to quite literally consider the dimensions of our **model domain**. For consistency, the spatial domain of any numerical model is typically defined by using an (x,y,z) coordinate system, where x represents the horizontal dimension in the east–west direction (**easting**), y represents the horizontal dimension in the north–south direction (**northing**), and z represents the vertical dimension (**Figure 11.1**). By using this uniform geometric convention, the model domain can always be described such that:

$+x$ = east direction (+ longitude)
$-x$ = west direction (– longitude)
$+y$ = north direction (+ latitude)
$-y$ = south direction (– latitude)
$+z$ = upward direction (elevation)
$-z$ = downward direction (depth)

Although we typically think of the ocean as a truly three-dimensional domain, the dynamics we may wish to model may not require a full three-dimensional simulation. Since three-dimensional simulations are extremely complex (and very difficult to perform), it is certainly in our best interest to consider whether the processes we wish to model can be simulated using a simpler domain. For the moment, let us ignore the temporal dimension (that is, the passage of time) and focus our attention solely on the spatial dimensions that prescribe our model domain. If we remind ourselves of how spatial dimensions are defined in mathematics, we can envision which geometries are best suited for our particular model (**Figure 11.2**).

Zero-Dimensional Models Simulate What Occurs at a Single Point

In mathematics, a single point in space has zero dimensions. That means that the value of that point, wherever it may be, is completely unaffected by spatial variability. That also means that the only other way for the value to change is as a function of time, $f(t)$, but not as a function of longitude (x), latitude (y), or depth (z).

For very small water masses that are assumed to be perfectly well mixed (that is, homogeneous in both the horizontal and vertical dimensions), the model solution at any point in the system is the model solution at every point in

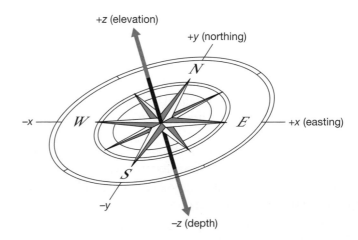

Figure 11.1 The Cartesian coordinate system most commonly adopted to define the spatial dimensions of ocean models, where the (x,y)-coordinates are defined by using the traditional compass directions, and the z-coordinate represents the vertical dimension.

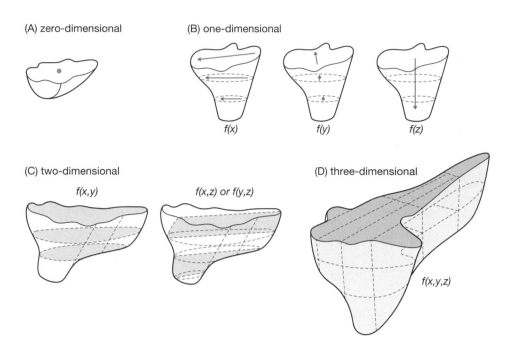

(A) zero-dimensional

(B) one-dimensional

$f(x)$ $f(y)$ $f(z)$

(C) two-dimensional

$f(x,y)$ $f(x,z)$ or $f(y,z)$

(D) three-dimensional

$f(x,y,z)$

Figure 11.2 The classic representations of zero-, one-, two-, and three-dimensional space, and how such geometries are best suited to represent natural aquatic systems that are expected to vary as a function of time (t) and spatial location (x,y,z).

the system. If this were the case, the model domain could be represented as a single, zero-dimensional point whose solution varies only as a function of time, as expressed in Equation 11.1:

$$\text{Zero-dimensional solution} = f(t) \qquad (11.1)$$

In the very special zero-dimensional case where the value remains unchanged throughout time, the entire mathematical model reduces to a simple constant. Of course, natural systems are almost never this simple, so it is highly unlikely that a zero-dimensional model could ever provide sufficient complexity to faithfully simulate the dynamics that require our attention.

One-Dimensional Models Simulate What Occurs Along a Line

One-dimensional models are those systems mathematically represented by a line, and whose time-dependent solution is also sensitive to a single spatial variable (x, y, or z). Using Equation 11.2, we can express this concept mathematically as:

$$\text{One-dimensional solution} = f(t,x) \text{ or } f(t,y) \text{ or } f(t,z) \qquad (11.2)$$

One-dimensional models are most appropriate for water masses that exhibit some gradient (or flow) along just one axis. This is not to say that only one variable is important. Quite the contrary: any number of variables may contribute to the gradient (or flow); but as long as all of those variables impart their influence along the same spatial dimension as the gradient (or flow), a one-dimensional model can be used.

For example, if we wished to develop a simple one-dimensional model to predict hydrostatic pressure in the ocean, we would be wise to consider the contributions of water temperature, salinity, and the compressibility of seawater as a function of depth (the vertical dimension). If we assume that the horizontal differences of water temperature, salinity, and compressibility within our

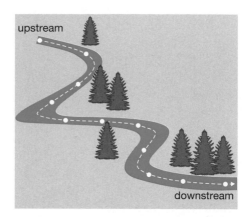

Figure 11.3 A one-dimensional representation of river flow, which is constrained along the single spatial dimension that defines the "upstream–downstream" direction. Note that despite all the twists and turns that create horizontal complexity in the river system, the use of the up-downstream dimension would allow the use of a much simpler one-dimensional model to simulate river dynamics. The length scale can then be easily divided into discrete sections (such as 5-km sections along the streamline).

domain are negligible, we can establish that all three vary only in the vertical dimension (that is, in the same direction as our pressure gradient).

In marine science, one-dimensional models are most commonly conducted in the vertical, because the natural stability of the stratified ocean will establish a persistent vertical gradient. Although horizontal gradients most certainly exist in the real ocean, it is quite rare to encounter water masses homogeneously mixed in one horizontal dimension but not the other. However, many freshwater systems (most notably in rivers) can also be simulated using one-dimensional models, if the spatial dimension is defined along the upstream–downstream axis (**Figure 11.3**).

Two-Dimensional Models Simulate What Occurs Within a Defined Area

Two-dimensional models are those systems that are geometrically represented by a plane (or area) and whose time-dependent solution is sensitive to some combination of two spatial variables (xy, xz, or yz). This concept can be written as a mathematical identity using Equation 11.3:

$$\text{Two-dimensional solution} = f(t,x,y) \text{ or } f(t,x,z) \text{ or } f(t,y,z) \qquad (11.3)$$

Two-dimensional models are most appropriate for water masses that exhibit some gradient (or flow) along two axes. Simulations of horizontal gradients or flows along some idealized surface (such as the ocean floor, the ocean surface, and virtually all terrestrial systems) are typically performed using a two-dimensional model.

For aquatic systems, it may be possible to perform simulations on various two-dimensional "slices" of the water column. If the water column exhibits no gradients or flow in the vertical dimension, it may be possible to simulate the system dynamics in **discrete** horizontal (xy) layers. If there are vertical gradients and/or flows, it may be advantageous to describe the horizontal dimension in the "along-transect" direction (**Figure 11.4**), thereby allowing the use of a simpler two-dimensional model to simulate the dynamics, represented as a vertical profile.

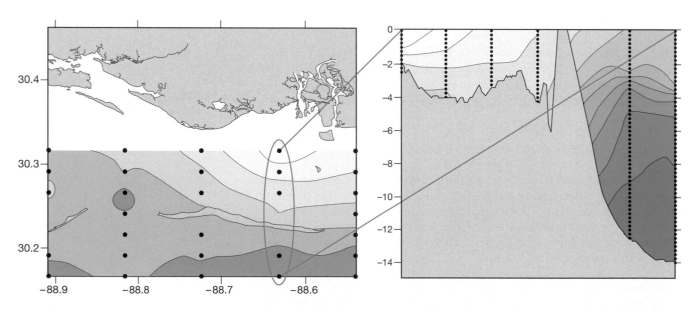

Figure 11.4 A two-dimensional representation of a spatial gradient within a plane, whether the plane is oriented as a horizontal layer (like the ocean surface) or as a vertical profile. The panel on the left depicts horizontal gradients of salinity as a two-dimensional surface layer. If it is more important to resolve vertical structure, data measured along a single transect (indicated by the black oval) can instead be depicted as a two-dimensional vertical profile, as shown in the panel on the right.

Three-Dimensional Models Simulate What Occurs Within a Fixed Volume

Three-dimensional models are those systems mathematically represented by a volume and whose time-dependent solution is sensitive to all three spatial variables (x, y, z), as Equation 11.4 indicates:

$$\text{Three-dimensional solution} = f(t,x,y,z) \qquad (11.4)$$

Three-dimensional simulations most accurately reflect real-world systems (which are also time-dependent in three dimensions), but three-dimensional models are notoriously complex and quite difficult to design. This is primarily because the ocean dynamics at work in the horizontal dimension typically occur at length scales much larger (103–105 m or more) than those in the vertical dimension (0.01–1 m). For example, the horizontal dimensions of the coastal area depicted in **Figure 11.5** are quite large (16.7 km latitudinal, 35.5 km longitudinal) compared to the vertical dimension, which spans only 14.4 m (a length scale more than three orders of magnitude smaller). This dramatic difference in length scales between the horizontal and vertical dimensions is actually quite typical of most marine science applications. To account for these differences in scale, three-dimensional models typically require the use of "mode-splitting" numerical schemes, where the horizontal and vertical gradients (or flows) are simulated separately from each other, using different spatial scales.

Spatial Dimensions Must Be Resolved Using a Model Grid

With the exception of a zero-dimensional model, all other models will possess geometries that include some sort of length scale. For example, a one-dimensional model oriented in the horizontal dimension can be used to represent the downstream distance along a river, as depicted earlier in Figure 11.3. If we wished to use a numerical model to simulate all aspects of the river's dynamics, it would be impractical for us to perform our model calculations at each infinitesimally small slice and integrate our results along the river's entire length. To avoid this, we would need to divide the length

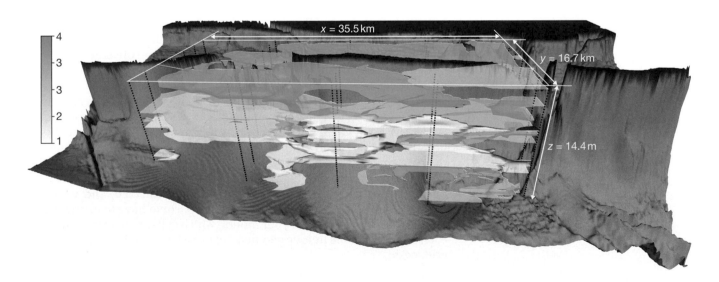

Figure 11.5 A three-dimensional representation of complex spatial gradients within a volume. Note the difference between the horizontal and vertical length scales within the study volume. When visualizing data (such as the three-dimensional chlorophyll concentrations depicted in µg L^{-1}), horizontal distances can be easily compressed a thousand-fold (in kilometers) relative to the vertical distances (in meters). In modeling applications, it is usually necessary to "split" the horizontal and vertical dimensions into different simulation modes because their length scales are so different.

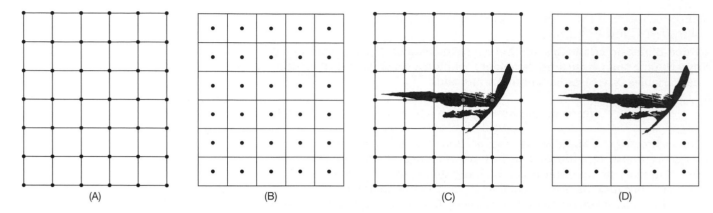

Figure 11.6 A geometrically spaced grid, where the length scales are divided equally to yield a regular two-dimensional mesh. The mesh can then be divided into compute nodes (points) where the numerical model will perform the calculations. Compute nodes can be located either at the corners of cells (A) or in the center of each cell (B). When a map of the simulation area is overlaid on the grid mesh, it is much easier to visualize whether the model resolution is sufficient, as well as the "connectivity" between adjacent cells. In this example, a hydrodynamic model of coastal circulation around an island (C, D) could not be performed at compute nodes that are occupied by the island itself (*blue*). The island would also represent a physical boundary to flow between certain adjacent cells (especially in the *y*-direction), but unless the island occupies a grid point in the mesh, it will be "invisible" to the model.

scale of the river into discrete sections and perform our calculations only so often, according to the **model resolution** we ultimately decide on.

For two- and three-dimensional models, the division of each length scale will ultimately produce a **grid mesh** of the area, which is represented by an array of cells, cell boundaries, and **compute nodes** (Figure 11.6). Using this strategy, the numerical model can then be used to perform calculations of the area only at each of the compute nodes, the results of which are used to represent the solution of the entire cell it represents.

Cartesian Grids

If the length scales are divided according to regular Cartesian coordinates, the resultant grid is more easily adapted to latitude–longitude coordinate systems. Although these coordinate systems may be convenient, they are not always best for the dynamics we wish to investigate. In the example given in Figure 11.6, the Cartesian grid represents an entire island simply as two or three compute nodes that are located over land and would therefore be "skipped" by a hydrodynamic model of coastal circulation around the island. Remember that our model is designed to simulate gradients (or flows) according to the spatial dimensions of the system. That means that the connectivity of the gradients (or flows) between adjacent cells will be affected by the physical boundary of the island itself. The model itself has no way of "knowing" which adjacent cells are connected, and which are not. Thus, if we used a rigid Cartesian grid, we would have to further define (within the computer code of our model) which cells are disconnected from each other due to the unique geometry of our island coastline relative to the grid mesh.

One common solution is to increase the spatial resolution of the model grid, so that the unique geometries of our system are better represented by a finer-grain mesh (Figure 11.7). Any increase in the grid resolution also means you are reducing the representative distance between compute nodes; as a result, the area of each grid cell becomes smaller. This will typically improve

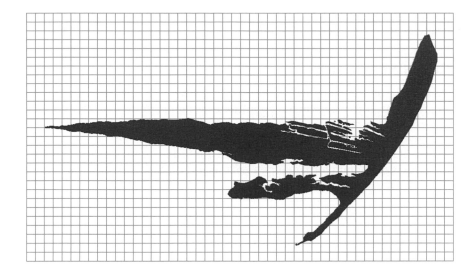

Figure 11.7 A Cartesian grid, similar to the one depicted in Figure 11.6, but with uniformly higher spatial resolution. An increase in grid resolution will always improve the spatial resolution of the model elements, but it also increases the number of compute nodes (which means more calculations are necessary to resolve the system).

how well the model simulates the true dynamics of the system, but it will also increase the number of calculations needed to resolve the system (which means your model will take longer to compute a solution).

Telescoping Grids

Telescoping grids are similar to Cartesian grids, except that they offer the ability for localized mesh refinement, where grid cells can be subdivided into four, thereby increasing grid resolution in regions of the model domain where fine-scale dynamics are important to resolve. If even finer-scale resolution is needed, a subdivided cell can be subdivided into four again, and so on (**Figure 11.8**).

Telescopic grids offer the best of both worlds: a simple, geometric grid with the flexibility to increase (or decrease) grid resolution to fit the complexity of

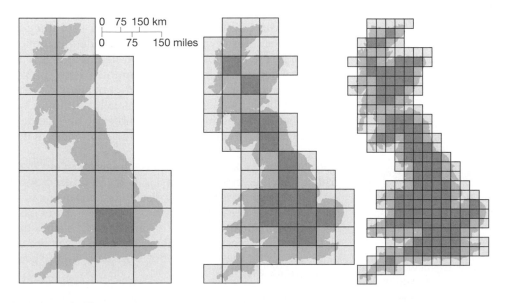

Figure 11.8 A grid mesh of Britain's coastline, where grid cells have been subdivided geometrically to produce a telescoping grid of varying resolution. (Courtesy of Alexis Monnerot-Dumaine / CC-BY-SA-3.0.)

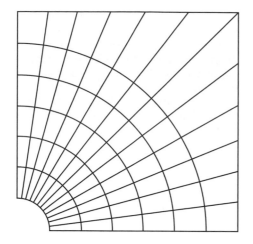

Figure 11.9 A curvilinear grid mesh, often used to represent lakes, embayments, curved surfaces, and rounded coastlines.

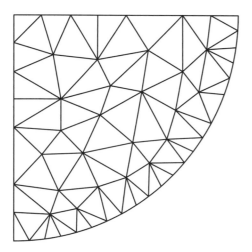

Figure 11.10 An unstructured grid mesh, produced by a patchwork of irregular triangles to serve as the grid cells, often used to represent highly complex river and coastal geometries.

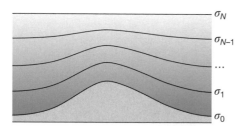

Figure 11.11 A σ-coordinate mesh is used to subdivide the vertical dimension into discrete layers using the bottom topography as the baseline (what one might call a "bottom-up" perspective). This scheme is particularly useful when modeling the vertical dimension within aquatic systems because the layers are free to follow the changing bathymetry while maintaining the relative "spacing" between the vertical layers.

the system. This also saves computational time, as the larger grid cells can be resolved with fewer compute nodes.

Curvilinear Grids

Depending on the complex geometry of the system, it may be better to use a curvilinear grid, whereby the coordinate system is translated according to some curve that represents a better "fit" to the coastal geometry (**Figure 11.9**). Although there are certain cases where the use of a curvilinear grid may better represent the overall spatial dimensions of the system, one of its major disadvantages is the nonuniform geometry of the grid cells, particularly with regard to the changing dimensions of the grid cell interfaces.

Also keep in mind that the resolution of the curvilinear grid gradually decreases as the grid extends outward. Sometimes this is precisely why the curvilinear grid is chosen. In most ocean models, the curvilinear grid is used so that grid spacing near the coast is very fine (where most of the important dynamics require high resolution computing). As the grid extends outward to the open ocean, the grid mesh becomes progressively coarser, thus requiring fewer compute nodes out in the open ocean (where the dynamics are considered to operate at much larger spatial scales anyway).

Unstructured Grids

Unstructured grids are reserved for highly complex geometries that cannot be sufficiently gridded using standard geometric or curvilinear grids. These typically require the use of complex software to map the surface of the area (or volume) to be gridded, and an irregular grid mesh is created from interconnected triangles (or tetrahedra), allowing for a much more complex grid structure (**Figure 11.10**). Although unstructured grids offer unbeatable "terrain-following" grids, the complexity of their grid-cell connections requires incredibly sophisticated computational schemes to simulate exchanges between irregularly shaped grid cells, with irregularly oriented interfaces.

Vertical Grids

Vertical grids are no different from other spatial grids, so they can use Cartesian, curvilinear, or unstructured grids also—they are simply oriented in the vertical. However, in most ocean modeling applications, the vertical dimension is defined on an entirely different length scale from that of the horizontal dimension. In fact, it is common to have compute nodes 1 to 2 km apart in the horizontal dimension, but <1 m apart in the vertical dimension. This large disparity in length scales usually requires a type of "mode splitting" in ocean models (as discussed earlier), where model simulations in the horizontal grid are conducted differently than those in the vertical.

One of the most common types of vertical grids in ocean models is the sigma grid, where the vertical dimension is divided into a set number of "sigma layers" (σ), which are vertically distributed according to some scaling function, where the ocean floor is always defined as the base sigma layer (σ_0). Because the bathymetry of the ocean floor establishes the base sigma layer, all of the other layers are distributed relative to the ocean floor. This results in a vertical grid that always follows the bathymetric contours of the ocean floor (**Figure 11.11**).

The Model Timestep Is Used to Resolve the Temporal Dimension

Just as we must divide the spatial dimensions of our model into discrete parcels, we must also do the same with respect to time. This means that at each compute node, our model must perform every calculation according to some predetermined timestep: an interval of time that "triggers" the model to calculate the dynamics that have occurred within each grid cell, across the entire model domain, over that particular time slice. For example, if we wished to model a particular current flowing at a constant velocity of 12 cm s^{-1}, and our model employed a 360-s timestep, our model would indicate that the initial water mass would move 43.2 m in a single timestep, as calculated in Equation 11.5:

$$\left(\frac{12 \text{ cm}}{\text{s}}\right) \cdot \left(360 \text{ s}\right) = 4320 \text{ cm} = 43.2 \text{ m} \tag{11.5}$$

If we had chosen a shorter timestep, the fundamental dynamics within our model would not change—our model would simply calculate the dynamics in smaller slices of time. In this case, a 60-s timestep would allow our modeled current to move that same water mass over a shorter distance of only 7.2 m.

Although this example is quite simple, it demonstrates a critical point: it is the combination of the model's length scale (spatial dimension) and time scale (temporal dimension) that will ultimately yield the solution to the model. Every conceivable dynamic in nature exerts its influence in terms of space and time. The fact that our model is designed to simulate space, as well as the passage of time, means that our model should be capable of resolving any of those dynamical processes.

The selection of an appropriate timestep is similar to our concerns with spatial resolution: the shorter the timestep, the better our model can resolve brief and transient events (but the greater the number of compute events). This is a particular concern in complex three-dimensional models, where fine-scale resolution in three dimensions typically involves thousands of compute nodes, which must all be solved at each timestep. Imagine 1 m^3 of water, with a grid node at each millimeter. Using a 10-s timestep, our model would require 8.64 trillion compute events per day! And what if we were simulating a multitude of physical and chemical properties of the water within that 1 m^3 volume at each compute event? You can see how quickly the number of compute events can skyrocket to an astounding (and impractical) magnitude.

Luckily, there are some guidelines when it comes to determining the appropriate level of spatial and temporal resolution for our model. As it turns out, any increase in your spatial or temporal resolution will generally improve model accuracy, but usually at the cost of computer processor time. For large, finely resolved model grids, it is possible for the compute time between timesteps to approach the duration of the timestep itself. After all, what value is a computer simulation if it takes 10 s of real time for your processor to compute a 10-s timestep? You might as well ditch the computer, kick back, and watch nature do its thing: Mother Nature requires exactly 10 s to execute a 10-s timestep.

Defining the lower limits of acceptable model resolution is another matter, as there are much nastier consequences if you choose a grid that is too coarse, or a timestep that is too large. If this is the case, the poor resolution will ultimately cause numerical instabilities within your model, such that the error associated with each calculation becomes larger than the "true" solution, and that error becomes compounded and propagates throughout your entire model. To avoid this, it is necessary that you carefully consider your chosen spatial and temporal resolutions to better ensure the numerical stability of your model.

Striking a Balance Between Model Resolution and Stability Is Paramount

Although there are several different methods available for analyzing the appropriateness of a model's resolution with respect to its stability, the most widely used is the Courant-Friedrichs-Lewy (CFL) stability criterion. It is mathematically defined in Equation 11.6 according to the logic:

$$\left| \frac{v_x \cdot \Delta t}{\Delta x} \right| \leq 1 \tag{11.6}$$

where v_x represents the largest magnitude of velocity along the x-dimension, Δt is the model timestep, and Δx is the greatest distance between any two points within the model grid, in the x-direction. In order to use the CFL criterion, something must be known about v_x in the system. If v_x is known, or can be safely estimated as a "maximum probable velocity," the spatial resolution (Δx) and temporal resolution (Δt) of the model can be entered, and the CFL criterion can be computed according to Equation 11.6. Any values that exceed unity are a violation of the CFL criterion, and require that either Δx or Δt (or both) be adjusted until Equation 11.6 yields a true statement.

Note that Equation 11.6 uses the general form of analyzing stability in the x-direction. For two- and three-dimensional simulations, the CFL criterion must be calculated for each of the (x,y,z) spatial dimensions and their respective maximum probable velocities. In most hydrodynamic applications, v_x is usually represented as the sum of advective transport velocity v_{adv} and the speed of wave propagation (c_{grp}), as expressed in Equation 11.7:

$$v_x = \left(v_{adv} + c_{grp} \right) \tag{11.7}$$

Essentially, the CFL criterion is used to guard against the possibility that information within the model can propagate faster than can be resolved by the model grid.

The CFL criterion is quite useful as a decision tool, so that we may establish the "safe" spatial and temporal limits of our model. As an example, let's assume we wished to resolve the horizontal "patchiness" of algal blooms in the coastal ocean, but did not know what spatial or temporal resolutions would be most appropriate for our model. In the course of our research and literature review, we discover that algal blooms in our region of interest can be as small as 75 m across, and can persist for hours or days. If we chose a horizontal grid spacing (Δx) of 75 m in order to resolve even the smallest algal bloom, we would also need to know something about the maximum probable velocities v_x that could potentially transport our algal patches throughout the model domain. If we have determined that v_x can be as high as 1.2 m s^{-1}, the solution to the CFL criterion (see Equation 11.6) is now expressed in Equation 11.8:

$$\Delta t \leq 1 \cdot \left| \frac{\Delta x}{v_x} \right| \Rightarrow \Delta t \leq \left(\frac{75 \text{ m}}{1.2 \text{ m/s}} \right), \quad \text{or} \quad \Delta t \leq 62.5 \text{ s} \tag{11.8}$$

Thus, for all $\Delta t \leq 62.5$ s, our model resolution meets the CFL criterion for stability.

Boundary Conditions

Now that we have determined the geographical area (or volume) that will serve as your model domain, it is easy to see that each of the compute nodes of the grid mesh must be located inside the model domain. Everything

PLAYING IT SAFE WITH THE CFL CRITERION

It is important to remember that the CFL criterion is most susceptible to our chosen value of v_x, because this value sets the limit as to how quickly information can propagate throughout our model. From a stability perspective, it is always better to overestimate the maximum probable velocities within the system.

beyond the grid mesh is the "rest of the universe," which is external to your model domain. Naturally, the grid points located along the perimeter of your model domain represent the **external boundary** of your model.

External boundaries are a very tricky thing. Fundamentally, we are delineating the boundaries between the interior of the model (where all of our compute nodes are located) and everything else that lies beyond the domain (which is essentially being ignored). This raises a couple of key issues that must be resolved:

1. How do we allow external (unmodeled) features to enter our model domain and affect the interior?
2. How do we allow internal (modeled) influences to radiate outside our model domain and completely leave the interior?

To answer these questions, we develop a special set of mathematical rules that we must apply only at the boundaries of our model domain. These rules represent the **boundary conditions** of our model, which essentially define how the gradients and flows within our model will behave when they reach the edge of our domain.

Because each model will have its own boundary conditions, it is difficult to offer uniform guidance as to how they should be defined. At the most fundamental level, it will require that you first conceptualize how the fluxes and flows will most appropriately (and most realistically) be transmitted across the boundaries of your domain, either in or out. Ultimately, you will be required to translate these concepts into various mathematical functions designed to impose a specific condition at a specific boundary.

In most ocean models, there are two general categories to consider: closed versus open boundaries. A **closed boundary** can also be considered to be a **physical boundary**, because it is represented as a real, physical impediment to fluxes and flows (so information is prevented from being translated across the boundary). By contrast, an **open boundary** is one that does allow fluxes and flows across the boundary. In ocean models, an open boundary can also be considered a **hydraulic boundary** because the fluxes and flows across the boundary are taking place largely within the fluid medium.

As a general rule, the conditions imposed at closed boundaries are assumed to be "no slip, no flow." This means that for all grid points along a closed boundary, all fluxes and flows that are parallel and perpendicular to the boundary are set to zero (**Figure 11.12**). This boundary condition leads to a dampening effect as fluxes and flows approach a coastline (or other physical boundary). For most ocean modeling applications, this is an appropriate assumption; however, there are conditions where it may be more appropriate to "reflect" the modeled phenomenon rather than reduce it to zero. An obvious example where this may be warranted is the reflection and refraction of various waveforms when they encounter a physical boundary.

Open boundaries are much more challenging, because it will be necessary to define where and how information outside of the model domain can appropriately enter the model interior. For example, if we wished to define our model domain as the west Florida shelf in the United States (**Figure 11.13**), the Loop Current (a transient yet significant influence on the hydrodynamics of the entire Gulf of Mexico) is exterior to our model domain and will therefore be excluded from our explicit modeling efforts. Yet its presence near our open boundary should be defined as a "special case," so that its influence can enter our model at the open boundary.

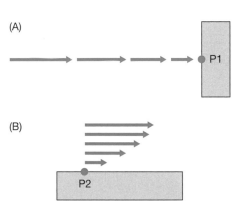

Figure 11.12 A typical boundary condition prevents flows across a physical boundary (*grey*), when the flow is normal (perpendicular) to the boundary (A). This can be done by establishing a boundary condition at point P1, where all flows at this grid point are simply set to zero (a "no flow" condition). When the flow is parallel to the boundary (B), a boundary condition can also be established at point P2, where all flows are set to zero as well (a "no slip" condition).

Figure 11.13 A representation of a typical coastal ocean model domain (represented using a curvilinear grid), bounded by the coastline (a closed boundary) and the ocean basin exterior to the domain (an open boundary). In this case, the Loop Current within the Gulf of Mexico lies external to the model domain, yet its significance and proximity to the open boundary would require the development of a special boundary condition so the Loop Current could influence the numerical solution of the model without actually being included in the model. (Courtesy of the National Oceanic and Atmospheric Administration, NOAA.)

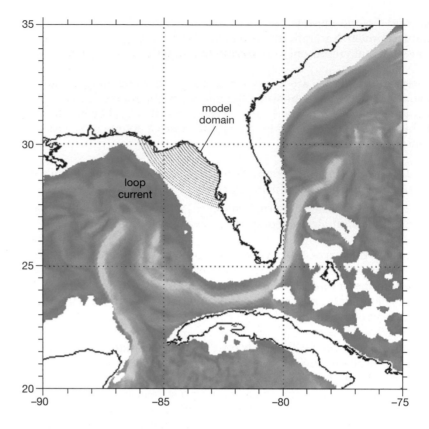

As already mentioned, it is impossible to explicitly define every possible open boundary condition, particularly when every model (and every open boundary therein) is unique. This chapter is intended to provide core guidance on the subject, but the reader is strongly encouraged to refer to the available literature that would more thoroughly define open boundary conditions for "typical" hydrodynamic models for the reader's region of interest. Of particular interest is the Orlanski (1976) treatment of open boundaries.

Computational Schemata

Beyond our concerns of the spatial and temporal resolution and the appropriate boundary conditions within our model, it is important to keep in mind that our final model will consist of many different calculations being performed at each compute node, all in an effort to resolve many different phenomena. Of course, all of these different phenomena are included within our model because we have some fundamental understanding that they are interrelated. After all, why would we include them as state variables within our model if they had absolutely no consequence to the model solution?

So by their very inclusion in our model, we recognize that all of our chosen state variables must be relatable to each other. That means we must decide on some common unit of measurement—something of a "common denominator" in all our calculations—so that the information contained within any one state variable can be easily translated to all of the state variables throughout the model. In modeling parlance, this grand unifying unit of measurement becomes the "common currency" of our model, so we can track its ebb and flow throughout the model domain and throughout time.

Model Currency Is the Unifying Medium of Exchange Between Variables Within the Model

When deciding on the appropriate currency for our model, it is best to choose a unit of measurement that can be uniformly applied to any numerical model. Luckily, Albert Einstein has defined for us humans what the natural world has known all along—that all masses (matter) in the universe are inexorably linked with, and can be translated into, energy according to Equation 11.9, his famous equivalence relation:

$$E = mc^2 \qquad (11.9)$$

The relevance of Equation 11.9 to the topic at hand is not trivial; there is sublime grandeur in the simplicity of that equation. Essentially, this relationship allows us the freedom to choose whether our model currency should be a measurement either of mass or of energy, and that our choice for one will not hinder the quantitation of the other. We could choose any other mathematical relationship between matter and energy (for example, potential energy or calorific energy), but our understanding of basic Newtonian physics tells us that mass and energy are always relatable to each other. It is by this relation that all of our state variables can be linked together and made consistent with each other, either as mass or as energy.

In most ecological models, biomass (as organic carbon, or ash-free dry weight) is most commonly used as the model currency, where important biological processes such as production, growth, reproduction, and mortality can be easily quantified by time-dependent shifts in the gain or loss of biomass within each grid cell (thereby including the spatial dimension of volume). This same relationship of mass per unit volume (that is, mass concentration) is also very useful in biogeochemical and toxicological models, where resolving the concentrations and fluxes of specific molecules is the primary goal.

Bioenergetic models (and many physical models) tend to use energy as the model currency instead. For instance, it may be a more intuitive choice to simulate energy flows for those physical phenomena that behave more as waves (energy specific) rather than as particles (mass specific). Physical models of wave propagation behavior and the radiative transfer of electromagnetic radiation (heat, light, etc.) are two obvious examples where it might make more sense to simulate energy flow rather than mass flow. That is not to say that energy flows should be avoided in biological models; on the contrary, bioenergetic models may be more interested in tracking energy flow (such as the caloric content of the biomass) across different trophic levels in a food web.

Mass and Energy Budgets Must Obey the Law of Conservation

Another benefit to using a mass-specific (or energy-specific) currency in your model is the fact that both follow the physical law of conservation: that matter (or energy) can be neither created nor destroyed, but must be conserved. Certainly, mass can be converted to energy, and energy to mass, but the fundamental quantities of each must be accounted for. That means we can easily create a mass budget (or energy budget) as a companion to our model, so we can verify that every bit of mass and energy in the simulation is conserved (and appropriately contained) within our model domain.

As a practical matter, adherence to the law of conservation is typically employed as a special boundary condition that applies to all grid cells, not just those at the boundary of the model domain. In order to impose this rule

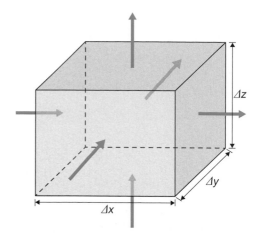

Figure 11.14 A representation of a model grid cell and the conceptual significance of the continuity equation, as it relates to mass and energy balance within the grid cell. The spatial dimensions of the grid cell (Δx, Δy, Δz) define its physical volume, where the arrows normal to each face of the grid cell are indicative of the fluxes into (or out of) the cell volume. Note that each face of the cell can have different fluxes: some coming in and some going out. But as a whole, all of the fluxes into the cell must be exactly equal to all of the fluxes out.

as a mathematical certainty, the **continuity equation** (see Equation 11.10) is applied to each grid cell (**Figure 11.14**):

$$\frac{\partial u}{\partial x} + \frac{\partial v}{\partial y} + \frac{\partial w}{\partial z} = 0 \tag{11.10}$$

where the "∂" term reflects the operation of a partial differential equation (PDE) that defines the rate of change in a particular variable (u, v, or w) with respect to the changing spatial dimension (x, y, or z). For example, the first term of Equation 11.10 is a mathematical expression of a more complex PDE that describes the rate of change in the flow in the x-direction (∂u) as the flow changes its position in the x-direction (∂x). Similar to the first, the second term describes the rate of change in the flow in the y-direction (∂v) as the flow changes its position in the y-direction (∂y), and the third term describes the rate of change in the flow in the z-direction (∂w) as the flow changes its position in the z-direction (∂z). Each of these terms are free to vary, but they are not free to vary independently of one another. In fact, the law of conservation dictates that they must all add to zero, if the mass (or energy) in the system is conserved.

Although the reader may be unfamiliar with the mathematical notation used in Equation 11.10, the continuity equation simply dictates that for any fixed volume of seawater, the balance of all fluxes and flows into (or out of) the control volume must be zero (using the safe assumption that seawater is an incompressible fluid). So, if we had an imaginary "box of water" that was completely full, the addition of any more water to our box would cause the same amount of water to pour out of the box somewhere else. In the context of our model grid cells, there are six faces to every cell where water (or anything else) is free to flow into or out of the box, but the net effect of all those flows must be zero. In this way, the volume of each grid cell remains unchanged—that's conservation of mass!

Since the mass and energy within our model domain must be conserved, we know that the solution to our model must contain the same amount of mass and energy at the end of the simulation as was present when the simulation began. This is a very important point, because it means that no matter how well (or how poorly) our model agrees with reality, we know that the final solution will, at the very least, conserve all of the mass and energy we started with. In other words, when your model is initialized, the finite quantity of mass and energy within the model domain is defined by that initialization, so all you really need to get your model started is one good dataset.

Model Initialization Requires That We Begin with a "Filled" Model Domain

As we have already discussed in this chapter, the spatial resolution of your model grid will be somewhat defined by the CFL stability criterion. In a perfect world, you would possess data for every explicit variable, and at every grid point in your model domain, with which to initialize your model. Of course, there is very little possibility that you will ever possess such a robust dataset. It is far more typical that you will have a handful of measurements scattered throughout your model domain, at locations that do not necessarily coincide with the compute nodes of your model.

The process of **interpolation** is a simple numeric method by which new data can be estimated from a discrete set of known measurements. Using interpolation, it is possible to populate an entire model grid using a handful of known measurements (so long as the measurements were gathered from an area consistent with your model domain). Although there are several different interpolation schemes one might use, the simplest and most common method is to utilize Equation 11.11, which defines the **finite difference** approximation, centered in space:

$$\frac{\Delta\varphi}{\Delta x} = \left(\frac{\varphi_{i+1} - \varphi_{i-1}}{2\Delta x}\right) + O(\Delta x^2) \tag{11.11}$$

where the change in φ (our variable of interest) can be defined spatially, relative to the change in the x-direction (Δx). If we use a **stencil** to help us visualize our grid (**Figure 11.15**), we can see that the value in the center of the stencil (φ_i) can be estimated using the values of the neighboring points on either side (φ_{i-1}, φ_{i+1}). Keep in mind that Equation 11.11 is not a perfect solution; rather, it is an estimation method that will always have inherent rounding errors (O), which are quite sensitive to the magnitude of Δx. As Δx approaches zero, the rounding errors are reduced; thus, our estimates are improved as Δx becomes smaller and smaller.

Now that we have calculated the rate of change in φ with respect to Δx in Equation 11.11, we can use that quantity to estimate the actual value of φ_i using Equation 11.12:

$$\varphi_i = \varphi_{i-1} + \left[\left(\frac{\Delta\varphi}{\Delta x}\right) \cdot \Delta x\right] \tag{11.12}$$

If we were using the simple grid depicted in Figure 11.15, we would repeat this exercise (using Equations 11.11 and 11.12) for each and every grid point inside the domain, shifting our stencil "downstream" after calculating each value of φ_i. After the first interpolation pass, sometimes called the first **iteration**, every single grid point within the domain will now have a value estimate (where before, many might have had none). If the interpolation scheme is begun again (for the second iteration), the estimated values from the previous iteration will autocorrect the estimated values in the second pass, creating a much more consistent field of data. For each iterative pass, the estimates of φ at each grid point will change less and less and eventually converge on a single solution where all of the values, throughout the grid mesh, are self-consistent and will no longer change. When this occurs, we can be confident that our entire grid has been populated with stable, consistent estimates based on the relatively few observations in our original dataset.

To demonstrate the usefulness of this method, let's assume our linear grid from Figure 11.15 represents a single cross-shelf transect, where each grid point represents an offshore station spaced 5 km apart ($\Delta x = 5$ km). Let's also assume that x_1 represents the open ocean boundary of our one-dimensional model, and x_5 represents the coastal boundary. If we had only a single measurement of salinity from station x_4, but we wished to estimate the salinity at all stations within our model domain, we could use Equations 11.11 and 11.12 to do so. Our data and boundary conditions are as follows:

x_1: Open ocean boundary condition (assume salinity of 36.00 PSU at ocean boundary)

x_2: Unknown salinity (can set to any value to start)

x_3: Unknown salinity (can set to any value to start)

x_4: Salinity measured as 24.27 PSU

x_5: Coastal boundary condition (assume salinity of 0.00 at land boundary)

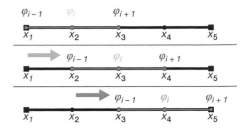

Figure 11.15 A finite differencing stencil can be used to visualize which points on the grid are used to estimate the value at the center of the stencil (φ_i) based on the value of the neighboring points on either side (φ_{i-1} and φ_{i+1}). In the simple grid depicted here, the point at x_1 defines an open ocean boundary condition and the point at x_5 defines the coastal boundary of the grid (where the values for φ at these points would be predetermined according to the chosen boundary conditions). Once the value of φ_i is determined at x_2, the stencil must shift to the right. In the second pass, φ_i can now be determined at x_3, incorporating our previous estimate of φ at x_2. In the third pass, the stencil shifts to the right once more, and φ_i can now be determined at x_4, incorporating our previous estimate of φ at x_3 and the boundary condition at x_5.

Since the values for x_1 will never change (as they are our open ocean boundary condition of 36.00 PSU), we start our calculations by estimating the salinity at x_2 using Equation 11.11 (and ignoring any truncation errors):

$$\frac{\Delta\varphi}{\Delta x} = \left(\frac{\varphi_{i+1} - \varphi_{i-1}}{2\Delta x}\right) = \left(\frac{0.00\ PSU_{x_3} - 36.00\ PSU_{x_1}}{2(5\ km)}\right)$$

$$= \frac{-36.00\ PSU}{10\ km} = -3.60\ PSU\ km^{-1}$$

Now we use Equation 11.12 to calculate φ_i, positioned at x_2:

$$\varphi_i = \varphi_{i-1} + \left[\left(\frac{\Delta\varphi}{\Delta x}\right)\cdot\Delta x\right] = 36.00\ PSU_{x_1} + \left[\left(\frac{-3.60\ PSU}{km}\right)\cdot 5\ km\right]$$

$$= 36.00\ PSU_{x_1} - 18.00\ PSU = 18.00\ PSU_{x_2}$$

Based on these calculations, our initial estimate of φ_i at x_2 is 18.00 PSU. Now we must shift our stencil so that φ_i is repositioned at x_3. Note that our estimate for φ_i located at x_3 will include the estimate we just calculated at x_2 (now designated as φ_{i-1}). At x_4 (φ_{i+1}), we already know what the salinity is, because it is the only station where salinity was actually measured. So for this calculation, our estimate at φ_{i-1} and our measurement at φ_{i+1} will give us:

$$\frac{\Delta\varphi}{\Delta x} = \left(\frac{\varphi_{i+1} - \varphi_{i-1}}{2\Delta x}\right) = \left(\frac{24.27\ PSU_{x_4} - 18.00\ PSU_{x_2}}{2(5\ km)}\right)$$

$$= \frac{6.27\ PSU}{10\ km} = 0.627\ PSU\ km^{-1}$$

$$\varphi_i = \varphi_{i-1} + \left[\left(\frac{\Delta\varphi}{\Delta x}\right)\cdot\Delta x\right] = 18.00\ PSU_{x_2} + \left[\left(\frac{0.627\ PSU}{km}\right)\cdot 5\ km\right]$$

$$= 18.00\ PSU_{x_2} + 3.135\ PSU = 21.14\ PSU_{x_3}$$

After our first iteration, we now have our first estimates of φ at all of our grid points where we were originally missing data:

   x_1: 36.00 PSU (ocean boundary condition)
   x_2: 18.00 PSU (estimated; iteration 1)
   x_3: 21.14 PSU (estimated; iteration 1)
   x_4: 24.27 PSU (measured)
   x_5: 0.00 PSU (coastal boundary condition)

Although it is highly unlikely that the salinity gradient would decrease from x_4 to x_2, and then reverse trend to increase from x_2 to x_1, keep in mind that this is just our first estimates of x_2 and x_3. With each successive iteration, our estimates of φ (in this case, salinity) will be dramatically improved. If we continued in this method, it would take only eight iterations before our φ values converged on the best estimates of salinity, as demonstrated in **Table 11.1** by the declining Δx_2 S and Δx_3 S (which represent the "corrections" made to our φ values with each new iteration). When your estimates

Table 11.1 Estimation of Unknown Salinity Values (blue), Based on an Iterative Scheme Using Known Salinity Values (black) to Interpolate and Improve on Previous Estimates ($\Delta x_i \to 0$) Using Finite Differencing

Iteration	$x_1\,S$	$x_2\,S$	$x_3\,S$	$x_4\,S$	$x_5\,S$	Δx_2	Δx_3
0	36.00	0.00	0.00	24.27	0.00	-	-
1	36.00	18.00	21.14	24.27	0.00	18.00	21.14
2	36.00	28.57	26.42	24.27	0.00	10.57	5.28
3	36.00	31.21	27.74	24.27	0.00	2.64	1.32
4	36.00	31.87	28.07	24.27	0.00	0.66	0.33
5	36.00	32.03	28.15	24.27	0.00	0.17	0.08
6	36.00	32.08	28.17	24.27	0.00	0.04	0.02
7	36.00	32.09	28.18	24.27	0.00	0.01	0.01
8	36.00	32.09	28.18	24.27	0.00	0.00	0.00

are no longer improved by new iterations, the interpolation scheme is complete.

Note that the linear interpolation method described in Equations 11.11 and 11.12 can only be used to interpolate an evenly spaced one-dimensional model "grid," as written. This method is appropriate for any one-dimensional "grid," regardless of whether the transect is oriented in the x-direction (east–west), in the y-direction (north–south), or in the z-direction (vertically). If you wished to interpolate a true two-dimensional (planar) grid, you could just as easily use Equations 11.11 and 11.12 to make a series of interpolation passes in the x-direction first, then again in the y-direction (**Figure 11.16A**).

To interpolate in both the x-direction and y-direction simultaneously, a **bilinear interpolation** scheme may be a better option, especially if you are using an unevenly spaced grid. Bilinear interpolation is quite diffcrent from finite differencing because it does not calculate the first derivative of φ as $\Delta\varphi/\Delta x$, or $\Delta\varphi/\Delta y$, or $\Delta\varphi/\Delta z$; it simply calculates an "average" value of φ, based on the values of φ at four neighboring nodes and the geometric distances to those four nodes (see Figure 11.16B). For unevenly spaced grids, the center node ($\varphi_{i,j}$) can be interpolated using Equations 11.13–11.15:

$$\varphi'_{j-1} = \left[\left(\frac{(i+1)-(i)}{(i+1)-(i-1)}\right)\cdot\varphi_{i-1,j-1}\right] + \left[\left(\frac{(i)-(i-1)}{(i+1)-(i-1)}\right)\cdot\varphi_{i+1,j-1}\right]$$

(11.13)

$$\varphi'_{j+1} = \left[\left(\frac{(i+1)-(i)}{(i+1)-(i-1)}\right)\cdot\varphi_{i-1,j+1}\right] + \left[\left(\frac{(i)-(i-1)}{(i+1)-(i-1)}\right)\cdot\varphi_{i+1,j+1}\right]$$

(11.14)

where Equations 11.13 and 11.14 provide temporary, intermediate values that are ultimately used to interpolate an average value for $\varphi_{i,j}$:

$$\varphi_{i,j} = \left[\left(\frac{(j+1)-(j)}{(j+1)-(j-1)}\right)\cdot\varphi'_{j-1}\right] + \left[\left(\frac{(j)-(j-1)}{(j+1)-(j-1)}\right)\cdot\varphi'_{j+1}\right]$$

(11.15)

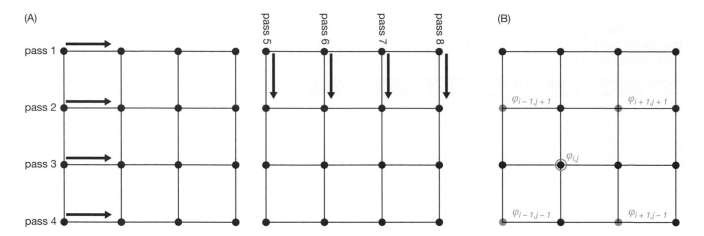

Figure 11.16 (A) Planar grids can be interpolated using finite differencing to calculate the first derivative of φ in the x-direction ($\Delta\varphi/\Delta x$) for each easting grid line, and then reoriented to calculate the first derivative of φ in the y-direction ($\Delta\varphi/\Delta y$) for each northing grid line. (B) As an alternative, bilinear interpolation uses a five-point "box stencil," where the average value of φ in the center of the box ($\varphi_{i,j}$) can be estimated as a geometric average of the φ values at the four corners of the box (blue).

If you are using an evenly spaced grid where $\Delta x = \Delta y$, Equations 11.13 and 11.14 become unnecessary and Equation 11.15 is simplified to become Equation 11.16:

$$\varphi_{i,j} = \frac{\varphi_{i-1,j-1} + \varphi_{i+1,j-1} + \varphi_{i-1,j+1} + \varphi_{i+1,j+1}}{4} \tag{11.16}$$

Now that you know how to interpolate, you can begin the iterations to dutifully ingest all of your initialization data and use those data to fill the model domain, so that each compute node contains the best possible "initial guess" for each of the state variables in your model. Keep in mind that the model will still require a significant amount of programming, so that the model can actually predict the future based on present values within the domain. But at least we now have a coherent model domain, fully populated with initial conditions, so we can begin our calculations forward in time.

In regard to that specific point, it should come as no surprise that many of the equations within complex aquatic models are differential equations that typically involve several different variables to determine their solution. Although differential equations are affected by the passage of time, it may not be necessary for us to explicitly solve each and every one of those equations. We can instead use a computer to provide an estimate of the solution at each compute node and then reanalyze the consistency of those numeric solutions at and between each compute node in our model domain, and at each timestep. This essentially boils down to a computerized guessing game, where each guess (at each compute node) is analyzed against the fundamental equations of motion, mass, energy, and so on. This is essentially what we have already seen: a method of spatial interpolation, where unknown values can be estimated by using the values at neighboring nodes. Believe it or not, this same concept can be used to estimate (that is, interpolate) what will happen at a particular node in the future ($t + \Delta t$), based on current conditions. But that concept shouldn't seem so foreign when you really think about it. After all, what ultimately happens in the future is necessarily driven by events in the present.

References

Carey GF (1997) Computational Grids: Generations, Adaptation, and Solution Strategies. Taylor & Francis.

Davis PJ (2014) Interpolation and Approximation. Dover Publications.

de Moura CA & Kubrusly CS (eds) (2011) The Courant-Friedrichs-Lewy (CFL) Condition: 80 Years After Its Discovery. Springer.

LeVeque R (2007) Finite Difference Methods for Ordinary and Partial Differential Equations: Steady-State and Time-Dependent Problems. Society for Industrial and Applied Mathematics.

Orlanski I (1976) A Simple Boundary Condition for Unbounded Hyperbolic Flows. *Journal of Computational Physics* 21(3):251–269.

Further Reading

Babuska I, Flaherty JE, Henshaw WD, et al. (eds) (1995) Modeling, Mesh Generation, and Adaptive Numerical Methods for Partial Differential Equations. Springer-Verlag.

Liseikin VD (2009) Grid Generation Methods. Springer.

Shashkov M (1995) Conservative Finite-Difference Methods on General Grids. CRC Press.

Thompson JF, Soni BK, & Weatherill NP (1998) Handbook of Grid Generation. CRC Press.

Chapter 12

Modeling Simple Dynamics

"Nature is pleased with simplicity. And nature is no dummy ..." – Sir Isaac Newton

Until now, our introduction to numerical modeling has focused on the establishment of model grid dimensions, boundary conditions, and constraints to numerical stability—all from a very general perspective. Now it's time to actually put some meat on those bones and begin the process of crafting the model as a cohesive assemblage of interrelated equations (which are essentially "submodels" within the larger whole).

To illustrate this process, it'll be helpful for us to use a common example, such as the conceptual diagram of the nitrogen cycle first introduced in Chapter 10. In this chapter, we will work from that conceptual model, in stepwise fashion, to demonstrate how models of simple dynamics (that is, models of individual forcing functions) are constructed. Since all models are different, there are no magic recipes—no "off-the-shelf" model that will meet your every need. So at some point, you'll have to write your own model—either in pieces or in whole.

Let's start by focusing on the process by which we might construct our mathematical models. And what better way to get our feet wet than to start with some simple examples?

Key Concepts

- The simplest models are little more than a single regression equation, called a "forcing function," which utilizes the value of some input variable to predict the value of an output variable.
- Any model, no matter how complex, can be constructed from a series of simpler interconnected submodels.
- Conceptual models provide an invaluable guide for the creation of numerical models, which is typically done in piecemeal for simplicity.
- Multidimensional models require multidimensional variables, which can be accomplished by using data arrays.
- State variables should have defined boundary conditions and a mass budget to prevent numerical instability within the model.

Constructing Your First Model

It is important to remember that there is nothing inherently mystical or difficult about modeling. You probably didn't realize it at the time, but every equation you've ever used to solve a math problem was a mathematical model. The fact that you were able to solve equations—any equations—means that you already have experience as a modeler.

In Chapter 9, we spent a great deal of time discussing the method by which we can use measurements of one (or more) independent variables to establish mathematical relationships with other variables (whose values are dependent on that relationship). In fact, in an earlier example we demonstrated how ocean density could be modeled (estimated) by using measurements of water temperature and salinity (see Equation 9.11). If we decided to use Equation 9.11 to estimate ocean density in a variety of locations and at various depths, we would have ourselves a three-dimensional model of ocean density!

Consider that the model is simply a mathematical representation of the process by which a measured variable can be used to predict the value of a different variable: a variable that is not measured, but is instead estimated by the model itself. The simplest models are those that require that we mathematically define a single forcing function or process (like Equation 9.11). However, if we wanted to make things a little more complicated, we could develop an entirely new model that uses the solution to Equation 9.11 (estimated density) to predict other physical features related to density, such as water pressure.

On and on we could go, stringing hundreds of individual models together, using the solution of one model to drive the solution of the next model, and then the next. Regardless of how complicated a model might look, all mathematical models are fundamentally constructed in the same fashion, where explicit variables (preferably field or laboratory measurements) are used by the model to provide an estimate of some other variable we are interested in and were unable to measure ourselves (**Figure 12.1**).

That's the good news—that means we can build very complex models simply by constructing each of the cross-linked models, one at a time.

The First Rule of Model Construction: Do Your Homework

If you've decided to embark on a modeling project as part of your research, you should already have a general idea of what you want your model to do. But in order for your model to faithfully represent the true dynamics of the natural world, it is absolutely imperative that you have a thorough understanding of all the different variables and forcing functions that are pertinent to the phenomena you wish to simulate with your model. It's OK if you don't know exactly how they are pertinent—for starters, it's enough to know that they're important enough to be included (somehow) in your model.

The importance of a thorough literature review cannot be overstated here. It is only through very diligent research that you will begin to feel confident that you have considered all of the important variables and forcing functions for your particular investigation. Once you are convinced that your laundry

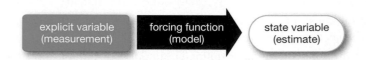

Figure 12.1 Conceptual diagram of the simple structure common to all mathematical models.

list of model elements is complete, you can turn your attention to which models already exist, and how you can incorporate those preexisting mathematical models into your own creation, with minor tweaking. There is no sense in reinventing the wheel—the whole point of the scientific literature is for you to build on the foundations that have been already built.

The Second Rule of Model Construction: Follow Your Conceptual Model

Having done your homework, you should now have a very good picture in your mind of how all the variables and forcing functions are (or should be) related to each other. Commit that picture in your mind to paper, and sketch out every one of those interrelationships to form a conceptual model of what will eventually become your mathematical model.

In Chapter 10, we used the conceptual model of the nitrogen cycle (see Figure 10.5) as an example. An exhaustive treatment of the nitrogen cycle might require such a detailed conceptual model (**Figure 12.2A**), but if your particular research interests were focused only on NH_3 concentrations and the effects of primary production, excretion, and bacterial nitrification on NH_3 concentrations, there would be no need to model the entire nitrogen cycle and the abridged version of the conceptual model would be more appropriate.

If you take a more careful look at the simplified version of our conceptual model for the nitrogen cycle (**Figure 12.2B**), you should notice that it is essentially three different submodels:

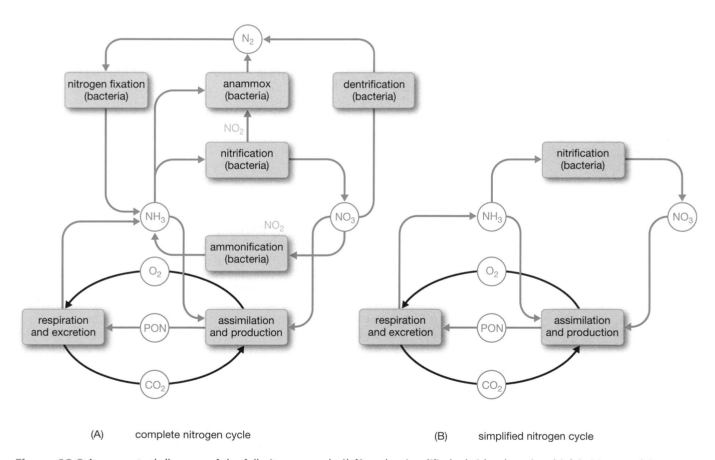

(A)   complete nitrogen cycle   (B)   simplified nitrogen cycle

Figure 12.2 A conceptual diagram of the full nitrogen cycle (*left*) and a simplified, abridged version (*right*). Measured (or model-estimated) concentrations of those molecules important to the nitrogen cycle are indicated in *blue* (with intermediate molecules in *light blue*). The biochemical processes that control the interconversions of nitrogen appear in boxes, while the forcing functions (models) that drive those processes appear as arrows.

Input variable(s)	Submodel	Output variable
NH_3	Nitrification	NO_3
PON	Excretion	NH_3
NH_3, NO_3	Production	PON

Each of these submodels can be represented by an equation, where the input variable is used to calculate the model's estimate of the output variable. Every model, no matter how complicated, can be broken down into several "simple" submodels, which are always bookended by an input variable and an output variable.

So let's take another look at the more complicated conceptual model of the nitrogen cycle (see Figure 12.2A). Every one of the boxes in the conceptual model can be represented by a simple submodel, with its own input variable(s) and output variable(s). A quick count of the number of boxes tells us that the more complicated model of the nitrogen cycle has to have seven submodels. It also tells us exactly which state variables we must include in our über-model:

N_2

NH_3

NO_3

NO_2

PON

CO_2

O_2

Of course, some of these variables must be used in more than one submodel, and some submodels have more than one input and output variable (for example, the assimilation and production submodel is driven by NH_3 and CO_2 inputs and calculates particulate organic nitrogen (PON) and O_2). Those are details to work out when you are crafting each of the constituent submodels; when all of the submodels are complete, you need only link them together (using your conceptual map as a guide) so that the output from one submodel can be used as an input variable for another submodel. When all the connections are complete, your model is ready for testing.

For each submodel, it is always best to have some real measurements of the input and output variables; that way, you can easily test the performance of your submodel by initializing it with your field or lab measurements (as the input variable) and allowing the model to perform its calculations and estimate the values of the output variable. If you also have field or lab measurements of the output variable, you can easily compare how well your modeled values agree with the observed data using a "goodness of fit" analysis. If your model results are way off base, it's time to take a very careful look at your submodel to make sure you didn't make any mistakes in its construction.

The Third Rule of Model Construction: Use the Appropriate Arrays

Keep in mind that the spatial and temporal resolution of your model will dictate how each of your variables must be set up, usually as an **array**. Since the value for each variable within your model is, by definition, allowed to vary in space and time, it will be necessary for you to define those dimensions of variability. For example, if you wanted to design a two-dimensional model of ocean salinities at the surface, you could define a single variable

S to represent the salinity value. Since it is likely that the surface salinity at one location (coordinates x_1,y_1) may be different at another location (x_2,y_2), you would need to define S as an array with the spatial dimensions x and y: $S(x,y)$. If you wanted to allow salinity to vary with time as well, you would need to add the temporal dimension t to the salinity array: $S(x,y,t)$.

Each of the variables within your model must be set up as its own array, but it is up to you to determine which spatial and temporal dimensions are relevant to each variable. Some may have several dimensions, while others may have as few as one dimension, but all will require some kind of array structure. The only exception to this rule is the use of mathematical constants within your models—since constants do not vary in space and time, they do not require an array.

The Fourth Rule of Model Construction: Define the Boundary Conditions

In Chapter 11, we first discussed the notion of boundary conditions in the context of the model domain; that is, defining the specific conditions we must impose at the edges of our domain to separate the dynamics our model will compute in the interior of the model (where all of our compute nodes are located) and what dynamics will essentially be ignored external to the domain. Here, we must also consider the appropriate boundary conditions for each of our variables. When you are setting up each of the variables in your model, it is important to remember that the model code will not be able to recognize what are "realistic" values for your variables—the computer will simply perform the calculations you have programmed and produce a numeric result. Thus, it is necessary for you to consider the realistic boundaries for the values of each of your model variables, and impose those boundaries in all your model's calculations.

For example, we can comfortably state that salinity could never be less than zero, but the computer will not impose that lower boundary to $S(x,y,t)$ unless you tell it to. This is as simple as inserting a line of computer code that establishes a default condition that if the calculated value $S(x,y,t) < 0$, then the calculated value should be replaced with $S(x,y,t) = 0$. Depending on the nature of your model, you may also wish to set a boundary condition that establishes an upper limit as well. In this fashion, the numeric values for all your variables are always constrained to a meaningful and realistic range.

You must be very deliberate and cautious in your choice of boundary conditions for each variable, because you are essentially defining when the results of the model should be overridden, according to your imposed rules. A poorly chosen boundary condition may also mask a larger problem in your model. For instance, if your model did indeed compute a negative value for salinity, it would indicate that the model possesses an error and is somehow computing a nonsensical result. By summarily setting all negative salinity values to zero, your boundary condition might be causing you to miss a very significant flaw in your model. So be warned: if the boundary conditions you choose to impose are too restrictive, or do not accurately reflect the natural range of your modeled variables, the solution to your model simulations will be more of a reflection of your own rule making rather than the natural dynamics you are seeking to model.

The Fifth Rule of Model Construction: Establish Your Refuges

Sometimes it is necessary to maintain a threshold value for a particular variable in your model. For example, if your ocean model included phytoplankton biomass being grazed by a variety of consumers, it would be numerically possible to graze the phytoplankton biomass down to zero. If such a case occurred, your model would never allow new phytoplankton production to occur again, because the phytoplankton population had been eradicated. Since this is unrealistic in natural systems, it may be necessary for you to establish a very low (nonzero) refuge value, so the model could never reduce those values below the refuge threshold. The establishment of any refuge within your model is essentially a special type of boundary condition and should be treated with the same care, especially since refuge populations can easily violate the continuity equation, as it relates to the conservation of mass in your mass budget (discussed in the next section).

The Sixth Rule of Model Construction: Stay Within Your Mass Budget

Whatever your model is designed to do, it will use some common type of model currency, which is tracked throughout the different processes of the model. If we were going to write a model to simulate the nitrogen cycle (see Figure 12.2), it would make the most sense to use atomic nitrogen (N) as our currency, because we are quantifying some form of N in every one of our variables. When the model is initialized, we will be defining exactly how much N is contained within our simulation at start-up, either as N_2, NH_3, NO_2, NO_3, or in the organic matrix as PON.

At any given time in our model calculations, we will always be able to quantify how much N is present in each "compartment." Since our conceptual model of the nitrogen cycle does not allow us to import or export N, there should be no way for our model to gain or lose N beyond start-up. That means we can establish a budget (a "nitrogen budget" in this case), and periodically add up all of the forms of N in our model. If that value ever departs from the original quantity of N we had at model start-up, we will know that our model has somehow created or destroyed N: a violation of the law of conservation and a strong indication that our model formulations are erroneous.

For models that do allow import and export, we can still use mass budgets, but we just have to make sure all our mass fluxes and mass reservoirs are balanced. This is similar to the approach we use to conceptualize the continuity equation (see Figure 11.10). From a "mass balance" perspective, we can imagine our model as a glass of water, full with the mass we used to initialize the model (**Figure 12.3**). If any mass is imported to the model, it will either be added to the mass held in reserve (that is, within the model itself), or it can be exported from the model.

Figure 12.3 Like a glass full of water at rest, a closed-system model (*left*) does not allow mass import or export, so the mass budget always stays the same. In an open-system model (*right*) that allows mass import and export, the total mass exported from the model can never exceed the total mass imported, plus the total mass held in reserve. In other words, whatever you pour into a full glass will overflow by the exact amount (influx = efflux).

As long as we keep track of all the mass being imported, exported, and held in reserve, we should still be able to perform our mass budget analyses to make sure our model is neither creating nor destroying mass. We could also do the same by using an "energy balance" approach, since energy is conserved similarly to mass.

The Unwritten Rule of Model Construction: Check Your Submodels for Goodness of Fit

For each submodel, it is always best to have some real measurements of both the input and output variables. Go ahead and initialize the submodel with your field or lab measurements (as the input variable) and allow the model to perform its calculations and estimate the values of the output variable.

If you also have field or lab measurements of the output variable, you can easily compare how well your model results agree with the observed data by using a "sum of squares" (*SS*) analysis to test what's called **goodness of fit**. The most common method for testing model performance is to start by simply plotting our observed data against the model's estimates of the same variable. For example, if we wished to test our nitrification submodel (**Figure 12.4**) it would be necessary for us to have both observed and modeled data for NO_3.

Ideally, our modeled NO_3 (*x*-axis) should exactly agree with the observed NO_3 (*y*-axis), giving us a simple linear regression of observed versus modeled (**Figure 12.5**), where $y_{OBS} = m \cdot x_{MOD} + b$. In the idealized case where our observations are perfectly predicted by our model, we would see that our modeled values would fall perfectly on the line prescribed by $y_{OBS} = 1 \cdot x_{MOD} + 0$ (see Figure 12.5A). In the more realistic case, our model will undoubtedly exhibit some variability with regard to our observed data: sometimes predicting higher (or lower) NO_3 values than what was observed. The imperfections of our model are reflected by the degree of "scatter" about the idealized trend line (see Figure 12.5B).

Fortunately, there is a more objective measure of model performance than simply relying on a visual analysis of "scatter." In essence, the degree of scatter is really a reflection of goodness of fit (or the lack of goodness of fit, depending on your perspective). Ultimately, we are interested in describing how well our modeled data fit the observed data, preferably by using a more objective method than simply eyeballing the degree of scatter in our plots of modeled versus observed. To some extent, the r^2 value of the linear regression (that is, correlation) between observed versus modeled results can serve as a simple quantitation of goodness of fit. But for a more thorough analysis, the *SS* method is the preferred method to analyze model fidelity.

Goodness of fit can be easily tested by first computing the sum of squares (*SS*) for the total body of observed data (SS_{tot}) by determining the relative departure of each observed datum (OBS_i) from the mean of all observed data ($\overline{OBS}$), as defined in Equation 12.1:

$$SS_{tot} = \sum (OBS_i - \overline{OBS})^2 \qquad (12.1)$$

Once SS_{tot} has been computed, it can be compared to the sum of squares of the residual error (SS_{res}) between each of the observed (OBS_i) and the modeled (MOD_i) values using Equation 12.2:

$$SS_{res} = \sum (OBS_i - MOD_i)^2 \qquad (12.2)$$

Figure 12.4 Conceptual diagram of nitrification, as a process.

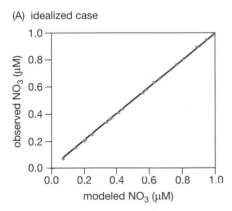

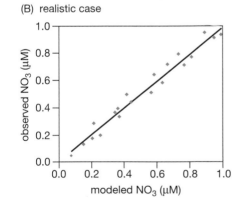

Figure 12.5 Visual comparison of measured versus modeled results for goodness of fit. In an idealized case, our modeled values of NO_3 should be exactly equal to our observed values of NO_3 (A). However, this level of agreement is almost impossible to achieve. It is more realistic to expect that the modeled values of NO_3 will deviate from our observed values of NO_3 but remain strongly correlated (B).

These quantities are then used to determine the coefficient of determination (R^2) as an estimate of the model's goodness of fit:

$$R^2 = 1 - \frac{SS_{res}}{SS_{tot}} \tag{12.3}$$

where R^2 ranges from 0.00 to 1.00. As $R^2 \rightarrow 1.00$, the model approaches perfection in its ability to predict the observed data. That means the higher the R^2 value, the better our goodness of fit. This is analogous to the correlation coefficients (r^2) we might calculate for a linear regression (see Chapter 9), but Equation 12.3 calculates the correlation coefficient specifically with regard to the SS.

Although the R^2 value provides us an objective measure by which to compare goodness of fit, the R^2 value can also be loosely interpreted to define the relative percentage of our observed data that are accurately predicted by our model. As an example, let's assume we were testing a nitrification model (and the model's estimates of NO_3) relative to our observed data in Table 12.1.

Table 12.1 Observed vs. Modeled Results of NO_3 Concentrations (μM), Including the Sum of Squares for the Total Body of Observed Data (SS_{tot}) and the Residual Error (SS_{res})

Observed NO_3 (μM)	Modeled NO_3 (μM)	SS_{tot}	SS_{res}
0.20	0.17	0.1011	0.0009
0.25	0.19	0.0718	0.0036
0.34	0.36	0.0317	0.0004
0.56	0.51	0.0018	0.0025
0.21	0.28	0.0949	0.0049
0.07	0.03	0.2007	0.0016
0.37	0.33	0.0219	0.0016
0.44	0.44	0.0061	0.0000
0.76	0.71	0.0586	0.0025
0.36	0.39	0.0250	0.0009
0.41	0.49	0.0117	0.0064
0.81	0.77	0.0853	0.0016
0.58	0.64	0.0038	0.0036
0.73	0.79	0.0449	0.0036
0.15	0.13	0.1354	0.0004
0.66	0.72	0.0202	0.0036
0.89	0.95	0.1384	0.0036
0.63	0.58	0.0125	0.0025
0.95	0.91	0.1866	0.0016
0.99	0.93	0.2228	0.0036
$\overline{OBS}$ = 0.52	$\overline{MOD}$ = 0.52	ΣSS_{tot} = 1.4751	ΣSS_{res} = 0.0494

$$R^2 = 1 - \frac{0.0494}{1.4751} = 1 - 0.0335 = 0.9665$$

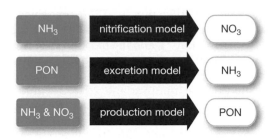

Figure 12.6 A conceptual diagram of the three submodels contained within the model of the nitrogen cycle, where the simplified nitrogen dynamics are subject only to the forcing functions that define nitrification, excretion, and production.

These data were used to generate the plot of modeled versus observed data in Figure 12.5B, where we can see that model agreement is generally good. However, if we wanted to use a sum of squares analysis to test goodness of fit, we would first use Equation 12.1 to compute the difference of each observation OBS_i from the mean ($\overline{OBS} = 0.52\ \mu\text{M NO}_3$), and then sum all of those values to determine $SS_{tot} = 1.4751$. Similarly, Equation 12.2 is used to compute the residual error of each modeled value MOD_i relative to the observed value OBS_i it was trying to predict, and then sum those values to determine $SS_{res} = 0.0494$. Now it is an elementary task to compute the goodness of fit of our model using Equation 12.3, where we find that $R^2 \approx 0.97$.

Quantifying the goodness of fit is a tremendously powerful tool in modeling, because it allows us to test how well our model comports with reality. Generally speaking, a low R^2 value would indicate that a given model either is formulated incorrectly or is too simple (that is, missing some key components). In our example, we used a simplified version of the nitrogen cycle (see Figure 12.2B), where we assumed that our nitrification model was the primary forcing function that predicted NO_3. If our R^2 value was low, it would indicate that our nitrification model was doing a poor job of predicting NO_3, which in turn would suggest that we must include more complex dynamics to account for other sources and removals of NO_3 from the system, in order for our model to better predict the observed data. However, if our R^2 value is high (for example, $R^2 \approx 0.97$), we can feel confident that the model in its current form is sufficient and does not require us to unnecessarily add more complexity to it.

Examples of Modeling Simple Dynamics

Now that we have a better idea of how to construct a model of simple dynamics, let's take a look at a few examples, using the three submodels (**Figure 12.6**) from our simplified conceptual model of the nitrogen cycle (see Figure 12.2B). For the sake of illustration, we shall assume that each of these forcing functions can be modeled simply, as a single equation. In the next chapter, we will explore how these same forcing functions might be more appropriately modeled as a complex dynamic. But for now, let's keep things simple until we're forced to do otherwise.

Example 1: Nitrogen Dynamics of Nitrification

Do Your Homework: Let's presume our thorough review of the literature indicates that bacteria are responsible for the conversion of NH_3 to NO_3 via nitrification (**Figure 12.7**), and that this bacterial process can be calculated using a modified **Michaelis–Menton** function, as described by Equation 12.4:

$$\frac{dNIT}{dt} = \left(\frac{r_{nit} \cdot NH_3}{k_{nit} + NH_3} \right) \tag{12.4}$$

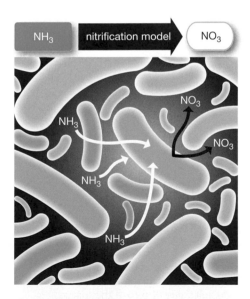

Figure 12.7 Nitrification is a bacterially mediated process whereby available NH_3 is ultimately oxidized to NO_3. Modeled changes in the nitrification rate will simultaneously define the rate of NH_3 removal from and NO_3 addition to the system.

where the rate of change in nitrification ($dNIT/dt$, μmol N m^{-3}) can be calculated using the *in situ* concentration of NH$_3$ (μmol N m^{-3}), the bacterial nitrification rate r_{nit} (μmol N m^{-3} day^{-1}), and the half-saturation constant for bacterial nitrification (k_{nit}, μmol N m^{-3}). Assuming both r_{nit} and k_{nit} are constant values that can be determined from the literature, the solution to Equation 12.4 will depend solely on the input variable NH$_3$. Of course, nitrification is a bacterially mediated process, so both r_{nit} and k_{nit} are likely to vary according to bacterial diversity, abundance, and/or changes to environmental parameters that affect bacterial metabolism (such as temperature and salinity). However, for the sake of this example, let us assume we are dealing with a generalized bacterial population in a controlled environment, such that $dNIT/dt$ is determined solely as a function of available NH$_3$ (as suggested in Equation 12.4).

Follow Your Conceptual Model: Recall that our original conceptual model of the nitrification submodel required NH$_3$ as the sole input variable and NO$_3$ as the output variable (see Figure 12.4). Our nitrification submodel (see Equation 12.4) certainly meets those requirements, provided that we can determine that the bacterial nitrification rate r_{nit} and half-saturation constant for bacterial nitrification (k_{nit}) are indeed constants.

Use the Appropriate Arrays: As our values for r_{nit} and k_{nit} are constant, they will not vary in space or time; therefore, they do not have to be defined according to an array. If you plan on using your submodel to calculate nitrification more than once, your values for NH$_3$ and NIT will most certainly vary with time t. Since we would expect NH$_3$ and NIT to vary in both the horizontal (x,y) and vertical (z) dimensions of the ocean, we would most likely define each variable as a four-dimensional array:

$$NIT(x,y,z,t)$$
$$NH_3\,(x,y,z,t)$$

That means our computer will record (and store in memory) a different value for NIT and NH$_3$ at each grid point (x,y,z) in our model and at each timestep t.

Define the Boundary Conditions: Although there is no inherent upper boundary to the concentration of nitrogen in the ocean, there is indeed a lower limit to any chemical concentration, and that limit is zero. Since NIT and NH$_3$ are both defined as the chemical concentration μmol N m^{-3}, we may wish to establish boundary conditions that prevent our model from calculating any concentrations less than zero. As a less invasive boundary condition, we may be better served to create a "flag condition" in our model code, so that any model calculation of a negative concentration will pause the model and provide an alert, indicating a nonsensical value has erroneously been calculated.

WATCH YOUR DENOMINATORS!

Be extremely cautious when defining boundary conditions that create a default value of zero. If that particular variable is used elsewhere in your model, make sure it is not in the denominator of any other equations. Otherwise, your boundary condition will cause division by zero and your entire model will crash.

According to our nitrification submodel (see Equation 12.4), there is no possible way to calculate a negative value for NIT, so we need not define any boundary conditions for NIT. Since nitrification represents a conversion of NH$_3$ to NO$_3$, any positive value of NIT will represent a loss from the NH$_3$ pool and a gain to the NO$_3$ pool. Based on our formulation of NIT in Equation 12.4, as the NH$_3$ pool is reduced, $dNIT/dt \rightarrow 0$. If NH$_3$ is actually reduced to zero, the numerator in Equation 12.4 becomes zero (thus, $dNIT/dt$ becomes zero) and the denominator simply becomes $k_{nit} + 0$. Since k_{nit} is always a positive, nonzero number, we will never be in danger of dividing by zero, even as NH$_3 \rightarrow 0$. Thus, our boundary conditions will essentially take care of themselves.

Establish Your Refuges: If you are concerned that a boundary condition of $NH_3 = 0$ is unrealistic (as trace concentrations of NH_3 always seem to be present), you may choose to establish a refuge concentration at a very low, but nonzero, value. Exactly what that value should be is a matter for careful consideration, and should depend on your research experience and familiarity with the natural dynamics you are attempting to simulate. Keep in mind that all methods and instruments of measure will possess some inherent limit of detection (LOD), below which the method or instrument will assign a measure of zero, simply because the actual value falls somewhere between true zero and the LOD. If you wish to be conservative in your treatment of measured zeroes, you may wish to use a value of "one-half LOD" to define the refuge concentration.

Stay Within Your Mass Budget: Since NIT represents the conversion of NH_3 to NO_3, NIT also defines the amount of nitrogen lost in the NH_3 pool, which will be exactly equal to the amount of nitrogen gained in the NO_3 pool. Thus, we must take care that our values for NH_3 and NO_3 are updated after each model calculation of NIT.

Keep in mind that the NH_3 and NO_3 pools will also be affected by the results of your other submodels as well (see Figure 12.2). Losses to the NH_3 and NO_3 pools will occur as a result of the primary production submodel, which will be further complicated by NH_3 gains due to the excretion submodel. As you link each of your submodels together, it will be increasingly important to maintain mass balance in all your calculations.

Example 2: Nitrogen Dynamics of Excretion

Do Your Homework: If we assume that the dominant source of NH_3 excretion shall come from the zooplankton grazers in the water column (**Figure 12.8**), our exhaustive research on the matter might reveal a relationship between the weight of the zooplankton grazer and the amount of NH_3 it is capable of excreting, such that Equation 12.5 can be used to estimate NH_3 excretion based on the biomass of the excretor:

$$\frac{dE_{NH_3}}{dt} = 0.1278W \qquad (12.5)$$

where the daily rate of NH_3 excretion (E_{NH_3}, μg atomic-N ind^{-1}) for each zooplankton individual can be estimated as a simple function of zooplankton dry weight W (μg ind^{-1}).

Follow Your Conceptual Model: Our original conceptual model of the excretion submodel called for the concentration of biogenous PON as our input variable and NH_3 as the output variable (see Figure 12.6). To stay consistent with our other submodels, we'll need to convert E_{NH_3} into units of μmol N m^{-3}; other than that, it looks like we're doing just fine with regard to our output variable. But Equation 12.5 (as currently written) does not follow our conceptual model (see Figure 12.2B) with regard to our input variable. That means we will have to figure out a (hopefully simple) way to convert zooplankton dry weight (μg ind^{-1}) into our required input variable, PON (μmol N m^{-3}).

Although this might at first seem like a difficult task, we just have to be a little clever about using unit conversion to get us where we need to be. If we were able to determine through additional literature research the average dry weight for herbivorous zooplankton ($\overline{W} = 16.4$ μg ind^{-1}), as well as an estimate for the typical population density of such zooplankton ($Z = 9000$ ind m^{-3}), we could use Equation 12.6 to determine their NH_3 excretion as a

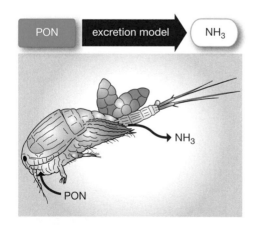

Figure 12.8 Nitrogen excretion is the process by which particulate organic nitrogen (PON) is metabolically converted to NH_3 and released to the system. Modeled changes in the excretion rate will simultaneously define the rate of PON removal from and the rate of NH_3 addition to the system.

function of their concentration rather than their dry weight. Thus, Equation 12.5 becomes

$$\frac{dE_{NH_3}}{dt} = 0.1278 \cdot \overline{W} \cdot Z \qquad (12.6)$$

$$\frac{dE_{NH_3}}{dt} = \left(\frac{0.1278 \, \mu g \, N}{\mu g \, dw} \right) \cdot \left(\frac{16.4 \, \mu g \, dw}{ind} \right) \cdot \left(\frac{9000 \, ind}{m^3} \right) = \frac{18,863.28 \, \mu g \, N}{m^3}$$

Converting E_{NH_3} into consistent units of $\mu mol \, N \, m^{-3}$ is just a simple matter of using Equation 12.7 to divide by the atomic weight of nitrogen, so Equation 12.6 ultimately becomes

$$\frac{dE_{NH_3}}{dt} = \left(\frac{18,863.28 \, \mu g \, N}{m^3} \right) \cdot \left(\frac{1 \, \mu mol \, N}{14.0067 \, \mu g \, N} \right) = \frac{1,346.73 \, \mu mol \, N}{m^3} \qquad (12.7)$$

If we choose to use this formulation, we would not need to have an input variable after all, since the solution to Equation 12.7 does not vary in space or time. This is probably a dubious oversimplification, especially since we assumed our zooplankton grazers (that is, our excretors) had a fixed biomass and a fixed abundance throughout the model domain. Since our zooplankton grazers will most likely exhibit variable grazing rates, variable abundances, and variable biomass, it is unrealistic for us to expect that Equation 12.7 will be an accurate submodel for excretion. It is far more likely that we will be forced to use a more complex formulation for excretion, but let's address that prospect in the next chapter and assume, for now, that Equation 12.7 is adequate.

Use the Appropriate Arrays: As we have demonstrated in our oversimplified submodel of excretion, the solution to Equation 12.7 is a fixed value and will not vary in space or time. Therefore, E_{NH_3} does not require an array. Since E_{NH_3} represents a constant gain to the NH_3 pool, and we have already established that there will be various other gains and losses to the NH_3 pool, it is clear that we will need to set up NH_3 as an array. In order to keep our variables consistent with all other submodels, we shall need to maintain NH_3 as a four-dimensional array, as we did in our earlier treatment of NH_3 in the nitrification submodel, where:

$$NH_3 \, (x,y,z,t)$$

Define the Boundary Conditions: Since our formulation of E_{NH_3} in Equation 12.7 will always yield a constant value, there is no need to establish a boundary condition for it.

Establish Your Refuges: Just as there was no need to establish a boundary condition for a value that is constant, there is no need to establish a refuge concentration for E_{NH_3} either.

Stay Within Your Mass Budget: Here's where we really run into trouble with our formulation for E_{NH_3}. Since Equation 12.7 is not in any way dependent on the nitrogen reservoir in our model, the constant source of excreted NH_3 represents a source of N that is in no way balanced by the removal of N. As a result, it will be impossible to maintain a mass budget, and the model will continue to accumulate nitrogen, perhaps to ridiculous proportions. If we had any misgivings about how oversimplified our excretion submodel was, our concerns about the mass budget for nitrogen would compel us to seek a new, more complex formulation of excretion.

Example 3: Nitrogen Dynamics of Production

Do Your Homework: In researching the common methods for modeling nitrogen assimilation among producers, we would discover that this is typically done using the classic exponential growth function, as defined in Equation 12.8:

$$B_t = B_0 e^{\mu t} \qquad (12.8)$$

where the producer biomass at some time in the future (B_t, μg C m^{-3}) can be estimated from some initial value of producer biomass prior to growth (B_0, μg C m^{-3}), the realized daily growth rate of the producer population (μ, day^{-1}), and time elapsed (t, day).

Although Equation 12.8 seems a rather straightforward equation, our continued research into the methods for calculating B_t will ultimately lead us to the realization that the parameter that mathematically describes the realized growth rate (μ) is deceptively challenging to quantify in a simple manner. We would prefer that only NH_3 and NO_3 concentrations serve as our input variables to determine nitrogen assimilation, but μ is in fact a complex function that is also dependent on (1) *in situ* temperature; (2) producer diversity; (3) Michaelis–Menton kinetics of nitrogen absorption; (4) availability of complimentary dissolved nutrients, such as phosphorus (P), silicon (Si), and iron (Fe); and the list goes on (**Figure 12.9**).

Based on our research on the topic, and the inherent complexity associated with μ, it is obvious that we won't be able to establish a simple model for the assimilation of nitrogen as a result of production. That doesn't mean we just give up—it simply means we have to create a more complex submodel than we had originally hoped. That shouldn't come as a tremendous surprise since the natural world is a pretty complicated place, and there are times when a simple model just won't do. In the next chapter, we will explore some of the more challenging aspects of modeling complex dynamics.

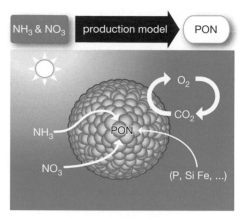

Figure 12.9 Nitrogen-specific production is the process by which dissolved inorganic nitrogen (as NO_3 or NH_3) is biologically converted to particulate organic nitrogen (PON, the nitrogen-specific portion of biomass) as a consequence of autotrophic growth. Modeled changes in the production of PON will simultaneously define the rate of nitrogen removal from the water column and the rate of PON increase in the photosynthetic biomass.

References

Carpenter EJ & Capone DG (eds) (1983) Nitrogen in the Marine Environment. Academic Press.

Fasham MJR, Ducklow HW, & McKelvie SM (1990) A nitrogen-based model of plankton dynamics in the oceanic mixed layer. *Journal of Marine Research* 48:591–639.

Fennell W & Neumann T (2014) Introduction to the Modelling of Marine Ecosystems, 2nd ed. Elsevier.

Hansell DA & Carlson CA (2014) Biogeochemistry of Marine Dissolved Organic Matter, 2nd ed. Academic Press.

Herbert RA (1999) Nitrogen cycling in coastal marine ecosystems. *FEMS Microbiology Reviews* 23:563–590.

Omori M & Ikeda T (1992) Methods in Marine Zooplankton Ecology. Krieger Publishing.

Pagano M, Gaudy R, Thibault D, & Lochet F (1993) Vertical migrations and feeding rhythms of mesozooplanktonic organisms in the Rhone River plume area (North-West Mediterranean Sea). *Estuarine, Coastal and Shelf Science* 37(3):251–269.

Wen YH & Peters RH (1994) Empirical models of phosphorus and nitrogen excretion rates by zooplankton. *Limnology and Oceanography* 39(7):1669–1679.

Further Reading

Chen S, Ling J, & Blancheton J (2006) Nitrification kinetics of biofilm as affected by water quality factors. *Aquacultural Engineering* 34:179–197.

Mayzaud P (1976) Respiration and nitrogen excretion of zooplankton: The influence of starvation on the metabolism and the biochemical composition of some species. *Marine Biology* 37(1):47–58.

Malone RF, Bergeron J, & Chad CM (2006) Linear versus monod representation of ammonia oxidation rates in oligotrophic recirculating aquacultutre system. *Aquacultural Engineering* 22(1–2):57–73.

Munn C (2011) Marine Microbiology, 2nd ed. Garland Science.

Sterner RW, Elser JJ, & Vitousek P (2002) Ecological Stoichiometry: The Biology of Elements from Molecules to the Biosphere. Princeton University Press.

Vallino JE, Hopkinson CS, & Hobbie JE (1996) Modeling bacterial utilization of dissolved organic matter: Optimization replaces Monod growth kinetics. *Limnology and Oceanography* 41(8):1591–1609.

Zhu S & Chen S (2002) Impact of temperature on nitrification rate in fixed film biofilters. *Aquacultural Engineering* 26:221–237.

Chapter 13

Modeling Complex Dynamics

"The complex is everywhere evolved out of the simple." – Thomas Henry Huxley

Much to our chagrin, the natural world can be quite complicated, so it should come as no surprise that our modeling efforts, on occasion, require a bit more forethought and insight. Of course, simplicity is the goal, but only when a simple model can faithfully simulate the dynamics we wish to explore. When the simple model fails, we must roll up our sleeves and build ourselves a better one, with all the necessary accoutrements to get the job done.

As we have already witnessed, the bulk of the hard work involved in modeling really does revolve around the scientific literature. The admonition to "do your homework" before embarking on a modeling project will resolve 80–90% of the fear and frustration you might otherwise feel. In preparation for this task, read many research manuscripts and keep in mind the conceptual model.

But above all, don't beat yourself up—this is supposed to be difficult! If modeling were easy, everyone would be doing it, and being a modeler would be nothing special. So let's work through a few examples of some complex modeling problems. After you finish your homework, of course.

Key Concepts

- Model solutions that cannot be accomplished using a single regression equation must instead make use of a solution cascade, whereby several equations are linked together and solved sequentially.

- Conceptual models are critical to the process of defining which state variables will drive the model (as input variables) and which will serve as the model solution (as output variables).

- Coupled models must possess some common element or currency that will enable the solution of one model to serve as the input for another.

- Great care must be taken when establishing boundary conditions and performing mass budgets for complex models, as their inherent complexity makes such tasks all the more challenging.

Embracing Complexity in Our Model(s)

In Chapter 12, we spent a good deal of time discussing the steps a modeler should always follow when constructing each and every submodel within the larger simulation:

1. Do your homework.
2. Follow your conceptual model.
3. Use the appropriate arrays.
4. Define the boundary conditions.
5. Establish your refuges.
6. Stay within your mass budget.
7. Check the model for goodness of fit.

Although we might prefer to use a simple formulation for each of our submodels, the examples we used in the previous chapter demonstrate that it is not always possible (or prudent) to model a particular dynamic in a simple fashion. Sometimes a more complex formulation is necessary (making these guidelines even more useful).

Recall from our earlier efforts in Chapter 12 to model an abridged version of the nitrogen cycle (see Figure 12.2), we were able to successfully craft a simple submodel to describe the nitrogen dynamics associated with nitrification (see Equation 12.4), but our formulation for excretion (see Equation 12.5) was grossly inadequate. When it came time to analyze our submodel for assimilation/production (see Equation 12.8), we were quick to recognize that the inherent dynamics were so complicated that a simple formulation of assimilation/production would be pure folly. And so, here we are, challenged with the task to develop a more complex formulation for the nitrogen dynamics associated with excretion and assimilation/production, where simpler models could not do.

Complex Processes Are More Easily Linked by Converting Your Conceptual Models to Conceptual Equations

Before embarking on this task, it is often a useful exercise to review the conceptual model and focus one's attention solely on the state variables that are germane to the submodel you are currently attempting to design. The direction of the arrows are also quite helpful, because they can help you visualize which variable(s) will be inputs and which will be output(s). Since our original attempts to simplify the excretion and assimilation/production models met with failure, we should take the training wheels off and take a closer look at each of the relevant submodels within the full (rather than the abridged) conceptual model of the nitrogen cycle (see Figure 12.2).

A Variable-Oriented Perspective Will Help You Focus on What Goes Into (and What Comes Out of) Each Modeled Process

Remember that each submodel represents a specific natural process, which is driven by the numerical values of a unique combination of input variables. The submodel then calculates a solution that represents a different set of numerical values as output variables (**Figure 13.1**). For each process being modeled, it is often helpful to create a punch list of all the input and output variables that will be necessary to include in the submodel.

This perspective is also helpful in defining the number of mathematical equations that will likely be used to represent the natural process being modeled. Remember that an output variable produced from a model is analogous to a dependent variable produced from a regression equation.

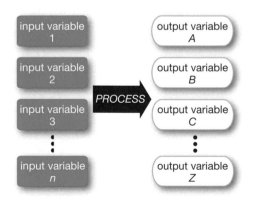

Figure 13.1 A variable-oriented perspective of submodels, used as a guide to help define each of the input and output variables that must be included in the modeled process. Blue boxes are used to represent input variables, while unshaded ovals represent the output variables.

Thus, a separate equation is usually required to calculate the solution for each output variable.

In those cases where only a single output variable is calculated from the submodel, the number of input variables can be helpful in establishing the type of mathematical equation that will be needed. For example, a solution driven by just a single input variable could potentially (but not necessarily) be a simple, linear relationship:

$$f(x) = mx + b \quad \text{Linear}$$
$$f(x) = e^x \qquad \text{Nonlinear}$$
$$f(x) = ax^b \qquad \text{Nonlinear}$$
$$f(x) = log_{10}^x \quad \text{Nonlinear}$$

However, any solution requiring two or more input variables will require a more complex, nonlinear mathematical relationship.

Recall from our modeling examples in the previous chapter that we were able to describe the nitrification model in relatively simple mathematical terms (see Equation 12.4) because we assumed the process was driven by a single input variable (NH_3) and produced a single output variable ($dNIT/dt$). This was not at all the case for the excretion (see Equation 12.5) and assimilation/production (see Equation 12.8) models, each of which was driven by multiple input variables and resulted in several output variables (**Figure 13.2**). From this perspective, it's no wonder that the excretion and assimilation/production models require a more complex formulation.

A Process-Oriented Perspective Will Help You Focus on Which Submodels Should Be Included

It is an equally important exercise to consider each of our model's state variables from a process-oriented perspective. Instead of focusing solely on the input and output variables of a single process, we consider the multitude of processes that are in some way tied to each state variable. Once more, let's take a closer look at the conceptual model of the nitrogen cycle (see Figure 12.2), using the NH_3 state variable as an example.

According to our conceptual model (and ignoring the more exotic anammox dynamics), there are a total of five different submodels that will all contribute, in some way, in the determination of the value of NH_3 at any given time (**Figure 13.3**). Three of these processes (N_2 fixation, ammonification, and excretion) will yield some estimate of NH_3 as their output variable, while two other processes (nitrification and assimilation/production)

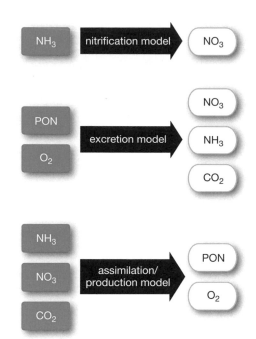

Figure 13.2 A variable-oriented perspective of the nitrification, excretion, and assimilation/production submodels within the nitrogen cycle.

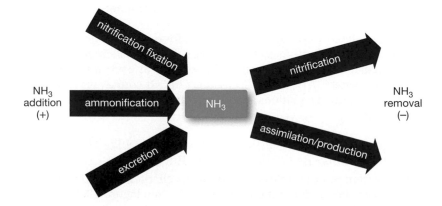

Figure 13.3 A process-oriented perspective, used as a guide to help define which (and how many) processes will affect the value of a particular state variable.

require NH_3 as an input variable. That means that the value of NH_3 within the model, at any given time, will be determined by the contributions from five different submodels simultaneously.

As complicated as that sounds, there is a nifty trick to making things much easier, if we use the process-oriented perspective to tackle this monster. The secret is to convert our conceptual model into a **conceptual equation**, where each process can be represented as a symbolic equation.

For example, we know that the concentration of NH_3 at any given time ($[NH_3]_t$) must be determined from the "standing stock" of available NH_3 in the system from the previous timestep ($[NH_3]_{t-1}$), where the magnitude of change in NH_3 will be affected by five different models (M_n). We can state this easily using Equation 13.1 as our conceptual equation of the various changes in NH_3:

$$\frac{dNH_3}{dt} = \left[NH_3 \right]_{t-1} + M_1 + M_2 + M_3 + M_4 + M_5 \quad (13.1)$$

Note that the d/dt term simply indicates a change of the pertinent quantity over some finite change in time t (such as the timestep of the model).

Thanks to our conceptual model, we already know exactly what each of these five models should be: (1) N_2 fixation (N_{fix}); (2) ammonification (N_{amm}); (3) excretion (N_{exc}); (4) nitrification (N_{nit}); and (5) assimilation/production (N_{prd}). We can then easily substitute these into Equation 13.1 to yield a more informative conceptual equation, now expressed in Equation 13.2, which describes the NH_3 state variable:

$$\frac{dNH_3}{dt} = \left[NH_3 \right]_{t-1} + \frac{dN_{fix}}{dt} + \frac{dN_{amm}}{dt} + \frac{dN_{exc}}{dt} + \frac{dN_{nit}}{dt} + \frac{dN_{prd}}{dt}$$

$$(13.2)$$

Note that the N_2 fixation, ammonification, and excretion processes all represent a potential gain (+) of NH_3, while nitrification and assimilation/production will represent a potential removal (−) of NH_3 from the system. Thus, it would be more accurate and informative to represent these features in our final conceptual equation, as Equation 13.3:

$$\frac{dNH_3}{dt} = \left[NH_3 \right]_{t-1} + \frac{dN_{fix}}{dt} + \frac{dN_{amm}}{dt} + \frac{dN_{exc}}{dt} - \frac{dN_{nit}}{dt} - \frac{dN_{prd}}{dt}$$

$$(13.3)$$

where all potential gains and losses are appropriately indicated in the conceptual equation. These conceptual equations are also tremendously helpful during mass budgeting as well (so you can immediately recognize which functions represent gains to the system, and which functions are losses).

And there you have it: a conceptual equation for our NH_3 state variable. We can now use the results from each of the five submodels defined in Equation 13.3 to calculate the instantaneous value of NH_3 at each timestep in our larger model, where $[NH_3]_t = [NH_3]_{t-1} + dNIT/dt$. Of course, we still have to define the conceptual equations for each of our other state variables, but you get the picture.

Conceptual Equations Are the Best Way to Visualize How Submodels Should Be Connected with Each Other

Remember from our examples in the previous chapter, we had used a simplified conceptual model of the nitrogen cycle (see Figure 12.2B) and had limited our interest to just three processes: nitrification, excretion, and assimilation/production. Based on this simplification, we can eliminate N_2 fixation and ammonification from Equation 13.3, so our conceptual equation can be rewritten as Equation 13.4:

$$\frac{dNH_3}{dt} = \left[NH_3 \right]_{t-1} + \frac{dN_{exc}}{dt} - \frac{dN_{nit}}{dt} - \frac{dN_{prd}}{dt} \qquad (13.4)$$

We can debate the prudence of eliminating N_2 fixation and ammonification from our model, but for the purposes of continuing our earlier example from Chapter 12, let's assume for now that Equation 13.4 is adequate.

Remember that among the nitrification, excretion, and assimilation/ production processes we explored in Chapter 12, we were able to create a simple model for nitrification using Equation 12.4, but the more complex models for excretion and production still elude us. We'll deal with that matter in just a bit; for now, let us define the conceptual equations for the other state variables in our example from Chapter 12, where we can use Equations 13.5 and 13.6 to define the d/dt for NO_3 and PON, respectively:

$$\frac{dNO_3}{dt} = \left[NO_3 \right]_{t-1} + \frac{dN_{nit}}{dt} - \frac{dN_{prd}}{dt} \qquad (13.5)$$

$$\frac{dPON}{dt} = \left[PON \right]_{t-1} + \frac{dN_{prd}}{dt} - \frac{dN_{exc}}{dt} \qquad (13.6)$$

Now we can pause to take stock of everything we've done, and what we have left to do. First, the good news: we have successfully defined all of the equations for each of our state variables using Equations 13.4–13.6, and we also have a simple model that describes nitrification (N_{nit}, using Equation 12.4). Let's stop a moment to really let that sink in. We have just defined the mathematical framework that will allow us to use our submodels to compute time-dependent changes in NH_3, NO_3, and PON concentrations, at any time and at every compute node in our model domain, using Equations 13.4–13.6. Even better, we already have a formulation for N_{nit}. All we need to do now is figure out a way to model the more complex dynamics of excretion (N_{exc}) and production (N_{prd}).

Examples of Modeling Complex Dynamics

Just because a natural process *can* be represented by a simple model doesn't necessarily mean that it should be. One of the more challenging aspects to modeling the natural world is knowing how much detail to include in your models. If a simple formulation performs quite well and reflects reality reasonably well, use it. But if a simple formulation yields results that do not reflect the real system, a more complex model will be necessary. As a practical matter, it is always preferable to attempt a simple formulation first and test it for goodness of fit. If the simple model doesn't perform to your expectations, complexity can always be added later (if and when it is necessary).

Let us continue the examples we began in Chapter 12, where we determined that we could adequately model the effects of nitrification (N_{nit}) using a simple formulation (Equation 12.4), but that the excretion and production models would require a more robust effort. Thus, we shall assume that Example 1 (the nitrification model) from the previous chapter does not require further attention, and we shall proceed directly to a more thorough treatment of Example 2 (excretion) and Example 3 (assimilation/production) from Chapter 12, as they relate to the nitrogen cycle.

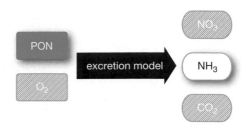

Figure 13.4 A variable-oriented perspective of the excretion model, indicating the necessary input variables used to drive the model, as well as the output variables produced from the model. Note that the hashed variables are ignored in the simplified excretion model, as per the current example.

Example 2 Revisited: Complex Nitrogen Dynamics of Excretion

Do Your Homework: As you may recall from Chapter 12, we originally made the assumption that the dominant source of NH_3 excretion was the result of zooplankton consumption. The function we ultimately used was a relation between NH_3 excreted, the average biomass of the zooplankton consumer, and the average abundance of zooplankton in the water column (Equation 12.6). Since our simple formulation used average values of zooplankton biomass and abundance, their magnitude did not vary in time; thus, the solution to Equation 12.6 was a constant.

According to the more complex conceptual model of the nitrogen cycle (see Figure 12.2A), the excretion model is driven by the following variables:

$$[PON], [O_2]: \quad \text{Input variables}$$
$$[NH_3], [NO_3], [CO_2]: \quad \text{Output variables}$$

However, for this example, let's limit our excretion model to simply include $[PON]$ as the sole input variable and $[NH_3]$ as the sole output variable (Figure 13.4), since we're interested only in the excretion of NH_3. If we wanted to include submodels to compute the respiration of O_2 to CO_2, or the nitrification of NH_3 to NO_3 we would need to include them as state variables in order to maintain our mass budgets. But since we're focusing solely on the excretion of NH_3 from PON in this example, let's keep it as simple as we can; therefore, our formulation of the excretion model must include only PON and NH_3 as the relevant state variables.

Since we must use PON to drive our excretion model, our literature review would require that we consider the many steps involved in rendering edible nitrogenous biomass (PON) into excreted ammonia (NH_3). If we are careful and thorough in our consideration of these steps, we can construct a **solution cascade**, where our initial input variable is used to solve one equation, whose solution is used to solve another then another, until we ultimately arrive at our destination: a value for the required output variable.

We start with the notion that the available edible nitrogenous biomass (as particulate organic nitrogen, or PON) will not likely escape the attention of zooplankton consumers, which will ingest PON according to some intrinsic feeding rate, which will depend on such factors as:

1. Concentration of available food, as $[PON]$
2. Species of zooplankton present in the grazer community (Z_s)
3. Population density (or volumetric abundance) of each zooplankton species (V_z)
4. Biomass (or weight) of each zooplankton species (W_z)
5. Water temperature, as it affects zooplankton metabolism (T)

Since our zooplankton would have to ingest the PON before they could excrete its nitrogen as NH_3 (Figure 13.5), we would be required to determine the amount of PON ingested (PON_i) as a function of those variables listed above. Thus, Equation 13.7 serves as our conceptual equation for determining PON_i:

$$PON_i = f(PON, Z_s, V_z, W_z, T) \tag{13.7}$$

The PON, once ingested, may either be metabolically assimilated or it can be egested as undigested PON in the fecal material of the consumer. The amount of assimilated nitrogen (AN) and that of the egested nitrogen (EN)

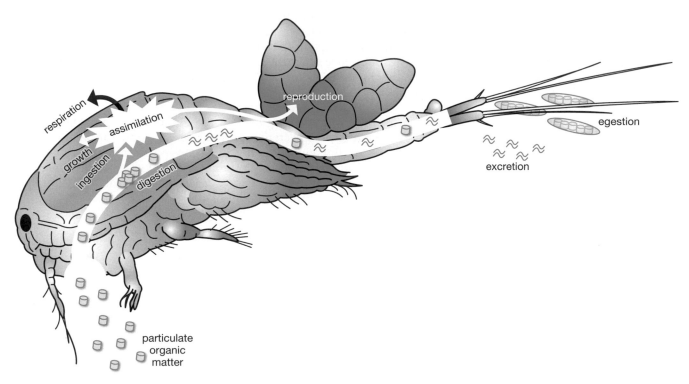

Figure 13.5 A graphical representation of the metabolic processes involved in the partitioning of ingested organic material. Particles must first be ingested before they can be assimilated. Assimilated biomass is either invested in the creation of gamete or larval biomass (reproduction), incorporated as new biomass (growth), "spent" on basic metabolic costs (respiration), or eliminated as metabolic waste (excretion). Particles that are ingested but never assimilated are simply eliminated as undigested fecal material (egestion).

will be proportional to the amount of PON ingested, which can be defined by Equation 13.8:

$$AN = (\%_A) \cdot (PON_i)$$
$$EN = (\%_E) \cdot (PON_i) \qquad (13.8)$$
$$100\% = (\%_A) + (\%_E)$$

where the assimilation efficiency of nitrogen ($\%_A$) is defined as the proportion of ingested nitrogen that becomes metabolically available to the consumer. Undigested (nonassimilated) nitrogen is simply passed through the consumer and eliminated as egested nitrogen. Since the proportion of assimilated ($\%_A$) and egested nitrogen ($\%_E$) must necessarily add to 100%, it is a simple matter to calculate $\%_E$ if we already know the value for $\%_A$.

Let's consider the egested nitrogen (EN) first. It is possible that some NH_3 could be released from the pool of egested fecal nitrogen (EN), either through direct **dissolution** (N_{dis}) of NH_3 or from bacterial **remineralization** (N_{min}) of the organic nitrogen in the fecal material. To include each of these possibilities, we can use Equation 13.9:

$$\frac{dN_E}{dt} = \left(\frac{dN_{dis}}{dt} + \frac{dN_{min}}{dt} \right) \cdot EN = (0.05) \cdot EN \qquad (13.9)$$

where the total amount of NH_3 released as a result of egestion (dN_E/dt) can be determined from the rates of nitrogen dissolution (dN_{dis}/dt) and remineralization (dN_{min}/dt). Based on preexisting models of such dynamics, prior research indicates that we may be able to simplify things a bit and assume the combined total of dN_{dis}/dt and dN_{min}/dt is ~5% of EN. Of course, if we find that this simplification does not provide adequate goodness of fit, we would have to create all new submodels to explicitly calculate the NH_3 dissolution (dN_{dis}/dt) and remineralization (dN_{min}/dt) rates included in Equation 13.9.

When we turn our attention back to the amount of assimilated nitrogen (AN) available for NH_3 excretion, we can use our earlier formulation of NH_3 excretion from Equation 12.6, but incorporate the variables zooplankton biomass W_Z (μg dw ind^{-1}) and abundance V_z (ind m^{-3}), summed across all of the zooplankton species in the community (Z_s) to yield a better, more realistic formulation using Equation 13.10:

$$\frac{dN_A}{dt} = \left(\frac{0.1278\,\mu\text{g}\,N}{\mu\text{g}\,\text{dw}} \right) \cdot \sum^{Z_s} \left(W_Z \cdot V_Z \right) \tag{13.10}$$

where dN_A/dt represents the rate of NH_3 excretion as a result of total zooplankton assimilation.

Now it is just a simple matter of adding our sources of NH_3, either from egestion (N_E) or from assimilation (N_A). Ultimately, we can use Equation 13.11 to yield our final estimate of total excreted NH_3 as dN_{exc}/dt:

$$\frac{dN_{exc}}{dt} = \frac{dN_A}{dt} + \frac{dN_E}{dt} \tag{13.11}$$

If we use direct substitution to redefine Equation 13.11, we can pull dN_A/dt from Equation 13.10 and dN_E/dt from Equation 13.9, which gives us an updated model (Equation 13.12) of dN_{exc}/dt:

$$\frac{dN_{exc}}{dt} = \left(\frac{0.1278\,\mu\text{g}\,N}{\mu\text{g}\,\text{dw}} \right) \cdot \sum^{Z_s} \left(W_Z \cdot V_Z \right) + 0.05\,EN \tag{13.12}$$

Once we calculate dN_{exc}/dt using Equation 13.12, we can use the result to help solve dNH_3/dt and $dPON/dt$ as defined in Equations 13.4 and 13.6, respectively.

Keep in mind that our more complex excretion model, as defined by Equation 13.12, could be easily made even more complex, depending on what other dynamics we wished to include. Conspicuously absent from Equation 13.12 is any term that would suggest that excretion (or ingestion, or assimilation) is heavily influenced by the surrounding water temperature. If we had done our homework, we would know that ambient temperatures have a huge impact on nearly all biologically mediated processes, so we should expect that water temperature should exert some influence on our zooplankton ingestion rates, which in turn will influence assimilation rates, which in turn will influence excretion rates. So it looks like we would be wise to figure out a way to make our excretion model in Equation 13.12 more temperature dependent.

Fortunately for us, Thornton and Lessem (1978) have already devised a very clever algorithm for this very purpose (**Figure 13.6**). If we assume Equation 13.12 represents the maximum theoretical rate of NH_3 excretion under ideal conditions, then we can introduce a separate **damping function** that will reduce our theoretical maximum to a rate that is more realistic (somewhere between 0 and 100% of the maximum rate) in response to the ambient temperature. At extremely low temperatures, biological activity will be exactly (or very

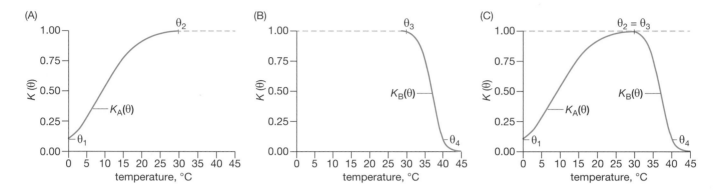

Figure 13.6 Generalized biological reaction rate curves, where the realized rate $K(\theta)$ is defined as a damping function that varies between 0 and 100% of the maximal theoretical rate, as a function of temperature. (A) The typical biological response to increasing temperatures is indicated as $K_A(\theta)$. (B) $K_B(\theta)$ reflects the negative consequences as temperatures become too great. (C) When the full range of temperature responses are combined, $K(\theta)$ describes the biological response curve across all temperatures. Note that the optimal temperature thresholds ($\theta_2 = \theta_3$) are used to define the temperature at which the maximal rate is achieved, while the minimum (θ_1) and maximum (θ_4) temperature thresholds are used to define the temperatures at which biological function has ceased (Adapted from Thornton KW & Lessem AS (1978) *Trans. Am. Fish Soc.* 107:284–287. With permission from Taylor and Francis Group.)

nearly) zero, but will typically increase in logistic fashion (approaching 100% of the maximal rate) as the temperature increases. This behavior is represented by the $K_A(\theta)$ curve in Figure 13.6A, and can be defined using Equation 13.13:

$$K_A(\theta) = \frac{K_1 e^{\gamma_1(\theta - \theta_1)}}{1 + K_1(e^{\gamma_1(\theta - \theta_1)} - 1)} \tag{13.13}$$

where

$K_A(\theta)$ = ascending reaction rate multiplier (unitless; ranges from 0.00 to 1.00)

K_1 = reaction rate multiplier near minimum temperature threshold (unitless; typically set between 0.01 and 0.10)

θ = ambient temperature (°C)

θ_1 = low-temperature threshold (°C; temperature below which activity ceases)

γ_1 = low-temperature specific rate coefficient (unitless), defined as

$$\gamma_1 = \left(\frac{1}{\theta_2 - \theta_1}\right) \cdot \ln\left(\frac{K_2 \langle 1 - K_1 \rangle}{K_1 \langle 1 - K_2 \rangle}\right) \tag{13.14}$$

θ_2 = optimum temperature (°C; temperature at which activity is maximal)

K_2 = reaction rate multiplier at optimal temperature (unitless; typically set to 0.99)

As temperatures increase beyond the optimal temperature, biological activity will become adversely affected and will decrease logistically (approaching 0% of the maximal rate at extreme temperatures). This behavior is represented by the $K_B(\theta)$ curve in Figure 13.6B, and is defined using Equation 13.15:

$$K_B(\theta) = \frac{K_4 e^{\gamma_2(\theta_4 - \theta)}}{1 + K_4(e^{\gamma_2(\theta_4 - \theta)} - 1)} \tag{13.15}$$

where

$K_B(\theta)$ = descending reaction rate multiplier (unitless; ranges from 1.00 to 0.00)

K_4 = reaction rate multiplier near maximum temperature threshold (unitless; typically set between 0.01 and 0.10)

θ = ambient temperature (°C)

θ_4 = high-temperature threshold (°C; temperature above which activity ceases)

γ_2 = high-temperature specific rate coefficient (unitless), defined as

$$\gamma_2 = \left(\frac{1}{\theta_4 - \theta_3}\right) \cdot \ln\left(\frac{K_3 \langle 1 - K_4 \rangle}{K_4 \langle 1 - K_3 \rangle}\right) \tag{13.16}$$

θ_3 = optimum temperature (°C; temperature at which activity is maximal)

K_3 = reaction rate multiplier at optimal temperature (unitless; typically set to 0.99)

Once the curves for $K_A(\theta)$ and $K_B(\theta)$ have been established, it is a simple task to combine them to yield the overall response curve $K(\theta)$ across all temperatures, as depicted in Figure 13.6C and calculated using Equation 13.17:

$$K(\theta) = K_A(\theta) \cdot K_B(\theta) \tag{13.17}$$

Now we can apply $K(\theta)$ as a reaction rate multiplier to our excretion model in order to make our zooplankton sensitive to temperature and therefore reduce their realized excretion rate from the theoretical maximum rate (described by Equation 13.10) to a more realistic rate, in response to ambient temperatures (which will vary throughout our model domain, and in time). So, after all that effort, we merely apply our calculated value of $K(\theta)$ from Equation 13.17 to our idealized model of zooplankton excretion (Equation 13.10) and redefine dN_A/dt as a more realistic excretion model now described by Equation 13.18:

$$\frac{dN_A}{dt} = \left\langle \left(\frac{0.1278\,\mu g\,N}{\mu g\,dw}\right) \cdot \sum^{Z_s}(W_Z \cdot V_Z) \right\rangle \cdot K(\theta) \tag{13.18}$$

If we plug this new formulation of dN_A/dt to update Equation 13.11, we now have a complete, temperature-dependent excretion model, defined in Equation 13.19:

$$\frac{dN_{exc}}{dt} = \left[\left\langle \left(\frac{0.1278\,\mu g\,N}{\mu g\,dw}\right) \cdot \sum^{Z_s}(W_Z \cdot V_Z) \right\rangle \cdot K(\theta) \right] + 0.05\,EN \tag{13.19}$$

Follow Your Conceptual Model: Our original conceptual model of the excretion submodel looks deceptively simple, requiring PON as our only input variable and NH_3 as the only output variable (see Figure 13.4). However, it was necessary to take multiple steps, using a solution cascade of models within models, to ultimately arrive at our solution (**Figure 13.7**). Fortunately, this solution can also be used in all the other submodels that require an estimate of dN_{exc}/dt.

Keep in mind that your conceptual model is meant to provide you with the "big picture" of how the general processes link the state variables together, and how those same state variables are used to drive all of the modeled processes. The solution cascades that are often necessary to perform (as we saw in the excretion model) involve the calculation of many different **implicit**

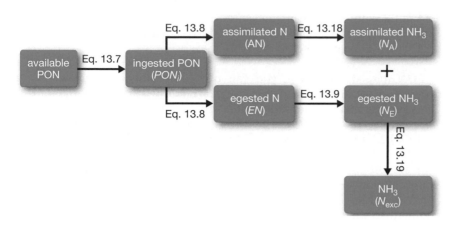

Figure 13.7 A graphical representation of the solution cascade within the excretion model, initially driven by the input variable PON, and used to calculate the output variable NH_3 (N_{exc}).

variables, which do not enjoy the same status as our state variables, but are critical to the model nonetheless. In other words, follow your conceptual model as a guide, but take care that you don't use it as a prescriptive limit on what computations are necessary to perform.

Use the Appropriate Arrays: Unlike our original formulation of excretion (see Equation 12.5), our new and improved version includes a wide variety of implicit variables in addition to our state variables. Since our input (PON) and output (NH_3) variables will likely vary in both space and time, they should each be established as a four-dimensional array. Since water temperature is also likely to vary in space and time (and is needed to calculate $K(\theta)$ within Equation 13.18), we must add it to the list of four-dimensional arrays:

$$\theta\,(x,y,z,t)$$
$$PON\,(x,y,z,t)$$
$$NH_3\,(x,y,z,t)$$

Since the values for intrinsic variables are used once, to calculate the instantaneous value of some other variable, their values are not typically stored in memory. Although they may vary in space (x,y,z), they are usually calculated anew at each timestep, so there is no need to maintain their temporal dimension.

Define the Boundary Conditions: You'll notice that Equation 13.10, which defines the rate of NH_3 excretion from assimilated nitrogen AN, does not incorporate the implicit variable AN; it is calculated solely as a function of zooplankton biomass W_z and abundance V_z for all zooplankton species Z_s in the community (which has the ingestion of PON_i already "built in" to the function described by Equation 13.18). That is an important point because we did not explicitly model the actual quantity of PON_i ingested by our zooplankton; we merely set the proportion of PON_i that could ultimately be assimilated (AN, in Equation 13.8) and later excreted as NH_3. Since our actual calculation of excreted NH_3 is decoupled from AN and PON_i, we establish a boundary condition that stipulates that the excretion of assimilated nitrogen N_A can never exceed the quantity of nitrogen actually ingested and assimilated (AN). This would involve a simple line of code in our model, which would look something like:

$$\text{if}\quad N_A > AN, \text{ let } N_A = AN$$

If we take a look at our numerical method for estimating NH_3 excretion as a result of egestion (N_E, in Equation 13.9), we can see that the equation includes an implicit variable that defines the quantity of egested nitrogen (EN). Since the solution of N_E is already dependent on the value of EN, there would be no need to establish a boundary condition for N_E.

Establish Your Refuges: It is probably an unrealistic condition that [NH_3] and/or [PON] would ever truly reach zero. In order to protect against this possibility within your model, you may wish to establish a refuge concentration for both. Assuming there are measurements of [NH_3], it may be advisable to set a NH_3 refuge to the lowest measured [NH_3] within the dataset; likewise for [PON] if the data exist.

If PON data do not exist, you can use your refuge population of the "prey" species in your model. If you can relate that refuge population to their carbon-specific biomass (that is, the amount of carbon in the refuge population), you can use the Redfield ratio (106 moles C: 16 moles N, or C:N = 6.625) to derive a first-order approximation of PON in that refuge.

Stay Within Your Mass Budget: Since the nitrogen in the nitrogen cycle comes in many forms (NH_3, NO_3, PON), it will be much easier to maintain a nitrogen budget using atomic nitrogen (at-N). Using molar quantities (rather than weights) will also make mass budgeting rather easy, because of the following relationships:

$$1 \text{ mole } NH_3 = 1 \text{ mole at-N } (17.031 \text{ g mol}^{-1})$$

$$1 \text{ mole } NO_3 = 1 \text{ mole at-N } (62.004 \text{ g mol}^{-1})$$

$$1 \text{ mole PON} = 1 \text{ mole at-N (highly variable g mol}^{-1})$$

Care should be taken to consider whether your model will include nitrogen import and export. If this is the case, the overall nitrogen budget will require you to track all import, export, and resident nitrogen throughout the duration of the model run in order to maintain mass balance. If the import and export of nitrogen is disallowed in your model domain, the amount of nitrogen in your model at the end of the run must be exactly equal to the amount of nitrogen in the domain when the model was initialized.

In our particular formulation of excretion, it is possible that a certain amount of egested nitrogen (*EN*) is neither dissolved nor remineralized (Equation 13.9). In those cases when $N_E < 0.05$ *EN*, a portion of the egested but unexcreted nitrogen could remain in the fecal material, representing a potential "loss" of available nitrogen in the model. Although this is a realistic condition (either as **refractory** PON or as PON sinking and becoming buried in the sediments), a balanced nitrogen budget would require that the unavailable PON be included in your calculations.

Likewise, it is also possible (in fact, expected) that the excretion of assimilated nitrogen (N_A) is far less than the total amount of assimilated nitrogen (*AN*). This is because some of the assimilated nitrogen is actually incorporated into the organic matrix of the consumer, either as increased biomass (growth) or invested in the production of eggs or larvae (reproduction). Any nitrogen invested in growth or reproduction will essentially remove that nitrogen from the pool of available nitrogen in the model domain, now that it is being stored in the consumer biomass.

Example 3 Revisited: Complex Nitrogen Dynamics of Assimilation/Production

Do Your Homework: As we first discussed in Chapter 12, the general equation that defines somatic growth (see Equation 12.8) is ultimately dependent on a very complex formulation that mathematically describes the organism's growth rate μ. Over a single timestep of the model, we can represent that exponential growth function as Equation 13.20:

$$B_t = B_{t-1} e^{\mu t} \tag{13.20}$$

where we can calculate the current biomass B_t as a result of growth, based on the biomass from the previous timestep (B_{t-1}). Of course, the amount of time elapsed in a single timestep (t) should be defined according to the Courant-Friedrichs-Lewy (CFL) stability criterion (see Equation 11.6).

In order to calculate the realized growth rate μ, the maximum theoretical growth rate μ_{max} for that particular organism must first be determined experimentally or gleaned from the literature. Since growth is a metabolic function, the temperature-dependent growth rate μ_θ for a particular species is naturally a function of that organism's μ_{max} and the ambient water temperature θ, which we can represent using Equation 13.21:

$$\mu_\theta = f\left(\mu_{max}, \theta\right) \tag{13.21}$$

As we saw in our earlier example in the excretion model, the Thornton and Lessem (1978) algorithm (Equations 13.13–13.17) can be used to introduce temperature dependence on all sorts of biological activity, and can be easily modified to yield a rate multiplier $K(\theta)$ that can then be applied to μ_{max} to yield Equation 13.22, which now defines the temperature-dependent growth rate μ_θ as

$$\mu_\theta = \mu_{max} \cdot K\left(\theta\right) \tag{13.22}$$

Within the **autotrophic** biomass, growth is limited by the least available growth factor, which is typically a toss-up between available light or nutrient(s). We can demonstrate this mathematically using Equation 13.23, where the realized growth rate μ is determined by the lesser value of the light-limited growth rate μ_{ll} or the nutrient-limited growth rate μ_{nl}, such that:

$$\mu = MIN \begin{cases} \mu_{ll} \\ \mu_{nl} \end{cases} \tag{13.23}$$

Since we are only considering nitrogen dynamics at the moment, we can ignore the effects of light limitation and focus solely on nutrient limitation (more specifically, nitrogen limitation). That being the case, $\mu = \mu_{nl}$.

When the full gamut of limiting nutrients is being explored, it is customary to consider $[NH_3]$, $[NO_3]$, and $[PO_4]$ as potentially limiting. Depending on the dynamics being explored, it may be necessary to include even more nutrients, such as $[NO_2]$ or others. For example, if the autotrophic biomass includes **diatoms**, we must also include dissolved silica as $[SiO(OH)_3]$. Remember that the autotrophic biomass is limited by the least available growth factor (in this case, the least available nutrient), so we can use Equation 13.24 to define the nutrient-limited growth rate μ_{nl} as:

$$\mu_{nl} = MIN \begin{cases} MAX \begin{cases} \mu_\theta \cdot \left(\dfrac{[NO_3]}{k_{NO_3} + [NO_3]} \right) \\[2em] \mu_\theta \cdot \left(\dfrac{[NH_3]}{k_{NH_3} + [NH_3]} \right) \end{cases} \\[4em] \mu_\theta \cdot \left(\dfrac{[PO_4]}{k_{PO_4} + [PO_4]} \right) \\[3em] \mu_\theta \cdot \left(\dfrac{[SiO(OH)_3]}{k_{SiO(OH)_3} + [SiO(OH)_3]} \right) \end{cases} \tag{13.24}$$

where the Michaelis–Menton kinetics of nutrient limitation are sensitive to the organism's half-saturation constant k for each nutrient in question. Note that the values for k are different for each species under investigation and must be determined experimentally (or gleaned from the literature). Since we are focusing our attention on nitrogen limitation, we can ignore phosphate and silica limitation, so Equation 13.24 can be rewritten as Equation 13.25:

$$\mu_{nl} = MAX \begin{cases} \mu_\theta \cdot \left(\dfrac{[NO_3]}{k_{NO_3} + [NO_3]} \right) \\[3em] \mu_\theta \cdot \left(\dfrac{[NH_3]}{k_{NH_3} + [NH_3]} \right) \end{cases} \qquad (13.25)$$

Now we have everything we need to calculate the nitrogen-specific growth of the autotrophic biomass in our model. We simply use $[NO_3]$, $[NH_3]$, and μ_θ in Equation 13.25 to calculate μ_{nl} (which, for our purposes, is the same as μ). Using the autotrophic biomass from the previous timestep (B_{t-1}), we can use Equation 13.20 to calculate the current biomass B_t.

In very simple terms, growth (dB/dt) is the amount of biomass added over one timestep. Using Equation 13.26, this is calculated as:

$$\frac{dB}{dt} = B_t - B_{t-1} \qquad (13.26)$$

If we are using carbon as our model currency, growth is expressed as some amount of organic carbon added per timestep. Since we are interested in the amount of nitrogen that is being removed from the $[NO_3]$ or $[NH_3]$ pools and added to the biomass pool, we will need to convert from carbon-specific growth to nitrogen-specific growth.

The most straightforward way to accomplish this is to assume that nitrogen is accumulated in the autotrophic biomass according to the Redfield ratio (C:N = 106 mol C : 16 mol N). Then our last step is a very simple one, using Equation 13.26 to compute the quantity of nitrogen associated with the growth of our autotrophic biomass, so that Equation 13.27 now defines dN_{prd}/dt as:

$$\frac{dN_{prd}}{dt} = \left(\frac{dB}{dt} \right) \cdot \left(\frac{N}{C} \right) \qquad (13.27)$$

Note that our solution for dN_{prd}/dt was not terribly complicated, but it was certainly not possible to compute with a single equation. As we can now see from the solution cascade within the assimilation/production model (**Figure 13.8**), our input variables (NO_3 and NH_3) are neatly coupled to the model solution: our output variable (PON).

Now we have a simple method to compute nitrification dynamics (as dN_{nit}/dt, in Equation 12.4) and more complex methods to compute excretion (as dN_{exc}/dt, in Equation 13.19) and production (as dN_{prd}/dt, in Equation 13.27). According to the master equations that define all of the dynamics we want to simulate in our simplified nitrogen cycle (Equations 13.4–13.6), we now have everything we need to run our first complex model.

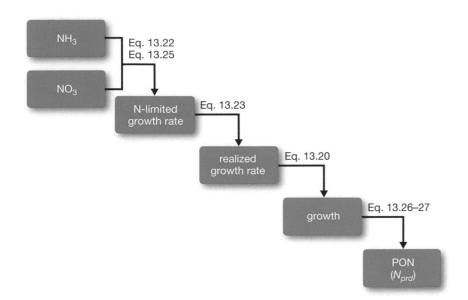

Figure 13.8 A graphical representation of the solution cascade within the assimilation/production model, initially driven by the input variables NH₃ and NO₃, and used to calculate the output variable PON (N_{prd}).

Follow Your Conceptual Model: If we take another look at our simplified conceptual model of assimilation/production (**Figure 13.9**), we can see that only two state variables are necessary as input variables: [NO_3] and [NH_3]. Our sole output variable is represented as organic nitrogen, or [PON]. These are consistent with our current formulations, where [NO_3] and [NH_3] are used to drive the function that defines the nitrogen-limited growth rate of organic material (μ_{nl}, in Equation 13.25), and the overall result of our production model (that is, the output variable) yields an estimate of growth in terms of organic nitrogen (dN_{prd}/dt, in Equation 13.27).

Use the Appropriate Arrays: Since our input (NH_3, NO_3) and output (PON) variables will likely vary in space and time, they should each be established as a four-dimensional array. Since water temperature is also likely to vary in space and time (and needed to calculate μ_θ), we must add it to the list of four-dimensional arrays:

$\theta\,(x,y,z,t)$
$NH_3\,(x,y,z,t)$
$NO_3\,(x,y,z,t)$
$PON\,(x,y,z,t)$

Since the values for the autotrophic biomass B also vary in space, and we have determined that we must be able to compare the biomass from one timestep to another, it too should be established as a four-dimensional array. However, if we wished to create an even more complicated model, one that contained multiple species (sp) of autotrophs, we would need to expand the biomass array to include a species dimension (yielding a five-dimensional array):

$B\,(x,y,z,t,sp)$

Keep in mind that the values for μ_θ and μ_{nl} are intrinsic variables that must be calculated anew at each grid point in the domain, and during each timestep; therefore, there is no need to keep track of their spatial or temporal dimensions. However, if our model contained more than one autotrophic species,

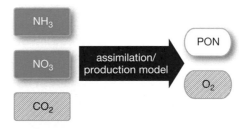

Figure 13.9 A variable-oriented perspective of the assimilation/production model, indicating the necessary input variables used to drive the model, as well as the output variables. Note that the hashed variables are ignored in the simplified model, as per the current example.

we would have to at least keep track of the different values of μ_θ and μ_{nl} for each species in our simulation:

$$\mu_\theta(sp)$$
$$\mu_{nl}(sp)$$

Although the half-saturation constants k cited in the nutrient limitation equations (Equations 13.24 and 13.25) are fixed values and therefore do not vary in space or time, they do vary according to species. Hence, we must add them to the list of one-dimensional, species-specific arrays:

$$k_{NO_3}(sp)$$
$$k_{NH_3}(sp)$$
$$k_{PO_4}(sp)$$
$$k_{SiO(OH)_3}(sp)$$

Define the Boundary Conditions: Since growth is driven by $[NO_3]$ and/or $[NH_3]$, the availability of those nutrients will automatically establish the boundary conditions that apply to nitrogen-specific growth. According to Equations 13.24 and 13.25, as nutrient concentrations approach zero, μ_{nl} will also approach zero. Conversely, even in the most nutrient-rich environments, μ_{nl} will never exceed μ_θ.

Since our assimilation/production model was focused on nitrogen dynamics, we ignored a significant source of variability in the realized growth rate μ of the autotrophic biomass: the role of light limitation. By ignoring light limitation, our current formulation of growth effectively assumes that the bulk of the autotrophic biomass is continually light-saturated—a highly unrealistic assumption. Therefore, the calculated growth rates μ will be much higher than most natural populations would ever enjoy. If the modeler wished to avoid inclusion of a detailed light-limitation scheme, it would be necessary to cap the realized growth rates μ at some value more consistent with natural environs, if such information is available in the scientific literature. A far better solution would be to include a light model, so the light-limited growth rate μ_{ll} could be calculated explicitly and then used in Equation 13.23 to determine the least-limiting growth factor within the autotrophic biomass.

Establish Your Refuges: As discussed earlier for the excretion model, it is probably unrealistic that $[NH_3]$, $[NO_3]$, and/or $[PON]$ would ever truly reach zero. Assuming there are measurements of $[NH_3]$ and $[NO_3]$, it may be advisable to set refuges to the lowest measured $[NH_3]$ and $[NO_3]$ within the dataset; likewise for $[PON]$ if the data exist.

Stay Within Your Mass Budget: The same concerns we expressed in the mass budgeting of nitrogen in the excretion model hold true for the assimilation/production model as well. With the completion of the assimilation/production model, we have closed the loop and all of our model linkages are complete, so our nitrogen has no choice but to be conserved in our model domain either as NH_3, NO_3, or PON (unless you choose to allow nitrogen flux at the boundaries of your model domain, in which case mass budgeting becomes much more challenging).

Moving On to Bigger and Better Models

So where do we go from here? So far we've been focusing on a relatively small portion of the overall modeling picture, using a simplified version of the nitrogen cycle to demonstrate how to construct simple and complex submodels alike. Depending on the grandiosity of your model, we still have a long way to go, as there are a multitude of dynamics far beyond the scope of the nitrogen cycle that may be important to your modeling efforts. Or perhaps the nitrogen cycle is the least of your concerns and you'd like to model something entirely different. Be that as it may, the next step is to take the skills you've learned thus far and apply them in the construction of a large system model, primarily by linking submodels to other submodels, and then linking those to others still; on and on it goes. The universe is a big place—even your tiny little piece of it—so we'll definitely have our work cut out for us as we venture forth into large-system modeling.

References

Bertalanffy L von (1957) Quantitative laws in metabolism and growth. *The Quarterly Review of Biology* 32(3):217–231.

Droop MR (1973) Some thoughts on nutrient limitation in algae. *Journal of Phycology* 9: 264–272.

Liu Y, Lin Y-M, & Yang S-F (2003) A thermodynamic interpretation of the Monod equation. *Current Microbiology* 46:233–234.

Milroy SP, Dieterle DA, He R, et al. (2008) A three-dimensional biophysical model of *Karenia brevis* dynamics on the west Florida shelf: A look at physical transport and potential zooplankton grazing controls. *Continental Shelf Research* 28(1):112–136.

Park RA & Clough JS (2009) AQUATOX: Modeling Environmental Fate and Ecological Effects in Aquatic Ecosystems, Volume 2: Technical Documentation. United States Environmental Protection Agency (US-EPA). Report EPA-823-R-09-004.

Thornton KW & Lessem AS (1978) A temperature algorithm for modifying biological rates. *Transactions of the American Fisheries Society* 107(2):284–287.

Walsh JJ, Weisberg RH, Dieterle DA, et al. (2003) Phytoplankton response to intrusions of slope water on the west Florida shelf: Models and observations. *Journal of Geophysical Research* 108(C6):3190–3208.

Wen YH & Peters RH (1994) Empirical models of phosphorus and nitrogen excretion rates by zooplankton. *Limnology and Oceanography* 39(7):1669–1679.

Further Reading

Bender EA (2000) An Introduction to Mathematical Modeling. Dover Publications.

Dale VH (ed) (2002) Ecological Modeling for Resource Management. Springer-Verlag.

Fennell W & Neumann T (2014) Introduction to the Modelling of Marine Ecosystems, 2nd ed. Elsevier.

Grant WE & Swannack TM (2007) Ecological Modeling: A Common-Sense Approach to Theory and Practice. Blackwell Publishing.

Haefner JW (1996) Modeling Biological Systems: Principles and Applications. Springer-Science.

Jopp F, Reuter H, & Breckling B (eds) (2011) Modelling Complex Ecological Dynamics: An Introduction into Ecological Modelling for Students, Teachers & Scientists. Springer-Verlag.

Jorgensen SE & Bendoricchio G (2001) Fundamentals of Ecological Modelling, 3rd ed. Elsevier Science.

Jorgensen SE, Chon T-S, & Recknagel FA (eds) (2009) Handbook of Ecological Modelling and Informatics. WIT Press.

Murray JD (2002) Mathematical Biology I: An Introduction, 3rd ed. Springer-Verlag.

Murray JD (2003) Mathematical Biology II: Spatial Models and Biomedical Applications, 3rd ed. Springer-Verlag.

Royle JA & Dorazio RM (2008) Hierarchical Modeling and Inference in Ecology. Elsevier.

Shiflet AB & Shiflet GW (2014) Introduction to Computational Science: Modeling and Simulation for the Sciences, 2nd ed. Princeton University Press.

Smith J & Smith P (2007) Environmental Modelling: An Introduction. Oxford University Press.

Chapter 14

Modeling Large System Dynamics

"Space and time are the framework within which the mind is constrained to construct its experience of reality."– Immanuel Kant

This chapter marks the culmination of our modeling discussions. Very soon you'll be designing your very own large system models. As we have demonstrated in earlier chapters, the ever-increasing complexity of our models simply reminds us to take one step at a time; that a complex model is merely an assemblage of much simpler models, linked together. The real challenge lies in preventing our linked models from becoming a tangled mess.

Toward that end, it is always best to sketch out a conceptual diagram of our model. In fact, a thorough and organized conceptual diagram of a large systems model is essential. But where to start?

Although large aquatic system models are so complex they defy simple generalization, more than likely they will incorporate at least four basic modeling elements: a hydrodynamic submodel to push all that water around, a biogeochemical submodel to keep track of vital chemical species and maintain mass balance of suspended and dissolved constituents, a submodel to elucidate the radiative transfer of heat and/or light into the ocean, and an ecological submodel to simulate the critical biological processes of biomass production. That's where we start.

Key Concepts

- Large aquatic system models are generally organized as a network of linked hydrodynamic, biogeochemical, radiative transfer, and ecological submodels.
- Hydrodynamic submodels are designed to simulate diffusive and advective transport (fluid dynamics), heat fluxes, and water column structure.
- Biogeochemical submodels are used to maintain mass balance and quantify the bacterial remineralization of vital chemical constituents.
- Radiative transfer submodels are used to determine thermal transfers to the ocean and the below-surface light field as it relates to biological production.
- Ecological submodels are used to explore paradigms of primary and secondary production, fisheries production, and food web connectivity.

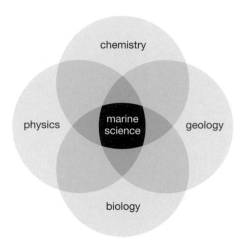

Figure 14.1 Marine science (oceanography) is a union of the four major disciplines of the natural sciences: biology, chemistry, geology, and physics.

Putting It All Together

As we have seen in our earlier discussions of modeling, there's actually a lot to consider when constructing a model: conceptual diagrams, spatial grids, timesteps, dimensional arrays, state variables, forcing functions, boundary conditions, mass budgets, and so on. The list can seem quite daunting, even for relatively simple models. For instance, the nitrogen cycle model we've been using as our example is but one small aspect of the larger reality of biogeochemical cycling. After all, there are other critical nutrients in aquatic systems beyond nitrogen. How should we include them in our model? And what else are we missing?

Before we head down that road, it's important to consider just how inclusive you want your model to be. If the model is meant to simulate large aquatic systems like the coastal or open ocean, then it will be necessary for you construct a highly complex model that integrates all of the major disciplines of the natural sciences: biology, chemistry, geology, and physics (**Figure 14.1**). Keep in mind that, in reality, the biology and chemistry and geology and physics of the ocean are not disconnected from each other; within the context of marine science, they are intimately coupled in very complex ways. That means your model will have to include all of the relevant forcing functions you can possibly think of, in each of these disciplines, and fully integrated with one another.

Let's pause for a moment to really emphasize the importance of model integration. Very often, models are constructed by professionals who have expertise in one particular emphasis area; rarely does one person possess complete knowledge and skill to build all the hydrodynamic, biogeochemical, radiative transfer, and ecological models. So practicality demands that we integrate our model (built from our own expertise) with preexisting models that were built by other experts. Resist the temptation to use the **multidisciplinary** approach of simply combining the results of several different and disconnected models. You will find that your results will be quite different if you instead take great care to employ an **interdisciplinary** method, where the models "talk to" and provide immediate feedback to each other.

Fortunately, the process of crafting (and integrating) a large system model is essentially the same as that for simpler models: all we really need is an input and a relevant mathematical forcing function to estimate (simulate) the model output (**Figure 14.2**). The only real difference in a large systems

Figure 14.2 Large system models are constructed by linking several submodels together, so that the solution of one submodel can be used to calculate the solution of another, then another. In this fashion, very large and complex models can be constructed by simply linking many individual forcing functions together. Dashed lines indicate where each submodel solution can be checked for goodness of fit, so all of the coupled submodels can be revised and improved (if necessary).

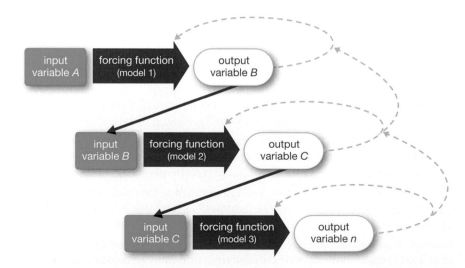

model is that we need a lot of forcing functions to simulate all the dynamics going on in our large system, and we need to allow the solution of one forcing function to provide feedback to the other forcing functions in our model, as appropriate.

That makes for a very complicated web of cross-connected forcing functions, each with its own input and output variables. Sometimes these variables provide feedback to a multitude of other forcing functions within the larger model, so these linkages can become overwhelming in a hurry. In order to make sense of it all, and to keep track of all the linked forcing functions within a large systems model, it is absolutely critical to maintain a detailed conceptual diagram of the overall model. Most large system models consist of four main elements (**Figure 14.3**):

1. Geophysical model (primarily for hydrodynamics)
2. Biogeochemical model (primarily for chemical fluxes)
3. Radiative model (primarily for light and heat fluxes)
4. Ecological model (primarily for biological responses)

It is not absolutely necessary that all of these elements be included in every model; rather, it is far more important that the model be constructed in keeping with the research interests of the model's creator. For example, hydrodynamic models of ocean circulation would not be significantly affected by biological components, so their inclusion in a physical model designed to simulate coastal ocean currents would be irrelevant. However, if we were more focused on the biology of the system, an ecological model might require the inclusion of a hydrodynamic model, as our biological

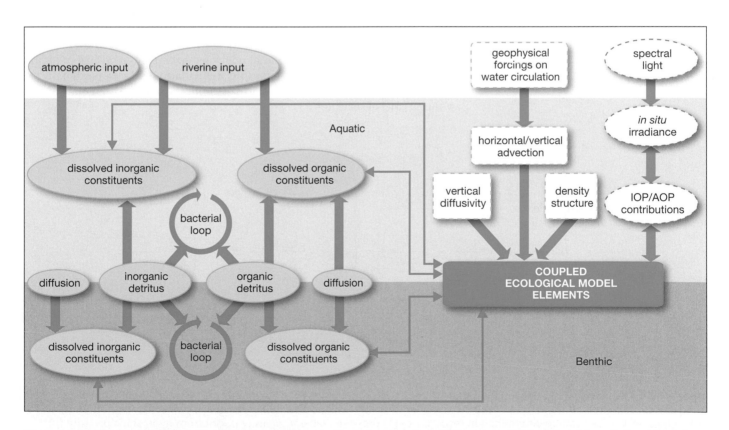

Figure 14.3 A conceptual diagram of the hydrodynamic (dotted rectangles), biogeochemical (solid ovals), radiative (dotted ovals), and ecological (solid rectangle) model elements that are present in most large system models.

components are likely to be very responsive to ocean circulation dynamics. Thus, it is important to customize the model elements to fit the purpose of the model.

In other words, you should focus your attention only on those model elements that are most relevant to the star of your show. That being said, it is helpful for us to briefly review some of the more common variables and forcing functions that serve as "minimum requirements" for most large system ocean models.

Typical Hydrodynamic Model Elements

Any geophysical model of hydrodynamics (or more specifically, of ocean circulation dynamics) will possess the same, fundamental model elements that define the physics of the ocean. Since most ocean circulation models are primarily interested in simulating velocity fields (to include the speed and direction of fluid motion), these models employ the classic **Navier–Stokes equations**: a collection of nonlinear partial differential equations that solve four-dimensional flows (three-dimensional in space + time) in the ocean by assuming the conservation of mass, energy, and momentum of an incompressible fluid (seawater) subject to variable forces (**Figure 14.4**).

Hydrodynamic Models Are Solved by Defining the Forces at Work Within the Spatial and Temporal Domain of Fluid Motion

In order for hydrodynamic models to solve the Navier–Stokes equations of fluid motion, it is necessary that the physical model contain a variety of physical elements, all of which are ultimately related to the balance of forces acting on the water, in space and time (according to our chosen frame of reference). The spatial scales of fluid motion are usually defined by the model grid, where the horizontal (x,y) and vertical (z) spatial scales within the model are fixed. This also allows the model to perform distance (one-dimensional), area (two-dimensional), and volume (three-dimensional) calculations, as these sorts of spatial relationships are often critical in defining other ocean properties. For example, a calculation of seawater density $(\rho,$ in units of kg m$^{-3})$ would be impossible without the ability to calculate a unit volume (m^3) of seawater.

$$
\begin{array}{l}
\text{easting} \\ \text{component}
\end{array}
\left\langle \frac{\partial u}{\partial t} \right\rangle + \left\{ u\frac{\partial u}{\partial x} + v\frac{\partial u}{\partial y} + w\frac{\partial u}{\partial z} \right\} = \left(-\frac{1}{\rho o} \cdot \frac{\partial p}{\partial x} \right) + (fv) + \left(\frac{1}{\rho_o} \cdot \frac{\partial \tau_x}{\partial z} \right) - Ju
$$

$$
\begin{array}{l}
\text{northing} \\ \text{component}
\end{array}
\left\langle \frac{\partial v}{\partial t} \right\rangle + \left\{ u\frac{\partial v}{\partial x} + v\frac{\partial v}{\partial y} + w\frac{\partial v}{\partial z} \right\} = \left(-\frac{1}{\rho o} \cdot \frac{\partial p}{\partial y} \right) + (fu) + \left(\frac{1}{\rho_o} \cdot \frac{\partial \tau_y}{\partial z} \right) - Jv
$$

$$
\begin{array}{l}
\text{vertical} \\ \text{component}
\end{array}
\left\langle \frac{\partial w}{\partial t} \right\rangle + \left\{ u\frac{\partial w}{\partial x} + v\frac{\partial w}{\partial y} + w\frac{\partial w}{\partial z} \right\} = \left(-\frac{1}{\rho o} \cdot \frac{\partial p}{\partial z} \right) + g_\Omega + \quad 0 \quad - Jw
$$

$$
\underbrace{}_{\begin{array}{c}\text{particle}\\\text{acceleration}\end{array}} \quad \underbrace{}_{\begin{array}{c}\text{field}\\\text{acceleration}\end{array}} \quad \underbrace{}_{\begin{array}{c}\text{pressure gradient}\\\text{force}\end{array}} \quad \underbrace{}_{\begin{array}{c}\text{Coriolis}\\\text{"force"}\end{array}} \quad \underbrace{}_{\begin{array}{c}\text{wind}\\\text{stress}\end{array}} \quad \underbrace{}_{\begin{array}{c}\text{miscellaneous}\\\text{friction}\end{array}}
$$

Figure 14.4 The Navier–Stokes equations are most commonly employed in hydrodynamic models to compute flow fields using the conservation of momentum. Vector components of flow in the x-direction (u, easting), y-direction (v, northing), and z-direction (w, vertical) can be determined in response to (1) pressure gradient forces wrought from hydrostatic pressure (p) and fluid density (ρ_0); (2) the latitudinally sensitive Coriolis "force" (f) and the Earth's gravitational force (g_Ω); and (3) wind stress (τ) and other transient frictional forces (J).

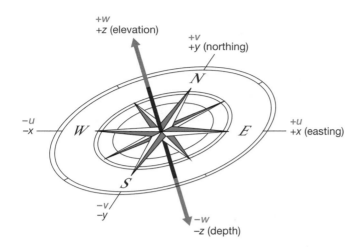

Figure 14.5 The Cartesian coordinate system most commonly adopted to define the spatial dimensions of ocean models, where the (x,y)-coordinates are defined by using the traditional compass directions, and the z-coordinate represents the vertical dimension. In similar fashion, the vector components of any three-dimensional flow can be defined using the same coordinate system.

The vector components of water currents (that is, current velocity) are also critical, because they describe both the direction and speed of flow, relative to the spatial domain of the model. These vector components are typically defined according to the cardinal direction of the flow ($+u$ = eastward flow, $+y$ = northward flow, $+w$ = upward flow), whereas the speed of the current is simply defined by the distance traveled per unit time (m s^{-1}, or perhaps cm s^{-1} for slower-moving currents). In this fashion, any three-dimensional flow field can be resolved by its vector components using Cartesian coordinates (**Figure 14.5**).

The temporal scales of water movement are usually defined according to the CFL stability criterion (see Equation 11.6) and serve as the model's timestep. Because the spatial scales in the horizontal are usually much larger than in the vertical, some physical models employ mode-splitting schemes so that horizontal phenomena are calculated using a different timestep than the vertical phenomena. For example, a shallow coastal sea may stretch for hundreds of kilometers in the horizontal dimension, but may be only 10 m (0.01 km) deep in the vertical. In this scenario, it may be necessary to have two completely different CFL stability criteria for water movement, because even slow-moving currents moving in the vertical direction can be significant with a spatial scale of only 0.01 km when compared to the horizontal scale of 100+ km.

Bathymetry and sea surface elevation are two sides of the same coin, but both are ultimately related to defining the vertical dimension of the spatial domain. Bathymetry essentially represents the distance to the bottom of the ocean, if we imagine ourselves on the surface of the ocean, looking down. From a modeling perspective, the depth of the water at any grid node is defined by the bathymetry at that specific location, according to a chosen bathymetric datum (for example, the depth relative to mean sea level). Spatially variable bathymetry is also used to define complex coastlines, where the height of the overlying water column becomes zero.

If we flip that perspective upside down, we might be able to imagine ourselves at the very bottom of the ocean, looking up. If we were concerned about the distance to the surface, we would in essence be defining the height of the water column that was sitting on top of us: this is exactly what is defined as sea surface elevation. Sometimes defining ocean depth by the sea surface elevation is a better representation of the true ocean because the bottom (that is, the bathymetry) of the ocean doesn't move, but the surface certainly does.

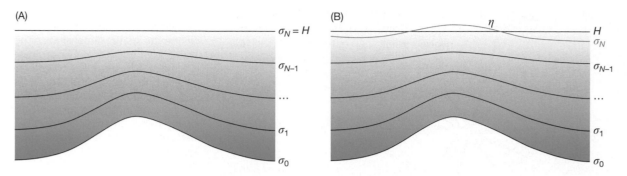

Figure 14.6 Sigma (σ) layers are often used in hydrodynamic models to subdivide the vertical dimension into discrete layers that follow the bottom topography. In certain instances, it may be possible to simplify the vertical dimension into fixed layers that assume that the ocean surface is rigid (A) and does not vary from the bathymetric datum (*H*). However, a more realistic condition (B) is one that allows the surface to freely respond to time-dependent height anomalies (η) above or below the bathymetric datum.

Some physical models use a "rigid lid" strategy where the ocean surface is not allowed to vary from the bathymetric datum (**Figure 14.6A**), so using the bathymetry to define the vertical dimension of the ocean may be a perfectly reasonable method. However, most modern ocean models employ a more realistic "free-surface" formulation (**Figure 14.6B**), where the momentary sea surface elevation (η) is calculated in terms of its +/– anomaly from the bathymetric datum (*H*). Most modern hydrographic models divide the vertical dimension into a number of sigma (σ) layers that are meant to replicate a free ocean surface but still follow the bottom topography, so each sigma layer can be defined using Equation 14.1:

$$\sigma_N = \frac{z_n - \eta}{H + \eta} \tag{14.1}$$

The Unequal Distribution of Mass, as It Relates to Water Density, Creates Pressure-Gradient Forces That Influence Water Movement

Without a doubt, the most critical variables in any physical model are θ (*in situ* temperature) and S (salinity). From these, the **equation of state** (see Equation 4.12) can be used to calculate the density of any water mass (ρ), which in turn can be used to calculate hydrostatic pressure and any pressure-gradient forces that are relevant to ocean circulation dynamics. Temperature and salinity are also necessary for calculations of entropy, water column stability, buoyancy, diffusion, mixing, and many other components critical to fluid dynamics.

In order to maintain mass balance within an oceanic model, it is often necessary to include precipitation fluxes (and other freshwater sourcings, such as river effluent), as well as evaporation. Each of these will affect the model's salt/water balance and can be critically important in investigating localized salt dilution and concentrating mechanisms. In coastal models, the influence of riverine and/or estuarine outfall can significantly impact coastal circulation regimes caused by the resultant density gradients, which in turn create pressure gradient forces that must be incorporated into the Navier–Stokes equations.

Although heat fluxes can influence evaporation rates, these concerns are usually subordinate to the bulk heat transfer to the ocean surface and the dissipation of heat throughout the water column. This ultimately affects water temperatures throughout the water column, and thus affects its entire density structure (again, affecting pressure gradient forces). Since the bulk circulation of the ocean is intimately tied to paradigms of heat transfer and its effects on fluid density, it should come as no surprise that heat fluxes are a critical component to any ocean circulation model.

Frictional Forces in the Ocean Play a Significant Role in Water Movement

Some of the primary forcings to initiate flow are ultimately related to the speed, direction, and **vorticity** of winds acting on the surface of the ocean. This **wind stress** enters the ocean due to the frictional interactions taking place only at the sea surface, so the vector components of friction (F_x, F_y) are ultimately what spawn the wind-driven currents.

The differential **viscosity** of seawater, in both the horizontal and vertical dimensions, is related to the capacity for turbulent mixing within a particular water mass. Although these terms are typically quite minor in comparison with other forcing functions, the long-term stability of ocean models often depends on a realistic formulation of the "effectiveness" of ocean mixing, which is ultimately related to viscosity as a type of frictional force included in the "catch-all" J-term in the Navier–Stokes equations (see Figure 14.4).

Fundamental to all models of large-scale fluid dynamics, the Coriolis "force" (f) is an apparent force that is imparted to all water masses that, over large spatial scales, will affect the direction and speed of the water parcel's motion because these flows are occurring on the surface of a rotating sphere. Although the angular velocity of the Earth (Ω) never changes, the rotational velocity of a fixed point on the Earth's surface will vary as a function of latitude (ϕ), where the rotational velocity of the Earth is zero at the poles and maximal at the equator. Thus, the apparent Coriolis force affects fluid movement as a function of latitude, so the vector components of flow must compensate for the Coriolis effect according to Equation 14.2:

$$f = 2\Omega \sin \phi \qquad (14.2)$$

where $\Omega = 7.29 \times 10^{-5} \text{ sec}^{-1}$.

As water masses flow past each other, frictional stresses due to current shear will result. Since the Navier–Stokes equations assume that all energy and momentum must be conserved, frictional losses due to current shear must also be included to compensate for these potential energy and momentum losses. This is particularly important at the solid boundaries of the ocean bottom and at the coastline, where most models assume a "no-slip" boundary (where velocities must be zero, either parallel with or perpendicular to a solid boundary such as the coastline or the ocean floor). Thus, solid boundaries represent a significant loss of momentum, for which there must be compensation elsewhere in the model if momentum is to be conserved.

The Influence of Tides Is as Varied as It Is Powerful

Depending on the system and the region under investigation, tidal forcings can be as significant to the hydrodynamics as wind forcings (if not more so). Generally speaking, tides will affect the sea surface elevation (η) as well as the horizontal vector components of velocity (u,v). Since tide-generating forces are caused as a result of the complex gravitational attractions between the Earth and other heavenly bodies in our solar system (most significantly the Sun and the Moon), hyperaccurate tidal models necessarily involve astrophysics and can be quite complex. For most modeling applications, a simplified tidal model (limited to diurnal and semidiurnal lunar and solar tide constituents) will usually suffice.

Typical Biogeochemical Model Elements

Biogeochemical models are used primarily to simulate the natural sourcing and recycling dynamics of various chemical constituents in seawater. Although strict geochemical models focus on the consequences of physical and chemical weathering and the **diagenesis** of crustal materials, the inclusion of biological weathering (as bacterial **remineralization**) is most prevalent in ecosystem models.

In most biogeochemical models, focus is most commonly placed on determining these chemical interconversions:

1. Particulate ↔ dissolved forms
2. Organic ↔ inorganic forms

In some specialized applications, various radioactive and stable isotopes can be used for additional analyses of biogeochemical dynamics within this context:

3. Selective isotopic enrichment ↔ impoverishment

The Fundamentals of Chemistry Are the Backbone of Biogeochemical Models

Not surprisingly, temperature θ and salinity S are among the most critical state variables in any biogeochemical model. Since temperature is so intimately tied to the thermodynamics of chemical reactions, it will also affect weathering and diagenetic conversion rates. Similarly, temperature also has a significant influence on determining the solubility and diffusion rates of dissolved molecules. Of course, the concentration and variety of dissolved salts in the ocean (what we would define as salinity) is also important in the calculation of solubilization, crystallization, ion activity, and conductance of ocean water.

Additionally, biogeochemical models can benefit from the inclusion of pH (H^+ ion activity), especially as it relates to ocean acidification dynamics and various weathering and diagenetic conversions that are accomplished via **hydrolysis** (acid/base reactions). Simulations that include dissolved oxygen (DO) are also quite valuable, as DO concentrations define the reduction/oxidation (**redox**) potential of ocean waters and can be used to define biochemically relevant conditions of hypoxia.

Inorganic Chemicals Define the Chemical Nature of Seawater, Divorced from Biology

Inorganic particulate matter is typically considered to be suspended sediments of **terrigenous** origin. These sediments are most often quantified as total suspended sediment (TSS, in units of mg L^{-1}). Within more complex biogeochemical models, it may be useful to characterize TSS fractions by their grain size, which can be an effective indicator of geologic origins and also help define particle sinking velocities (when sediment accumulation rates are important). Beyond tracking the geochemical fate of suspended particulates, sediment loads will also affect the absorption and scattering of light within the water column (as discussed later in the chapter).

In most biogeochemical applications, the relative content and solubility of various important ions within the particulate mineral matrix (such as

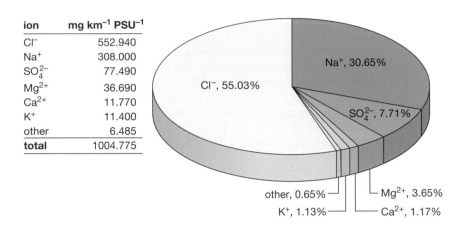

ion	mg km^{-1} PSU^{-1}
Cl$^-$	552.940
Na$^+$	308.000
SO$_4^{2-}$	77.490
Mg^{2+}	36.690
Ca^{2+}	11.770
K$^+$	11.400
other	6.485
total	1004.775

Figure 14.7 Relative abundance of conservative ions in seawater (mg kg^{-1} seawater), as a function of salinity (PSU). Note that while the total salt content will vary with salinity, the relative proportions of conservative ions within the salt mixture will remain the same.

phosphate, silicate, and heavy metals like iron) can be used to estimate ion release from the particulate inorganic fraction into the dissolved inorganic fraction. Occasionally the reverse occurs, whereby dissolved materials are adsorbed to suspended sediment particles, removing them from the water column and relocating them in the deposited sediments (due to sinking). In coastal regions heavily influenced by suspended sediment loads, it is often necessary to include these dynamics of sediment diagenesis and fate.

Since the relative proportions of individual salt ions in the world ocean are constant regardless of salinity (Figure 14.7), it is not often necessary to model the ion concentrations of mundane marine salts. Beyond the explicit modeling of salinity, the following dissolved inorganic constituents of seawater are quite commonly included in biogeochemical models, primarily for their significance in coupled ecological models:

H$^+$, to define pH

O$_2$ (dissolved oxygen gas)

CO$_2$ (dissolved carbon dioxide gas)

H$_2$CO$_3$ (carbonic acid)

HCO$_3^-$ (bicarbonate ion)

CO$_3^{2-}$ (carbonate ion)

DIC (dissolved inorganic carbon, as CO$_2$ + H$_2$CO$_3$ + HCO$_3^-$ + CO$_3^{2-}$)

NH$_4^+$ (ammonium ion)

NO$_2^-$ (nitrite ion)

NO$_3^-$ (nitrate ion)

DIN (dissolved inorganic nitrogen, as NH$_4^+$ + NO$_2^-$ + NO$_3^-$)

PO$_4^{3-}$ (phosphate ion)

DIP (dissolved inorganic phosphorus, as PO$_4^{3-}$)

H$_3$SiO$_4^-$ (orthosilicic acid)

DSi (dissolved silica, as SiO$_2$(OH)$_2^{2-}$ + SiO(OH)$_3^-$ + Si(OH)$_4$)

In recent years, the dynamics associated with iron availability (Fe^{2+}/Fe^{3+}) and its significance on nitrogen fixation (N$_2$ → DIN) have become increasingly favored for inclusion in modern biogeochemical models. Likewise, the increasing significance of ocean acidification and the effects on calcium carbonate hydrolysis might beg the inclusion of Ca^{2+}.

Organic Chemicals in Seawater Are the Calling Cards of Life

Organic particulates are those particles in the ocean that are, or recently once were, associated with living organisms. Living organic particulates can range from the smallest biological units (like viruses and bacteria) to the largest (such as whales), and everything in between. As a practical matter, living organic particulates are usually handled by a coupled ecological model rather than a biogeochemical model, since biological processes still determine the fate of the organism's chemical constituents while it is alive.

However, all of the nonliving organic particulates (collectively called **detritus**) behave somewhat similarly to inorganic particulates and are generally included in a biogeochemical model to determine the leaching and adsorption dynamics of their various chemical constituents. Additionally, detrital particles are subject to bacterial remineralization, a microbial process by which bacteria are capable of partially metabolizing the organic material within the detritus and thereby recycle a portion of the detrital constituents back into the dissolved inorganic fraction.

Within detrital particles, the particulate organic fractions of carbon, nitrogen, and phosphorus (POC, PON, and POP, respectively) are the focus of most biogeochemical models, simply because these are readily remineralized back to dissolved inorganic carbon (DIC, usually as CO_2), nitrogen (DIN, typically as NH_4^+), and phosphorus (DIP, as PO_4^{3-}) by naturally occurring bacteria. Of course, not all POC/PON/POP can be remineralized. Generally speaking, fresh detritus is quite reactive (or **labile**) and will be more easily converted back to DIC/DIN/DIP. In contrast, the more aged detrital particles possess organics that are far less reactive (or **relict**) and are more likely to retain that POC/PON/POP until it is removed from the system due to sinking or burial.

Not all organic molecules remain as particulates; the more soluble forms (such as carbohydrates and amino acids) can easily be leached from detrital particles and enter the dissolved organic fraction. Since most soluble organic molecules are also quite labile, they are swiftly consumed by other biologics (primarily bacteria) within the water column. Thus, it is usually necessary to link both the dissolved organic and the dissolved inorganic constituents from the biogeochemical model to a coupled ecological model.

As an added complication, some of these soluble organic molecules are strongly colored and can affect water clarity. Colored (or chromophoric) dissolved organic material (CDOM) is particularly significant in coastal waters, which are heavily influenced by riverine or estuarine effluent. Since much of the riverine CDOM is also relict, it can be quite persistent and can dramatically affect light penetration in coastal waters. Hence, the CDOM fractions within the dissolved organic constituents should also be coupled to the radiative model to help determine light transmissivity.

Bacteria Provide a Fundamental Link Between the Organic and Inorganic Chemistry of the Sea

Often referred to as the "microbial loop," all biogeochemical models should possess some capability to simulate the microbial consumption of particulate and dissolved organic material and their capacity to recycle organics back to their dissolved inorganic forms. This can be done as a simple recycling rate within the biogeochemical model, or it can be explicitly modeled using bacterial cells and their metabolic activities within a coupled ecological model.

Since bacterial abundances are not routinely measured, it is often difficult to explicitly model bacterial populations (unless that is the preferred purpose of the model). In most cases, bacterial remineralization rates are intimately tied to (1) *in situ* temperatures (θ) and (2) the availability of particulate and dissolved organic molecules as "food stocks" for the bacteria. Since these variables are required elements within the biogeochemical model, it is often a more practical solution to simply estimate remineralization rates as a function of ambient temperature and/or the concentrations of available organics (similar to the method we used when working through our nitrification example in Chapter 12).

Of course, it is certainly possible to explicitly model bacteria and their metabolic processes, but such an exhaustive treatment of bacterial processes would be more fitting within a coupled ecological model. Should bacterial populations be included in an ecological model, their remineralization rates would obviously require a feedback link to the biogeochemical model. In either case, as long as bacteria have a home somewhere in your model, it really just comes down to your preference as to where that home should be.

Diagenesis Represents the Fusion of Geology and Chemistry

Generally speaking, the **abiotic** processes that will define the geochemical dynamics wrought from weathering and diagenesis can be grouped into four general categories:

1. Dissolution (simple solubility)
2. Hydrolysis reactions (acid/base)
3. Reduction/oxidation reactions
4. Ion exchange reactions

In most cases, the simple dissolution of particulate matter into their dissolved ionic constituents grants sufficient detail for most ecological and/or physical models. However, if the thrust of your model is decidedly more geochemical in nature, it will be necessary to include all possible modes of diagenetic conversion (to include hydrolysis, redox, and ion exchange dynamics).

Typical Radiative Model Elements

Radiative models are used to determine two of the most important forcing functions within the larger model: (1) heat fluxes (as they relate to thermal energy transfer and their consequence on water temperatures in the hydrodynamic model) and (2) transmission of **spectral** light throughout the water column (as it relates to primary production in the ecological model). Since we have already discussed the inclusion of heat flux as a critical component in hydrodynamic models, let us turn our attention instead to the transmission of visible and ultraviolet light into the ocean, as these sources of radiation are most relevant to the biological processes within the ecological model.

Light and Heat Must Penetrate the Atmosphere Before They Reach the Ocean

The intensity and spectral quality of the solar radiation reaching the outer atmosphere of the Earth represents the upper limit of the radiation that is available to penetrate the atmosphere and ultimately reach the surface of the Earth. Hence, it is often necessary to quantify the radiant power of each wavelength λ reaching the outer limits of the Earth (**Figure 14.8**). Fortunately, these are measures that have been conducted by a variety of Earth-orbiting satellites and are relatively constant over subdecadal time scales.

The orbital position of the Earth relative to the Sun (which corresponds to the day of the year), the latitude, and the time of day are all used to define the

Figure 14.8 The solar radiation spectrum defines both the quantity and quality of radiation that reaches the upper atmosphere and the surface of the Earth. Note that the reduction of solar energy reaching the surface (dark blue), compared to the amount of energy actually reaching the earth itself (light blue), is primarily due to atmospheric gases like ozone (O_3), oxygen (O_2), carbon dioxide (CO_2), and water vapor (H_2O), all of which are very effective at absorbing and/or reflecting radiation at different wavelengths. (Courtesy of Robert A. Rohde / CC-BY-SA-3.0.)

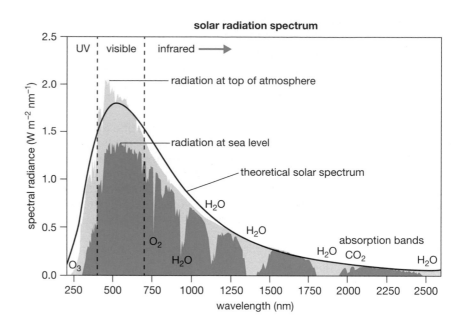

Sun's position in the sky at any given moment, which also defines the effective "thickness" of the Earth's atmosphere, through which the Sun's photons must penetrate. The thicker the layer of atmosphere through which a photon must travel, the greater its chances of being absorbed, scattered, or reflected. Thus, any model of light transmission must quantify the atmospheric path length in order to determine the length scale through which the photons are subject to atmospheric effects.

The concentration of gas molecules in the atmosphere (which can be estimated based on measures of barometric pressure) defines the number of particles that may lie in the path of each photon traveling through the atmosphere. Any interaction between a photon and an atmospheric molecule could result in scattering; that is, a change in direction of the photon's flight. Photons that are scattered forward may still reach the surface of the Earth, but their direction will be different from the main solar beam. Photons that are scattered backward are essentially reflected back out into space and never reach the Earth's surface.

Molecules of ozone (O_3) in the atmosphere are extremely effective absorbers of solar energy, particularly at ultraviolet wavelengths (200–400 nm). Thus, any radiative model that seeks to quantify the amount of UV radiation reaching the Earth's surface must incorporate some measure of atmospheric ozone and its likelihood of encountering and absorbing UV photons.

Like ozone, O_2 gas and H_2O vapor are very effective absorbers of solar radiation, although their concentrations in the atmosphere are much higher than those of ozone, and they absorb more effectively at infrared wavelengths. Since infrared radiation affects the heat flux to the Earth's atmosphere and surface (and into the ocean), these effects are more important to physical models of heat propagation; ecological models are rarely concerned with radiative transfer of infrared light.

The quantity and type of **aerosols** (large molecules and dust particles suspended in the atmosphere) can be quite variable and affect photon absorption and scattering in very complex ways. Although maritime atmospheres are typically less "polluted" with aerosols than terrestrial atmospheres, they are still a significant source of **attenuation** of light at UV, visible, and infrared wavelengths.

The Radiation Reaching the Ocean Surface Is Quickly Attenuated as It Travels Through the Water

If measures of light intensity and spectral quality are collected at sea level, there is no need to model the radiative transfer through the atmosphere: the radiation measured at sea level represents the quantity and quality of light actually available to enter the ocean (which is likely our primary interest). As a practical matter, most light instruments are designed to measure the radiant flux (or **irradiance**) at the surface, just above water. These instruments typically measure total downward irradiance at depth (E_{dz}) by orienting the light collector in such a fashion that only the downward vector component of the light field is measured. Since E_{d0-} is defined as the downward spectral irradiance in the infinitely thin layer just beneath the surface of the ocean (which represents the actual radiant flux entering the ocean in the downward direction), it is usually necessary to calculate E_{d0-} using measurements of E_{dz} and compensating for how much light was actually reflected from the ocean surface in relation to the light that actually entered the water (E_{d0-}).

Since the energy associated with light is dependent on its wavelength λ, we must take care to remember that both E_{d0-} and E_{dz} are dependent on wavelength, which we indicate as $E_{d0-}(\lambda)$ and $E_{dz}(\lambda)$. Consider Equation 14.3, which defines photon energy q:

$$q = \frac{hc}{\lambda} \tag{14.3}$$

where Planck's constant ($h = 6.626 \times 10^{-34}$ J s) and the speed of light ($c = 2.998 \times 10^8$ m s^{-1}) can be used to calculate the inherent energy of any photon as a function of its wavelength λ. We can easily demonstrate the different energies associated with violet light (400 nm) and red light (700 nm) at the extreme boundaries of the visible spectrum:

$$q_V = \frac{\left(6.626 \times 10^{-34} \text{ J s}\right) \cdot \left(2.998 \times 10^8 \text{ m s}^{-1}\right)}{\left(400 \times 10^{-9} \text{ m}\right)} = 4.966 \times 10^{-19} \text{ J}$$

$$q_R = \frac{\left(6.626 \times 10^{-34} \text{ J s}\right) \cdot \left(2.998 \times 10^8 \text{ m s}^{-1}\right)}{\left(700 \times 10^{-9} \text{ m}\right)} = 2.838 \times 10^{-19} \text{ J}$$

Because photons of different wavelengths also possess different energies, different colors of light will respond differently when they enter the ocean (**Figure 14.9**). But as long as we know how much and what kind of light enters the ocean as $E_{d0-}(\lambda)$, we can predict how those photons will behave as they penetrate deeper into the water column.

Assuming $E_{d0-}(\lambda)$ data exist, modeling radiative transfer within the water column is a relatively simple task using Beer's law (see Equation 14.4), which dictates:

$$E_{dz}(\lambda) = E_{d0-}(\lambda) \cdot e^{-K_d(\lambda) \cdot z} \tag{14.4}$$

where $E_{dz}(\lambda)$ represents the downward irradiance of a specific wavelength of light at some depth z. As long as we know $E_{d0-}(\lambda)$ and the **diffuse attenuation**

Figure 14.9 Visible light is actually comprised of many different wavelengths (colors) of light, ranging from 400 nm to 700 nm. Because photon energy is dependent on its wavelength, different colors of light will behave differently when they encounter an absorbing and/or scattering medium like the ocean. (From Moran MA & Miller WL, *Nat. Rev. Microbiology* 5:792–800, 2007. With permission from Nature Publishing Group.)

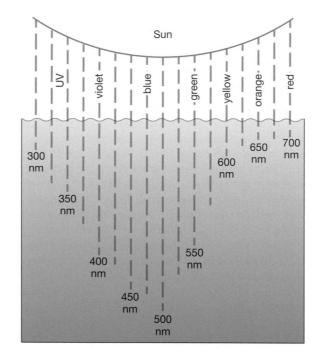

coefficient K_d (also as a function of wavelength), we can calculate the amount of light reaching any depth as E_{dz} (**Figure 14.10**).

The diffuse attenuation coefficient $K_d(\lambda)$ looks deceptively simple. In reality, it is quite difficult to parameterize this variable because there are so many ocean constituents that will influence the value of K_d (given in units of m^{-1}). Its calculation depends on all the different absorbing (a_{total}) and scattering (b_{total}) properties of seawater (and anything else that may be dissolved or suspended in that seawater), as well as the geometry of the light field ($\bar{\mu}$), where:

$$K_d(\lambda) = \frac{a_{total}(\lambda) + b_{total}(\lambda)}{\bar{\mu}} \tag{14.5}$$

The average downwelling angle of light ($\bar{\mu}$, sometimes called the "average cosine") is time and latitude dependent because it represents the angle of the direct solar beam once it enters the water column. K_d and $\bar{\mu}$ are both considered to be **apparent optical properties** (AOPs), because they are dependent on the geometry of the light field.

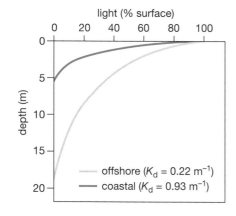

Figure 14.10 As light enters the ocean, it is quickly attenuated as a function of depth and K_d. Even the clearest waters ($K_d = 0.02$ m^{-1}) will cause an exponential decline in the amount of ambient light witnessed at depth when compared to surface light intensities. As concentrations of chlorophyll, CDOM, suspended sediments, and detritus increase (as commonly witnessed in coastal waters), the magnitude of K_d will also increase. As a result, light is more rapidly attenuated and cannot reach as deeply as in clear, open ocean waters.

In contrast, the total collection of absorbing (a_{total}) and scattering (b_{total}) constituents of and within seawater are considered to be **inherent optical properties** (IOPs) because they do not depend on the geometry of the light field. Each represents the intrinsic tendency for the total absorption (a_{total}; Equation 14.6) or scattering (b_{total}; Equation 14.7) of spectral light as a consequence of photon interaction with seawater (H$_2$O), pigmented biogenic particulates (*chl*), colored dissolved organic material (*cdom*), detritus (*det*), and/or suspended sediments (*ss*), so that:

$$a_{total}(\lambda) = a_{H_2O}(\lambda) + a_{chl}(\lambda) + a_{cdom}(\lambda) + a_{det}(\lambda) + a_{ss}(\lambda) \tag{14.6}$$

$$b_{total}(\lambda) = b_{H_2O}(\lambda) + b_{chl}(\lambda) + b_{det}(\lambda) + b_{ss}(\lambda) \tag{14.7}$$

Unless the nature of your radiative model requires an exhaustive determination of each IOP, it is generally acceptable (and certainly easier) to use modern satellite products, which provide reliable estimates of K_d(490 nm) over very broad geographic and temporal scales. In some applications where only Secchi depth (*SD*; in meters) data are available, a reasonable approximation of K_d across all visible wavelengths (400–700 nm) can be derived using Equation 14.8:

$$K_d(400-700\text{nm}) = \frac{\chi_{SD}}{SD} \qquad (14.8)$$

where χ_{SD} is an empirically derived correction factor (typically ranging from 1.7 to 2.5) that allows for local influences to the IOPs unique to the location where the SD data were collected.

Typical Ecological Model Elements

Although ecological models can be designed to focus on any biological processes relevant to ecology, most are used to investigate the dynamics of primary production, secondary production, fisheries production, or trophic (food web) connectivity. In most cases, ecological models are designed from a "functional group" perspective, where individual species are chosen to represent a larger group of organisms that share functional similarity within the ecosystem. For example, instead of modeling 15 different species of filter-feeding clams, it is usually better to choose just one or two species to serve as representatives for the larger diversity of clams in the natural environment.

Since most ecological models simulate organisms as biomass, the main "currency" in such models is organic carbon. Other important organic constituents (such as organic nitrogen, phosphorus, and chlorophyll) are easily related to the carbon-specific biomass using either Redfield ratios for C:N:P **stoichiometry** or carbon:chlorophyll ratios specifically for the autotrophic biomass. For ecological models that focus more on metabolic processes and energy transfer, it may be preferable to use caloric content as the model currency instead of carbon.

Although ecological models can be as diverse in their construction as the organisms populating the ocean, conservation of mass and energy dictates that the biomass (or bioenergy, as *B*) within the ecological model must be conserved among the following partitions:

$$\frac{\partial B}{\partial t} = \frac{\partial G}{\partial t} \pm \frac{\partial X}{\partial t} \pm \frac{\partial Z}{\partial t} \pm \frac{\partial Y}{\partial t} - \frac{\partial R}{\partial t} - \frac{\partial E}{\partial t} - \frac{\partial F}{\partial t} - \frac{\partial M}{\partial t} \qquad (14.9)$$

where time-dependent additions to biomass *B* due to growth (*G*) are affected by diffusive and advective transport (*X*), vertical migration/sinking (*Z*), as well as reproduction (*Y*), and are subject to localized losses as a result of respiration (*R*), excretion (*E*) of dissolved metabolic wastes, the egestion (*F*) of solid fecal material, and mortality (*M*; either as grazing, predation, harvest, or **senescence**).

Growth (∂G) Describes the Positive Increment of Biomass Either from Production Among Autotrophs or Assimilation Among Heterotrophs

Within the autotrophic biomass, growth and production are one and the same, where the idealized growth rates for each species are set by the ambient water temperature θ and are modulated either by light limitation (via the radiative transfer model) or nutrient limitation (via the biogeochemical model). Recall that examples of these functions were treated earlier in Equations 13.20–13.25.

For the heterotrophic biomass, growth and reproduction represent two different reservoirs of assimilated organic carbon (see Figure 13.5). Since there is a significant investment of biomass and bioenergy in the capture and ingestion of food within the heterotrophic biomass, the metabolic processes associated with assimilation (growth + reproduction) require a far more exhaustive treatment for heterotrophs within the ecological model. In Equation 14.9, reproduction is handled separately in the ∂Y term.

Diffusive and Advective Transport (∂X) Move Biomass Throughout the Domain in the Horizontal Dimension

The apparent gain or loss of biomass due to diffusive (slow, small spatial scales) and advective (fast, large spatial scales) transport must be computed as a local phenomenon because the compute nodes within the model grid (where these calculations are taking place) are fixed in space. Immigration of biomass into a specific compute node will be an apparent gain of biomass, while emigration from that same compute node will represent an apparent loss of biomass. Of course, the horizontal transport of biomass out of one compute node necessarily requires that the biomass will be transported into another (so the overall biomass throughout the entire model domain is conserved).

Vertical Migration and Sinking (∂Z) Move Biomass Throughout the Domain in the Vertical Dimension

Just as the diffuse and advective transport of biomass represents an apparent gain or loss in the horizontal dimension, organisms that exhibit some kind of vertical migration behavior will cause an apparent gain or loss of biomass in the vertical dimension of the model grid. Although vertical swimming behaviors are quite common within the heterotrophic biomass, the vertical migration behaviors of some planktonic autotrophs cannot be ignored (especially within the dinoflagellates, which exhibit strong vertical migration patterns). Of course, any transport of biomass between vertical layers (whether due to swimming or sinking) requires that we ensure all biomass fluxes are balanced.

Reproduction (∂Y) Represents the Fraction of Growth That Is Invested in the Next Generation

True reproduction within the heterotrophic biomass can represent either a gain or a loss to the carbon-specific biomass, depending on your perspective of carbon flow. Adult species that invest a certain portion of their assimilated carbon towards the production of gametes (egg and sperm) or their young will experience that investment as a loss of carbon biomass; however, if you are explicitly modeling larval and/or juvenile forms in your model, those organisms will experience that reproductive investment from their parents as a gain in their own carbon biomass. Obviously, any biomass gains in the gamete or larval pools must be exactly equal to the biomass losses experienced by the adults engaged in reproduction.

Respiration (∂R) Is the Loss of Biomass Attributed to Gaseous Wastes

Respiration is a significant contributor to biomass loss among autotrophs and heterotrophs alike. Although the extent of biomass loss attributed to respiration is largely influenced by ambient temperatures (θ), respiratory losses can also be impacted by complex locomotory behaviors (such as migration, foraging, or predator avoidance) that consume a great deal of assimilated carbon. In some bioenergetics models, the thermic effect of food (or **specific dynamic action**) being ingested and assimilated is sometimes significant enough to cause disproportionate effects on the respiration or modeled organisms. Osmotic stress (due to significant salinity changes) and hypoxia can also cause dramatic shifts in the respiratory losses of modeled organisms in coastal and estuarine systems.

Typically, respiration losses are modeled as a conversion of POC to DIC (specifically as CO_2). Of course, the respiratory conversion of POC to CO_2 will significantly impact any associated biogeochemical models, as O_2 is consumed (removed) from the water column as CO_2 is produced through respiration. Depending on the sophistication of your coupled biogeochemical model, additions of CO_2 will not only force changes to the carbonate equilibrium ($CO_2 + H_2O \leftrightarrow H_2CO_3 \leftrightarrow H^+ + HCO_3^- \leftrightarrow 2H^+ + CO_3^{2-}$), the associated liberation of H^+ may also affect the pH, ocean acidification, and some important anaerobic bacterial processes related to nutrient recycling.

Excretion (∂E) Is the Loss of Biomass Attributed to Liquid Wastes

Excretion within the autotrophic biomass is not excretion per se; rather, it is often a simple parameterization of the "leakiness" of autotrophic cells, which can lose small soluble molecules because of passive diffusion through their cell membrane. However, in most ecological applications, "excretion" from the autotrophic biomass can largely be ignored.

Within the heterotrophic biomass, excretion is a significant mechanism for the release of dissolved inorganic molecules as metabolic wastes wrought from assimilation. Although there are several forms of excreted materials, most ecological models focus primarily on the excretion of nitrogenous wastes (as NH_3, which forms NH_4^+ and OH^- in water) and secondarily on the excretion of phophogenous wastes (as PO_4^{3-}). These excretion products must then be linked to the biogeochemical model, where the total DIN and DIP stocks can be "updated" with the excreted nitrogen and phosphorus. As you may recall from Equations 13.7–13.12, we have already discussed an example of how the excretion of nitrogenous waste can be modeled; a similar methodology would be required for the excretion of PO_4^{3-} as well.

Egestion (∂F) Is the Loss of Biomass Attributed to Solid Wastes

Since egestion is technically defined as the process of eliminating undigested food as feces, egestion can be ignored within the autotrophic biomass. For heterotrophs, egestion represents the particulate organic material (detritus) that has escaped assimilation and must be linked to the biogeochemical model for potential bacterial remineralization of the organic carbon, nitrogen and phosphorus, therein.

Mortality (∂M) Is Always Represented as a Loss of Biomass, but It Comes in Many Forms

Within the autotrophic biomass, mortality is primarily a function of grazing (although senescence and cell lysis can be significant contributing factors and may require separate treatment). For heterotrophs, mortality is almost exclusively related to natural predation or anthropogenic harvest (where senescence and cell lysis can largely be ignored, unless they are used as a proxy for losses due to disease and/or exposure to toxic substances).

Since the behaviors surrounding grazing and predation are highly specific to the consumer as well as the type and availability of prey, your ecological model will likely require several different grazing or predation schemes to simulate those complex behaviors. Losses due to disease and/or toxic exposures are also very specific to the type of mortal exposure and the particular species being affected.

Closing Thoughts

As you can imagine, it would be impossible to fully discuss all of the elements of a large systems model in a single chapter, but we have covered the very basics—enough to get you on your way. A lot has been covered; from the statistical framework of data analysis, to the various field methods of data acquisition, and ultimately to the creation of mathematical models to simulate those variables that you cannot (or did not) measure directly. But more than that, we've been able to demonstrate that models are a great way to conduct very complex thought experiments, where you can perturb a marine system and investigate the response in simulation space.

That's important because there's a lot of experimentation you can do with a model that would be too expensive (or too dangerous) to perform in a real system. Additionally, there are an infinite number of "what if" scenarios you can explore using a model, whether it's to understand ocean dynamics at a much deeper level or to plan a response to an event that has yet to occur (but for which your simulation has a solution). So why wait for tomorrow when you can simulate it today?

References

Aumont O (2004) PISCES (Pelagic Iteraction Scheme for Carbon and Ecosystem Studies) Biogeochemical Model. http://www.lodyc.jussieu.fr/~aumont/pisces.html; Laboratoire d'Océanographie Dynamique et de Climatologie.

Bird R & Riordan C (1984) Simple Solar Spectral Model for Direct and Diffuse Irradiance on Horizontal and Tilted Planes at the Earth's Surface for Cloudless Atmospheres. Solar Energy Research Institute (SERI), Technical Report 2436.

Changsheng Chen C, Beardsley RC, Cowles G, et al. (2011) An Unstructured Grid, Finite-Volume Coastal Ocean Model (FVCOM) User Manual, 3rd ed. http://fvcom.smast.umassd.edu/fvcom; University of Massachusetts-Dartmouth.

Christensen V, Walters CJ, Pauly D, & Forrest R (2008) Ecopath with Ecosim (v6) User Guide. http://sources.ecopath.org/trac/Ecopath/wiki/UsersGuide; Ecopath Research and Development Consortium.

Gregg WW & Carder KL (1990) A simple spectral solar irradiance model for cloudless maritime atmospheres. *Limnology and Oceanography* 35(8):1657–1675.

Griffies SM, Harrison MJ, Pacanowski RC, & Rosati A (2008) A Technical Guide to MOM4 (Modular Ocean Model 4). GFDL Ocean Group, Technical Report No. 5; http://www.mom-ocean.org/web/docs/project/MOM4_guide.pdf; US National Oceanic and Atmospheric Administration (NOAA) Geophysical Fluid Dynamics Laboratory (GFDL).

Mellor GL (2002) Users Guide for a Three-Dimensional, Primitive Equation, Numerical Ocean Model (Princeton Ocean Model). http://www.ccpo.odu.edu/POMWEB/index.html; Princeton University.

Park RA & Clough JS (2009) AQUATOX: Modeling Environmental Fate and Ecological Effects in Aquatic Ecosystems, Volume 2: Technical Documentation. United States Environmental Protection Agency (US-EPA), Report EPA-823-R-09-004.

Moran MA & Miller WL (2007) Resourceful heterotrophs make the most of light in the coastal ocean. *Nature Reviews Microbiology* 5:792–800.

APPENDIX A

Table of Symbols

(occurrence indicates the Equation or Table where first defined)

symbol	definition	occurrence
‰	parts per thousand	4.15
$\%_A$	assimilation efficiency of nitrogen	13.8
$\%_E$	proportion of nitrogen egested	13.8
α	probability level specific volume	2.13 4.5
a	Haversine parameter	4.3
$\bar{a}$	density correction for temperature, equation of state	4.11
a_{cdom}	absorption coefficient for colored dissolved organic material	14.6
a_{chl}	absorption coefficient for pigmented biogenic particles	14.6
a_{det}	absorption coefficient for detrital particles	14.6
a_{H_2O}	absorption coefficient for pure water	14.6
a_{ss}	absorption coefficient for suspended sediments	14.6
a_{total}	total absorption coefficient	14.5
A	area Schnabel numerator	3.6 6.11
AN	assimilated nitrogen	13.8
AU	absolute uncertainty	2.1
$\bar{b}$	density correction for salinity, equation of state	4.11
$\hat{b}$	statistical estimate of the intercept	9.4
b_{chl}	scattering coefficient for pigmented biogenic particles	14.7
b_{det}	scattering coefficient for detrital particles	14.7
b_{H_2O}	scattering coefficient for pure water	14.7
b_{ss}	scattering coefficient for suspended sediments	14.7

symbol	definition	occurrence
b_{total}	total scattering coefficient	14.5
B	Schnabel denominator biomass	6.11 14.9
B_0	initial biomass	12.8
B_t	biomass at time t	12.8
χ_{SD}	correction factor for Secchi depth	14.8
c	speed of light	14.3
c_{grp}	speed of wave propagation	11.7
C	Fleiss parameter	3.3
CI	confidence interval	2.13
$CPUE$	catch per unit effort	7.10
cv	coefficient of variation	2.6
Δ	arithmetic difference, large degree of change	4.1
δ	arithmetic difference, small degree of change	Table 11.1
∂	rate of change; partial differential	11.10
d	arithmetic difference distance depth	2.3 4.4 4.9
$\bar{d}$	circular quadrat ideal diameter	3.9
d°	calibration distance	7.6
d_α	chosen significance level	2.13
$^{1,2,3}d_n$	rank ordered distance between two observations	3.13
$\dfrac{d}{dt}, \dfrac{\partial}{\partial t}$	rate of change	12.4
D	population density	3.13
D_{max}	theoretical maximum of Simpson diversity	8.12
D_S	Simpson index of diversity	8.10
ε	experimentwise error rate random error	8.5 9.2
η	momentary sea surface elevation	14.1
e	Euler's number; base of the natural logarithm	9.19
E	relative error effort model representation of excretion	3.1 7.11 14.9
E_{NH_3}	excretion of ammonia	12.5
E_S	Simpson index of species evenness	8.14
E_z	downward irradiance of light at depth z	14.4
E_{0-}	downward irradiance of light entering the ocean	14.4
EN	egested nitrogen	13.8
f	frequency of occurrence Coriolis "force"	6.4 14.2

symbol	definition	occurrence
F	filtration efficiency model representation of egestion	7.6 14.9
f_v	volume scaling factor	7.8
$f(var)$	any function of the variable var	11.1
$\gamma_{1,2}$	Thornton-Lessem temperature coefficients	13.13
g	mean constant of gravitational acceleration	4.9
G	model representation of growth	14.9
h	height Planck's constant	Table 7.2.1 14.3
H	water height according to the bathymetric datum	14.1
H'	Shannon-Weaver index of diversity	8.8
H'_{max}	theoretical maximum of Shannon-Weaver diversity	8.11
H_a	alternate hypothesis	Table 2.1
H_o	null hypothesis	Table 2.1
i	any integer, sometimes specific to the x-direction	6.1
j	any integer, sometimes specific to the y-direction	11.13
J'	Shannon-Weaver index of species evenness	8.13
$\bar{k}$	density correction for pressure, equation of state	4.11
k_{NH_3}	half-saturation constant for ammonia	13.24
k_{nit}	half-saturation constant for bacterial nitrification	12.4
k_{NO_3}	half-saturation constant for nitrate	13.24
k_{PO_4}	half-saturation constant for phosphate	13.24
$k_{SiO(OH)_3}$	half-saturation constant for silica	13.24
K	secant bulk modulus, equation of state	4.14
$K_{1,2,3,4}$	Thornton-Lessem reaction rate multipliers	13.13
$K(\theta)$	Thornton-Lessem rate multiplier	13.13
$K_A(\theta)$	Thornton-Lessem ascending rate multiplier	13.13
$K_B(\theta)$	Thornton-Lessem descending rate multiplier	13.13
K_d	diffuse attenuation coefficient	14.4
λ	wavelength	14.3
ℓ	Simpson index of species dominance	8.9
l	length (distance) of transect or belt	3.19
L	length	3.6
μ	realized growth rate	12.8
$\bar{\mu}$	average downward angle of light; average cosine	14.5
μ_{ll}	light-limited growth rate	13.23
μ_{max}	maximum theoretical growth rate	13.21
μ_{nl}	nutrient-limited growth rate	13.23
μ_θ	temperature-dependent growth rate	13.21

symbol	definition	occurrence
m	population density mass	3.9 4.5
$\hat{m}$	statistical estimate of the slope	9.4
m_i	number of marked specimens in the ith collection	6.11
M	number of marked specimens in trap model representation of mortality	6.9 14.9
M_i	accumulation of marked specimens after i collections biomass concentration of the ith species	6.11 7.2
MAX	maximum value	13.24
MIN	minimum value	13.23
MOD_i	value of an individual model estimate	12.2
n	any number number of occurrences	2.9 6.4
n_i	number of total specimens in the ith collection	6.11
n_o	minimum sample size	3.1
n_S	number of specimens in sample	7.9
N	total number of observations	2.2
N_A	assimilated nitrogen	13.10
N_{amm}	nitrogen associated with ammonification	13.2
N_{dis}	nitrogen associated with dissolution	13.9
N_E	egested nitrogen	13.9
N_{exc}	nitrogen associated with excretion	13.2
N_{fix}	nitrogen associated with nitrogen fixation	13.2
N_i	numeric concentration of the ith species	7.1
N_{min}	nitrogen associated with remineralization	13.9
N_{MR}	number of specimens in area from mark & recapture	6.7
N_{nit}	nitrogen associated with nitrification	13.2
N_{prd}	nitrogen associated with production	13.2
N_0	total population size estimated from mark & recapture	6.8
NIT	nitrification	12.4
Ω	angular velocity of the Earth	14.2
$\overline{OBS}$	mean of all observed data	12.1
OBS_i	value of an individual observation	12.1
ϕ	any value determined through finite differencing latitude	11.11 14.2
π	3.14159265358979…	3.7
p	probability value	Technical Box 8.2
p_c	proportion belonging to control group	3.3
p_e	proportion belonging to experimental group	3.3

symbol	definition	occurrence
P	perimeter pressure	3.6 4.10
P_h	hydrostatic pressure	4.9
PON	particulate organic nitrogen	13.6
PON_i	ingested PON	13.7
q	radiative energy	14.3
q_i	proportion of catchability for the ith species	7.10
Q	discharge rate	7.3
ρ	density of water	4.5
ρ'	regression estimate of density	9.10
ρ_0	standard density, equation of state	4.11
r	radius Pearson's coefficient of correlation	3.7 9.3
r_{nit}	bacterial nitrification rate	12.4
R	range model representation of respiration	2.7 14.9
R^2	coefficient of determination for goodness of fit	11.3
RHO	density of water, dependent regression variable	9.7
RU	relative uncertainty	2.1
Σ	arithmetic sum	2.2
σ	sigma as $\rho - 1000$	4.10
σ_n	sigma layer n	14.1
s	standard deviation	2.3
s^2	variance	2.8
s_P	pooled standard deviation	8.3
S	salinity species richness	4.10 8.11
S_0	standard salinity, equation of state	4.11
SAL	salinity of water, independent regression variable	9.7
SD	standard depth biometric Secchi depth	Table 8.1 14.8
SE	standard error	2.4
SL	standard length biometric	Table 8.1
SS_{res}	sum of squares of the residual error	12.2
SS_{tot}	sum of squares for the total body of data	12.1
θ	angle Thornton–Lessem ambient temperature	Technical Box 7.1 13.13
$\theta_{1,2,3,4}$	Thornton–Lessem temperature thresholds	13.13
t	time elapsed	7.3

symbol	definition	occurrence
t'	test statistic	6.2
T	temperature	4.10
T_0	standard temperature, equation of state	4.11
$TEMP$	temperature of water, independent regression variable	9.8
u	velocity component in x (east) direction	11.10
u_i	number of unmarked specimens in the ith collection	6.11
U	number of unmarked specimens in trap wind speed	6.9 Table 9.1
v	velocity component in y (north) direction	11.10
$\bar{v}$	mean velocity	7.5
v_{adv}	advective transport velocity	11.7
v_x	largest magnitude of velocity in the x-direction	11.6
V	mean swimming speed volume	3.21 4.5
V_s	sample volume	7.8
V_z	volumetric abundance of zooplankton species	13.7
VMR	variance-to-mean ratio	3.8
w	width of transect or belt velocity component in z (up) direction	3.19 11.10
w_n	maximum orthogonal width	3.20
W	width dry weight	3.6 12.5
$\bar{W}$	mean dry weight	12.6
W_z	dry weight of zooplankton species	13.7
x	horizontal easting dimension	11.10
x_i	distance from starting point to nearest target	6.1
X	model representation of advective transport	14.9
$\bar{X}$	arithmetic mean of X	2.2
y	horizontal northing dimension	11.10
Y	model representation of reproduction	14.9
$\bar{Y}$	arithmetic mean of Y	9.3
Y'	regression estimate of Y	9.9
z	z-score vertical dimension	2.5 11.10
z_i	distance from target to nearest neighbor	6.1
z_n	depth layer n	14.1
Z	encounter rate zooplankton density model representation of vertical migration or sinking	3.21 12.6 14.9
Z_s	zooplankton species s	13.7

APPENDIX B

Constants, Equivalents, and Common Units of Measure

Table B1. Scientific Notation.

Exponent	Common Name	Common Prefix
10^{15} = 1,000,000,000,000,000	quadrillion	peta-
10^{12} = 1,000,000,000,000	trillion	tera-
10^{9} = 1,000,000,000	billion	giga-
10^{6} = 1,000,000	million	mega-
10^{3} = 1,000	thousand	kilo-
10^{2} = 100	hundred	hecto-
10^{1} = 10	ten	deca-
10^{-1} = 0.1	tenth	deci-
10^{-2} = 0.01	hundredth	centi-
10^{-3} = 0.001	thousandth	milli-
10^{-6} = 0.000001	millionth	micro-
10^{-9} = 0.000000001	billionth	nano-
10^{-12} = 0.000000000001	trillionth	pico-

Table B2. Useful Constants and Unit Measurements.

Constant	Unit
angstrom (Å)	1.0×10^{-10} m
bar	105 Pa
day	86,400 s
dyne	$g \cdot cm \ s^{-2}$
erg	$g \cdot cm^2 \ s^{-2}$
gravitational acceleration (g)	$9.80665 \ m \ s^{-2}$
hour	3600 s
Joule (J)	$kg \cdot m^2 \ s^{-2}$
langley (ly)	$41,840 \ J \ m^{-2}$
Newton (N)	$kg \cdot m \ s^{-2}$
Pascal (Pa)	$kg \ m^{-1} \ s^{-2}$
Planck's constant (h)	$6.626 \times 10^{-34} \ J \cdot s$
speed of light (c)	$2.9979 \times 10^8 \ m \ s^{-1}$
Sverdrup (Sv)	$1.0 \times 10^6 \ m^3 \ s^{-1}$
Watt (W)	$J \ s^{-1}$
year	3.156×10^7 s

Table B3. Common Units of Conversion.

Length or Distance	
1 statute mile (mi)	0.868976 nmi 1.609344 km
1 nautical mile (nmi)	1.150779 mi 1.852000 km 1 min of latitude
1 kilometer (km)	1×10^3 m 0.621371 mi 0.539957 nmi
1 meter (m)	1×10^{-3} km 39.370079 in 3.280840 ft 1.093613 yard (yd)
1 centimeter (cm)	1×10^{-2} m 0.393701 in 0.32808 ft
1 millimeter (mm)	1×10^{-3} m 0.039370 in 0.003281 ft
1 inch (in)	2.540000 cm 25.400000 mm
1 foot (ft)	30.480000 cm 304.800000 mm

Table B3 (*Continued*). Common Units of Conversion.

Area	
1 hectare	2.471054 acres 0.01 km^2
1 mi^2	0.755120 nmi^2 2.589988 km^2 640.000000 acres
1 nmi^2	1.324293 mi^2 3.429904 km^2 847.547736 acres
1 km^2	0.386102 mi^2 0.291553 nmi^2 247.105381 acres
1 m^2	1 x 10^{-6} km^2 10.763910 ft^2 1.195990 yd^2
1 cm^2	1 x 10^{-4} m^2 0.155000 in^2 0.001076 ft^2
1 mm^2	1 x 10^{-6} m^2 0.001550 in^2
1 in^2	6.451600 cm^2 0.000645 m^2
1 ft^2	929.030400 cm^2 0.092903 m^2
Volume	
1 km^3	1 x 10^9 m^3 1 x 10^{12} L
1 m^3	1 x 10^3 L 35.314667 ft^3 264.172052 gallons (gal)
1 L	1 x 10^{-3} m^3 1 x 10^3 cm^3 1000 milliliter (mL) 0.035315 ft^3 0.264172 gal
1 cm^3	1 x 10^{-3} L 1 mL
Mass	
1 metric ton (MT)	1 x 10^3 kg 2,205 lb
1 ton	2,000 lb 907.184740 kg 0.907185 MT
1 kilogram (kg)	1 x 10^3 g 2.204620 lb 35.273962 oz
1 gram (g)	1 x 10^{-3} kg 1 x 10^3 milligrams (mg) 0.002205 lb 0.035274 oz

Table B3. *Common Units of Conversion.*

Mass	
1 stone	224 oz 14 lb 6.350293 kg
1 pound (lb)	16 oz 0.453592 kg 0.071429 stone
1 ounce (oz)	0.062500 lb 28.349523 g 0.004464 stone
Velocity	
1 mi hr^{-1}	1.609344 km hr^{-1} 0.447040 m s^{-1}
1 km hr^{-1}	0.621371 mi hr^{-1} 0.277778 m s^{-1}
1 nmi hr^{-1}	1 knot (kt) 1.852000 km hr^{-1} 1.150779 mi hr^{-1} 0.514444 m s^{-1}
1 m s^{-1}	3.6 km hr^{-1} 2.236936 mi hr^{-1} 1.943844 kt
Pressure	
1 atmosphere (atm)	1.01325 bar 1.01325 x 10^5 Pa 760 Torr 14.69595 lb in^{-2}
1 Pascal (Pa)	10 dynes cm^{-2} 1 x 10^{-5} bar 9.869233 x 10^{-6} atm
Density	
1 kg m^{-3}	1 g L^{-1}
1 g cm^{-3}	1 g mL^{-1}
Energy	
1 Joule (J)	1 x 10^7 ergs 0.239006 cal 9.478171 x 10^{-4} British thermal units (BTU)
1 calorie (cal)	4.184000 J 3.965667 x 10^{-3} BTU
Force	
1 Newton (N)	1 x 10^5 dynes

Table B4. Ocean Measures.

Area	
Atlantic Ocean	$106{,}463 \times 10^6$ km^2
Indian Ocean	$74{,}917 \times 10^6$ km^2
Pacific Ocean	$179{,}679 \times 10^6$ km^2
Mean Depth	
Atlantic Ocean	3,332 m
Indian Ocean	3,897 m
Pacific Ocean	4,028 m
Volume	
Atlantic Ocean	$354{,}679 \times 10^6$ km^3
Indian Ocean	$291{,}945 \times 10^6$ km^3
Pacific Ocean	$723{,}699 \times 10^6$ km^3

Table B5. Planetary Measures.

Measure	Value
area of the earth	5.10×10^8 km^2
area of the earth occupied by land	1.49×10^8 km^2
area of the earth occupied by water	3.61×10^8 km^2
highest land elevation	8.848 km
greatest ocean depth	11.035 km
angular velocity of the earth	7.292×10^{-5} s^{-1}
radius of earth at equator	6378 km
radius of earth at poles	6357 km
radius of earth (mean value)	6367.5 km
mass of the earth	5.97×10^{24} kg
mass of the sun	1.99×10^{30} kg
mass of the moon	7.35×10^{23} kg
mean distance from sun	149.7×10^6 km
mean distance from moon	384.4×10^3 km

Table B6. Typical Seawater Values.

Measure	Value
latent heat of fusion (liquid → solid)	3.3×10^5 J kg^{-1}
latent heat of vaporization (liquid → gas)	2.45×10^6 J kg^{-1}
molecular diffusion of heat	1.4×10^{-7} m^2 s^{-1}
molecular diffusion of salt	1.5×10^{-9} m^2 s^{-1}
molecular viscosity	1.0×10^{-6} m^2 s^{-1}
specific heat (at constant pressure)	0.4 J kg^{-1}°C^{-1}
surface tension	0.08 N m^{-1}

Common Algorithms for the Estimation of Important Water Properties

The algorithms presented here are based on the Practical Salinity Scale (PSS-78) and the International Equation of State (IES 80; Technical Box 4.1) for seawater. They are intended for use in "typical" marine waters that are within nominal ranges for salinity ($2 < S < 42$), temperature ($-2°C < T < 35°C$), and pressure (1,722 decibar $< P <$ 10,334 decibar), which would include the bulk of the world ocean.

Salinity (from Conductivity)

Although salinity can be determined by either a hydrometer or a refractometer, neither of these is as accurate or precise as a conductivity meter, which directly measures the specific conductance of electricity (which is directly proportional to the total dissolved salt content) within a water sample (Figure C.01). Conductivity measures are typically given in siemens per meter ($S\ m^{-1}$) or millisiemens per centimeter ($mS\ cm^{-1}$).

For any measured value of conductivity C (in $mS\ cm^{-1}$), you must first calculate the conductivity ratio R_C according to Equation C.01:

$$R_C = \frac{C(S,T,P)}{C(35,15,0)} = \frac{C}{42.9140\ \mathrm{mS\ cm^{-1}}} \qquad (\text{C.01})$$

where your measured value of conductivity C will vary as a function of salinity S, temperature T, and pressure P. In Equation C.01, the value for C (35, 15, 0) simply represents the electrical conductance of a standard sample at 35 PSU and 15°C, located at the surface (0 pressure).

Before we can calculate salinity values from our conductivity measures, we must first estimate the contributions to the conductivity ratio RC caused by

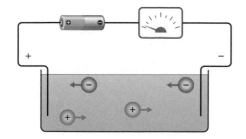

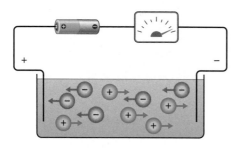

Figure C.01 Seawater contains a variety of dissolved cations (+) and anions (–), which are present in concentrations that are proportional to salinity. At low salinities *(top)*, there are fewer ions available to conduct electricity, so measures of specific conductance will be low. At higher salinities *(bottom)*, there are more dissolved ions to conduct electricity, so measures of specific conductance will be higher. It is precisely this relationship between conductance and salinity that allows oceanographers to indirectly measure salinity by measuring conductance instead.

temperature effects (r_t in Equation C.02) and pressure effects (r_p in Equation C.03), where

$$r_t = a_0 + a_1 T + a_2 T^2 + a_3 T^3 + a_4 T^4 \tag{C.02}$$

$$
\begin{aligned}
a_0 &= +0.6766097 \\
a_1 &= +2.00564 \times 10^{-2} \\
a_2 &= +1.104259 \times 10^{-4} \\
a_3 &= -6.9698 \times 10^{-7} \\
a_4 &= +1.0031 \times 10^{-9}
\end{aligned}
$$

$$r_p = 1 + \frac{P\left(b_1 + b_2 P + b_3 P^2\right)}{1 + c_1 T + c_2 T^2 + R_C\left(c_3 + c_4 T\right)} \tag{C.03}$$

$$
\begin{aligned}
b_1 &= +2.070 \times 10^{-5} & c_1 &= +3.426 \times 10^{-2} \\
b_2 &= -6.370 \times 10^{-10} & c_2 &= +4.464 \times 10^{-4} \\
b_3 &= +3.989 \times 10^{-15} & c_3 &= +4.215 \times 10^{-1} \\
           &   & c_4 &= -3.107 \times 10^{-3}
\end{aligned}
$$

Then it is a relatively simple matter to calculate the true conductivity ratio R_T using Equation C.04:

$$R_T = \frac{R_C}{r_P \cdot r_t} \tag{C.04}$$

This computed value for R_T is then used in Equation C.05 to calculate the salinity correction factor ΔS as

$$\Delta S = \frac{(T - 15)}{1 + 0.0162(T - 15)} \cdot \left(d_0 + d_1 R_T^{1/2} + d_2 R_T + d_3 R_T^{3/2} + d_4 R_T^2 + d_5 R_T^{5/2}\right) \tag{C.05}$$

$$
\begin{aligned}
d_0 &= +0.0005 \\
d_1 &= -0.0056 \\
d_2 &= -0.0066 \\
d_3 &= -0.0375 \\
d_4 &= +0.0636 \\
d_5 &= -0.0144
\end{aligned}
$$

Now the salinity S can finally be calculated using Equation C.06:

$$S = e_0 + e_1 R_T^{1/2} + e_2 R_T + e_3 R_T^{3/2} + e_4 R_T^2 + e_5 R_T^{5/2} + \Delta S \tag{C.06}$$

$$
\begin{aligned}
e_0 &= +0.0080 \\
e_1 &= -0.1692 \\
e_2 &= +25.3851 \\
e_3 &= +14.0941 \\
e_4 &= -7.0261 \\
e_5 &= +2.7081
\end{aligned}
$$

Depth (from Pressure)

As we saw earlier in Equation 4.09, the hydrostatic pressure P_h is calculated as a function of the density ρ, the mean constant of gravitational

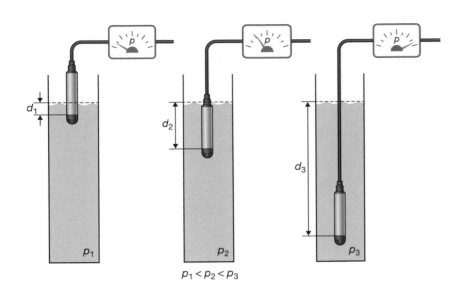

Figure C.02 A pressure sensor will experience an increase in hydrostatic pressure as it is lowered deeper and deeper into the ocean. As long as we know something about the density (ρ) of the water column above the sensor, we can use the measured pressure value to estimate the depth of the sensor.

acceleration (g), and the depth d at which we are attempting to measure P_h. If we were in possession of a sensor that could measure the hydrostatic pressure P_h *in situ* (**Figure C.02**), we would be able to determine the depth of our sensor simply by rearranging Equation 4.9 to solve for d, as demonstrated in Equation C.07:

$$d = \frac{P_h}{\rho \cdot g} \tag{C.07}$$

If this were the case, it would be simple to calculate water depth by measuring the hydrostatic pressure and using fixed values for ρ and g. In reality, ρ is almost never constant with depth, and even g will vary slightly as a function of the latitude ϕ. For the exact solution of d, we would need to use a latitudinally corrected value for g, as well as solve the integral of ρ, simultaneously.

As you can imagine, the exact solution of d would be impractical for routine use; fortunately, a very accurate method of estimating depth does exist, without requiring integral calculus. This is done first by calculating the variation of gravity with latitude (g_ϕ) using Equation C.08:

$$g_\phi = 9.780318 \cdot (1 + [(5.2788 \times 10^{-3} + 2.36 \times 10^{-5} X^2) \cdot X^2]) + 1.092 \times 10^{-6} P_h \tag{C.08}$$

$$X = sin\left(\frac{\phi}{57.29578}\right)$$

where P_h must be measured in decibars. Once g_ϕ has been determined, a very accurate estimation of the water depth d (in meters) can be obtained from P_h measured *in situ*, using Equation C.09:

$$d = \frac{\left(\left[\left\langle 2.279 \times 10^{-10} - 1.82 \times 10^{-15} P_h \right\rangle \cdot P_h - 2.2512 \times 10^{-5}\right] \cdot P_h + 9.72659\right) \cdot P_h}{g_\phi} \tag{C.09}$$

Euphotic Depth (from Secchi Depth)

As we were able to demonstrate in Equation 14.4, the amount of light reaching a specific depth (E_z) can be determined if we also know the total amount of light entering the ocean at the surface (E_{0^-}) and the diffuse attenuation

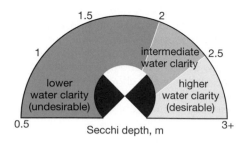

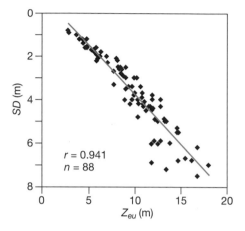

Figure C.03 Using something as simple as a Secchi disk, it is possible to quantify the Secchi depth SD as a simple measure of water clarity *(top)*. Measures of water clarity can then be used to determine the depth of light penetration (that is, the depth of the euphotic zone, z_{eu}). When SD is compared against z_{eu} *(bottom)*, the slope of the resultant line can be used to derive the appropriate correction factor χ_{SD}, which can ultimately be used to derive the diffuse attenuation coefficient for visible light, $K_{d(PAR)}$. (From Luhtala H & Tolvanen H [2013] *ISPRS Int J Geoinf* 2:1153–1168. CC-BY-SA-3.0.)

coefficient K_d. Each of these is a function of the wavelength λ, but it is occasionally more convenient to use values for E_z, E_{0^-}, and K_d that have been averaged across all photosynthetically active radiation (PAR) wavelengths (400–700 nm) to represent the behavior of all visible light.

In Equation 14.8, we also learned that a Secchi disk could be used to determine the Secchi depth (SD), which in turn could be used to provide an estimate of $K_{d(PAR)}$ using an empirically derived correction factor χ_{SD} to represent water clarity and to characterize the attenuation of light with depth (**Figure C.03**). If we combine Equations 14.4 and 14.8, we now have a more practical means of calculating the subsurface light at any depth, or $E_{z(PAR)}$, using Equation C.10:

$$E_{z(PAR)} = E_{0^-(PAR)} \cdot e^{\left(\chi_{SD}/SD\right)\cdot z} \tag{C.10}$$

If we want to determine the depth z at which $E_{z(PAR)}$ is reduced to a specific percentage of $E_{0^-(PAR)}$, Equation C.10 becomes much easier to solve. Since $E_{0^-(PAR)}$ represents the total amount of visible light entering the ocean, it is always 100%. The bottom of the euphotic zone is traditionally defined as the depth where the ambient light is ~1% of the surface light, so we simply set $E_{z(PAR)} = 1\%$ in order to determine the depth of the euphotic zone (z_{eu}) in Equation C.11:

$$1 = 100 \cdot e^{\left(\chi_{SD}/SD\right)\cdot z_{eu}} \tag{C.11}$$

If we solve for z_{eu}, we arrive at Equation C.12:

$$z_{eu} = -\left(\frac{SD}{\chi_{SD}}\right) \cdot \ln\left(\frac{1}{100}\right) \tag{C.12}$$

where the bottom of the euphotic zone (z_{eu}) can be determined by using the Secchi depth SD and the appropriate correction factor (χ_{SD}) for the region (in most cases, $1.7 < \chi_{SD} < 2.5$).

Adiabatic Lapse Rate

The adiabatic lapse rate for seawater (Γ) represents the change in temperature relative to the change in pressure experienced by a parcel of water, assuming there is no exchange of heat or salt with the surrounding water mass. The adiabatic lapse rate is a significant concept in physical oceanography, and is necessary for understanding the concept of potential temperature (discussed later). Γ (in °C decibar^{-1}) is a function of temperature T (in °C), salinity S (in PSU), and pressure P (in decibars), and can be estimated using Equation C.13:

$$\Gamma = a_0 + a_1 T + a_2 T^2 + a_3 T^3 + \left\langle (b_0 + b_1 T) \cdot (S - 35) \right\rangle$$
$$+ \left[c_0 + c_1 T + c_2 T^2 + c_3 T^3 + \left\langle (d_0 + d_1 T) \cdot (S - 35) \right\rangle \right] \cdot P \tag{C.13}$$
$$+ \left[e_0 + e_1 T + e_2 T^2 \right] \cdot P^2$$

$$a_0 = +3.5803 \times 10^{-5} \qquad c_2 = +8.7330 \times 10^{-12}$$
$$a_1 = +8.5258 \times 10^{-6} \qquad c_3 = -5.4481 \times 10^{-14}$$
$$a_2 = -6.8360 \times 10^{-8} \qquad d_0 = -1.1351 \times 10^{-10}$$
$$a_3 = +6.6228 \times 10^{-10} \qquad d_1 = +2.7759 \times 10^{-12}$$
$$b_0 = +1.8932 \times 10^{-6} \qquad e_0 = -4.6206 \times 10^{-13}$$
$$b_1 = -4.2393 \times 10^{-8} \qquad e_1 = +1.8676 \times 10^{-14}$$
$$c_0 = +1.8741 \times 10^{-8} \qquad e_2 = -2.1687 \times 10^{-16}$$
$$c_1 = -6.7795 \times 10^{-10}$$

Potential Temperature

The potential temperature θ is traditionally defined as the temperature a parcel of water would have if it were moved from its initial location with ocean pressure P to some reference location with pressure P_r that may be greater or less than P, although P_r is most commonly set as the pressure at the ocean surface ($P_r = 0$). As a result, potential temperatures in the deep ocean are slightly colder than *in situ* temperatures at the same depth, because any parcel of water from the deep ocean, if brought to the surface, would expand ever so slightly and subsequently cool from that expansion (**Figure C.04**).

Computed from the adiabatic lapse rate Γ as discussed earlier, θ (in °C) can be estimated using a solution cascade where we must first define the change in pressure (ΔP, in decibars) using Equation C.14:

$$\Delta P = P - P_r \qquad (C.14)$$

Before we can assess θ, we must first determine the adiabatic lapse rate Γ in Equation C.13 using the *in situ* values for salinity (S_0, in PSU), temperature (T_0, in °C), and pressure (P_0, in decibars). The first approximation of θ (θ_1) is calculated using Equation C.15:

$$\Delta\theta_1 = \Delta P \cdot \Gamma\left(S_0, T_0, P_0\right),$$
$$\theta_1 = T_0 + \frac{\Delta\theta_1}{2} \qquad (C.15)$$

Γ is recalculated (Equation C.13) using θ_1 as the new temperature parameter T, and we calculate our second approximation of θ (θ_2) using Equation C.16:

$$\Delta\theta_2 = \Delta P \cdot \Gamma\left(S_0, \theta_1, P_0 + 0.5\Delta P\right),$$
$$q_1 = \Delta\theta_1,$$
$$\theta_2 = \theta_1 + \left(1 - 1/\sqrt{2}\right) \cdot \left(\Delta\theta_2 - q_1\right) \qquad (C.16)$$

Again, Equation C.13 is used to recalculate Γ, this time using θ_2. We calculate our third approximation for θ (θ_3) using Equation C.17:

$$\Delta\theta_3 = \Delta P \cdot \Gamma\left(S_0, \theta_2, P_0 + 0.5\Delta P\right),$$
$$q_2 = \left(2 - \sqrt{2}\right) \cdot \Delta\theta_2 + \left(-2 + 3/\sqrt{2}\right) \cdot q_1,$$
$$\theta_3 = \theta_2 + \left(1 + 1/\sqrt{2}\right) \cdot \left(\Delta\theta_3 - q_2\right) \qquad (C.17)$$

Finally, Equation C.13 is used one last time to recalculate Γ using θ_3. The fourth and final approximation for θ (θ_4) is computed using Equation C.18:

$$\Delta\theta_4 = \Delta P \cdot \Gamma\left(S_0, \theta_3, P_0 + \Delta P\right),$$
$$q_3 = \left(2 + \sqrt{2}\right) \cdot \Delta\theta_3 + \left(-2 - 3/\sqrt{2}\right) \cdot q_2,$$
$$\theta_4 = \theta_3 + \frac{\left(\Delta\theta_4 - 2q_3\right)}{6} \qquad (C.18)$$

Now, we have our estimate for potential temperature ($\theta = \theta_4$).

Freezing Temperature (from Salinity)

Although the salt content of marine waters has the effect of lowering the freezing point of water, there are temperatures on the Earth's surface that are certainly cold enough to freeze seawater. Interestingly, the higher pressures

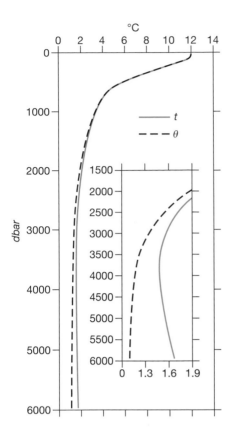

Figure C.04 At shallow depths (where pressure effects are minimal), the *in situ* temperature (*t*) and the potential temperature (θ) are essentially the same. However, as the pressure effects increase with depth, we see that $\theta < t$. This phenomenon is related to the adiabatic lapse rate, whereby deep water parcels are hypercondensed because of the great pressures at depth. If those parcels are brought to the surface, the release of pressure will cause the volume of those water parcels to expand, thereby inducing a cooling effect from that expansion. Inset shows a closer view of the departure between *t* and θ in the deep ocean. (From Pawlowicz, R [2013] *Nature Education Knowledge* 4[4]:13.)

of the deep ocean also lower the freezing point of water. As a function of both salinity S (in PSU) and pressure P (in decibars), Equation C.19 can be used to calculate the freezing temperature T_f (in °C) of marine waters within a salinity range of $4 < S < 40$:

$$T_f = -0.0575S + 1.710523 \times 10^{-3} S^{3/2} - 2.154996 \times 10^{-4} S^2 - 7.53 \times 10^{-4} P$$

(C.19)

For all surface waters, $P = 0$.

Speed of Sound in Seawater

Estimating the behavior of sound waves in the ocean, particularly the direction and attenuation of wave propagation, is incredibly complex. Fortunately, in the ocean it is much easier to predict the speed of sound (c_s), which varies as a function of salinity, temperature, and pressure (like most other variables important to physical oceanography). Because the world ocean is somewhat uniform in its salt content, the speed of sound is actually far more sensitive to temperature and pressure effects (**Figure C.05**).

In most applications where the speed of sound in seawater is needed, the Coppens equation is often preferred because of its relative simplicity. Valid for all waters from 0 to 35°C, salinity of 0 to 45 PSU, and depths from 0 to 4,000 m, the Coppens equation calculates the speed of sound (c_s, in m s^{-1}) as a function of depth d (in meters), salinity S (in PSU), and temperature T (in °C), as defined in Equation C.20:

$$t = T/10,$$
$$c_0 = 1449.05 + 45.7t - 5.21t^2 + 0.23t^3$$
$$+ (1.333 - 0.126t + 0.009t^2)(S - 35),$$
$$c_s = c_0 + (16.23 + 0.253)d + (0.213 - 0.1t)d^2$$
$$+ [0.016 + 0.0002(S - 35)](S - 35)td$$

(C.20)

For calculating the speed of sound in deeper waters (0 to 8,000 m), the Mackenzie equation is preferred, although its use is restricted to oceanic

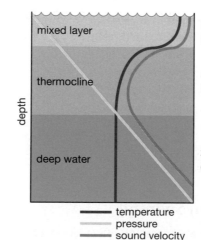

Figure C.05 The speed of sound in seawater (c_s) is most significantly affected by temperature and pressure, and as a result, the vertical ocean can be divided into three "zones" of sound speed. In the mixed layer (Zone 1), pressure effects are minimal and the warm ocean temperatures generally lead to maximal sound speeds. In the intermediate depths of the thermocline (Zone 2), the dramatically reduced water temperatures will slow the propagation of sound, leading to minimal sound speeds. In the deeper ocean where pressure effects become significant (Zone 3), the speed of sound will increase with greater depths, yielding maximal sound speeds in the deepest reaches of the ocean.

mixed layer

thermocline

depth

deep water

zone 1: maximum c_s
(dominated by warm temperature effects)

zone 2: minimum c_s
(dominated by cold temperature effects)

zone 3: minimum c_s
(dominated by pressure effects)

— temperature
— pressure
— sound velocity

waters with a salinity range of 25 to 40 PSU and a temperature range of 2 to 30°C. The Mackenzie equation is calculated using Equation C.21:

$$\begin{aligned}
c = {}& 1448.96 + 4.591T - 5.304 \times 10^{-2}T^2 \\
& + 2.374 \times 10^{-4}T^3 + 1.340(S - 35) \\
& + 1.630 \times 10^{-2}d + 1.675 \times 10^{-7}d^2 \\
& - 1.025 \times 10^{-2}T(S - 35) - 7.139 \times 10^{-13}Td^3
\end{aligned} \tag{C.21}$$

The most accurate (and most rigorous) equation for estimating c_s is the UNESCO algorithm, which uses pressure P (in bar) rather than depth. Valid for all waters with a temperature range from 0 to 40°C, salinity from 0 to 40 PSU, and pressures from 0 to 1,000 bar, the UNESCO algorithm uses four intermediate calculations to derive the final calculation of c_s. These intermediate steps (A, B, C, D) are defined by Equation C.22:

$$\begin{aligned}
A(T,P) = {}& \left(A_{00} + A_{01}T + A_{02}T^2 + A_{03}T^3 + A_{04}T^4 \right) \\
& + \left(A_{10} + A_{11}T + A_{12}T^2 + A_{13}T^3 + A_{14}T^4 \right)P \\
& + \left(A_{20} + A_{21}T + A_{22}T^2 + A_{23}T^3 \right)P^2 \\
& + \left(A_{30} + A_{31}T + A_{32}T^2 \right)P^3
\end{aligned}$$

$$B(T,P) = B_{00} + B_{01}T + \left(B_{10} + B_{11}T \right)P$$

$$\begin{aligned}
C(T,P) = {}& \left(C_{00} + C_{01}T + C_{02}T^2 + C_{03}T^3 + C_{04}T^4 + C_{05}T^5 \right) \\
& + \left(C_{10} + C_{11}T + C_{12}T^2 + C_{13}T^3 + C_{14}T^4 \right)P \\
& + \left(C_{20} + C_{21}T + C_{22}T^2 + C_{23}T^3 + C_{24}T^4 \right)P^2 \\
& + \left(C_{30} + C_{31}T + C_{32}T^2 \right)P^3
\end{aligned} \tag{C.22}$$

$$D(T,P) = D_{00} + D_{10}P$$

$A_{02} = +7.166 \times 10^{-5}$	$A_{30} = +1.100 \times 10^{-10}$	$C_{11} = +6.8999 \times 10^{-4}$
$A_{01} = -1.262 \times 10^{-2}$	$A_{31} = +6.651 \times 10^{-12}$	$C_{12} = -8.1829 \times 10^{-6}$
$A_{00} = +1.389$	$A_{32} = -3.391 \times 10^{-13}$	$C_{13} = +1.3632 \times 10^{-7}$
$A_{03} = +2.008 \times 10^{-6}$	$B_{00} = -1.922 \times 10^{-2}$	$C_{14} = -6.126 \times 10^{-10}$
$A_{04} = -3.21 \times 10^{-8}$	$B_{01} = -4.42 \times 10^{-5}$	$C_{20} = +3.126 \times 10^{-5}$
$A_{10} = +9.4742 \times 10^{-5}$	$B_{10} = +7.3637 \times 10^{-5}$	$C_{21} = -1.7111 \times 10^{-6}$
$A_{11} = -1.2583 \times 10^{-5}$	$B_{11} = +1.7950 \times 10^{-7}$	$C_{22} = +2.5986 \times 10^{-8}$
$A_{12} = -6.4928 \times 10^{-8}$	$C_{00} = +1402.388$	$C_{23} = -2.5353 \times 10^{-10}$
$A_{13} = +1.0515 \times 10^{-8}$	$C_{01} = +5.03830$	$C_{24} = +1.0415 \times 10^{-12}$
$A_{14} = -2.0142 \times 10^{-10}$	$C_{02} = -5.8109 \times 10^{-2}$	$C_{30} = -9.7729 \times 10^{-9}$
$A_{20} = -3.9064 \times 10^{-7}$	$C_{03} = +3.3432 \times 10^{-4}$	$C_{31} = +3.8513 \times 10^{-10}$
$A_{21} = +9.1061 \times 10^{-9}$	$C_{04} = -1.47797 \times 10^{-6}$	$C_{32} = -2.3654 \times 10^{-12}$
$A_{22} = -1.6009 \times 10^{-10}$	$C_{05} = +3.1419 \times 10^{-9}$	$D_{00} = +1.727 \times 10^{-3}$
$A_{23} = +7.994 \times 10^{-12}$	$C_{10} = +0.153563$	$D_{01} = -7.9836 \times 10^{-6}$

Ultimately, the solutions to Equation C.22 are combined to yield the speed of sound in seawater (c_s), as defined by the UNESCO algorithm in Equation C.23:

$$c_s = A(T,P) \cdot S + B(T,P) \cdot S^{3/2} + C(T,P) + D(T,P) \cdot S^2 \qquad \text{(C.23)}$$

References

Chen C-T & Millero FJ (1977) Speed of sound in seawater at high pressures. *Journal of the Acoustical Society of America* 62(5):1129–1135.

Coppens AB (1981) Simple equations for the speed of sound in Neptunian waters. *Journal of the Acoustical Society of America* 69(3):862–863.

Doherty BT & Kester DR (1974) Freezing point of seawater. *Journal of Marine Research* 32:285–300.

Fofonoff NP (1977) Computation of potential temperature of seawater for an arbitrary reference pressure. *Deep Sea Research* 24:489–491.

Fofonoff NP & Millard Jr. RC (1983) Algorithms for Computation of Fundamental Properties of Seawater. UNESCO Technical Papers in Marine Science No. 44, UNESCO.

Fujino K, Lewis EL, & Perkin RG (1974) The freezing point of seawater at pressures up to 100 bars. *Journal of Geophysical Research* 79:1792–1797.

Luhtala H & Tolvanen H (2013) Optimizing the use of Secchi depth as a proxy for euphotic depth in coastal waters: an empirical study from the Baltic Sea. *ISPRS International Journal of Geo-Informatics* 2(4):1153–1168.

Mackenzie KV (1981) Nine-term equation for the sound speed in the oceans. *Journal of the Acoustical Society of America* 70(3):807–812.

Pawlowicz R (2013) Key physical variables in the ocean: temperature, salinity, and density. *Nature Education Knowledge* 4(4):13.

Saunders PM (1981) Practical conversion of pressure to depth. *Journal of Physical Oceanography* 11:573–574.

Saunders PM & Fofonoff NP (1976) Conversion of pressure to depth in the ocean. *Deep Sea Research* 23:109–111.

Tyler JE (1968) The Secchi disk. *Limnology and Oceanography* 13(1):1–6.

Wong GSK & Zhu S (1995) Speed of sound in seawater as a function of salinity, temperature, and pressure. *Journal of the Acoustical Society of America* 97(3):1732–1736.

Further Reading

Bryden HL (1973) New polynomials for thermal expansion, adiabatic temperature gradient, and potential temperature of sea water. *Deep Sea Research* 20:401–408.

Cox RA, McCartney MJ, & Culkin F (1970) The specific gravity / salinity / temperature relationship in natural sea water. *Deep Sea Research* 17:679–689.

IOC, SCOR, & IAPSO (2010) The International Thermodynamic Equation of Seawater – 2010: Calculation and Use of Thermodynamic Properties. Intergovernmental Oceanographic Commission, Manuals and Guides No. 56, UNESCO.

Lewis EL & Perkin RG (1981) The Practical Salinity Scale 1978: conversion of existing data. *Deep Sea Research* 28A:307–328.

Lugo-Fernandez A, Gravois M, & Montgomery T (2008) Analysis of Secchi depths and light attenuation coefficients in the Louisiana-Texas shelf, Northern Gulf of Mexico. *Gulf of Mexico Science* 26(1):14–27.

Millero FJ & Leung WH (1976) The thermodynamics of seawater at one atmosphere. *The American Journal of Science* 276:1035–1077.

Perkin RG & Lewis EL (1980) The Practical Salinity Scale 1978: fitting the data. *IEEE Journal of Oceanic Engineering* OE-5(1):9–16.

UNESCO (1981) Background Papers and Supporting Data on the Practical Salinity Scale 1978. UNESCO Technical Papers in Marine Science No. 37, UNESCO.

UNESCO (1981) Background papers and Supporting Data on the International Equation of State of Sea Water 1980. UNESCO Technical Papers in Marine Science No. 38, UNESCO.

Wilson WD (1960) Speed of sound in seawater as a function of temperature, pressure, and salinity. *Journal of the Acoustical Society of America* 32:641–644.

Glossary

a priori relating to information that is known (or knowable) through logic and deductive reasoning alone, not through experimentation.

abiotic not derived from or related to living organisms.

absolute uncertainty the total uncertainty of any measurement taken with a particular instrument (typically defined as half the resolution of the measuring device).

abyssal zone the deepest stratum of the ocean (defined at 1,500 m and beyond), where the waters are characterized by crushing pressures, frigid temperatures, and a complete absence of light (see *aphotic zone*).

accuracy an assessment of how well a measurement (or estimate) comports with reality; in statistical applications, represented by the degree of agreement between a measured value and the true value.

advection the transfer of mass (or energy) by the flow of a fluid.

aerosol an extremely fine particle dispersed in the atmosphere (or any other gas).

aliquot a portion of liquid, taken from a larger sample.

alkalinity the measure of a liquid's capacity to neutralize acid.

along-shelf oriented parallel to the continental shelf or coastline.

ambient of or relating to the immediate surroundings.

aphotic zone the deepest stratum of the ocean that receives no sunlight and is therefore subjected to perpetual darkness.

apparent optical properties properties that attenuate light in relation to substances comprising the aquatic medium as well as the geometric structure of the light field.

array a single variable that may possess multiple dimensions.

artifact an inaccurate result that was caused by an imperfect instrument or method of measure.

ash-free dry weight any measure of dry weight biomass that represents the amount of metabolically active organic material and does not include mineralogical compounds (such as bone).

attenuation the reduction of spectral intensity as light passes through an absorbing and/or scattering medium (such as the atmosphere or the ocean).

autotrophic relating to organisms capable of producing their own food using inorganic substances, typically through photosynthesis.

axiom a statement or mathematical equation that is regarded as being well-established or self-evidently true.

bacterioplankton plankton that is exclusively comprised of bacteria.

bathyal zone A deep stratum of the ocean (defined between 200–1,500 m) which is also known as the "midnight zone" (see *disphotic zone*).

bathymetry the measurement of water depth.

benthic from or related to the benthos.

benthos a biogeographic region or aquatic habitat that is restricted to the bottom.

bias an unintended and unwanted influence on the results of an experiment.

bilinear interpolation a method of interpolation that seeks to derive an unknown value using a two-dimensional stencil.

biodiversity a group of mathematical indices (or concept), related to species diversity, but inclusive of intraspecies and ecological variability .

biogeochemistry a scientific discipline that requires the combined study of biology, geology, and chemistry.

biomass any measure of biological matter.

biometric any number of measurements that can be taken to represent biological parameters.

biotic habit a regular tendency or usual way of behaving among living organisms.

boundary conditions rules that specifically dictate how information is computed and/or exchanged at defined boundaries within the model domain.

canonical correlation a statistical procedure in which information is successively added to assess the relative importance of each new addition of information to the overall relationship between variables.

categorical data qualitative data that are grouped according to shared characteristics, defined subjectively.

causation an act or process that produces an effect.

central limit theorem a statistical theorem that states that the sampling distribution of any descriptive statistic can be considered normal if the sample size is large enough.

central tendency a simple numerical descriptor used to generalize the totality of data within a statistical distribution.

chi-square in reference either to the distribution of data, or the geometric curve that represents that distribution, the chi-square curve is a special probability distribution where the mean is equal to the degrees of freedom, and as degrees of freedom increase, the chi-square distribution approaches the Gaussian (normal) distribution.

chlorinity a measure of the dissolved chloride (Cl^-) content in a body of water.

chlorophyll *a* a universal biological pigment used in photosynthesis.

closed boundary any boundary within (or at the edge of) the model domain, through which no information computed by the model is allowed to pass.

coefficient of variation a measure used as a type of descriptive statistic to quantify the degree of dispersion relative to the central tendency.

colorimetric a method of quantification that requires the use of a colored reagent, such that the intensity of color is proportional to the amount of the compound being analyzed.

compute node any point within the model domain where the model calculations are conducted, usually located in the grid mesh where the boundaries of contiguous cells intersect, or in the geometric centers of the grid cells.

conceptual equation a logical arrangement of interrelated processes and concepts, using simple arithmetic relationships to visualize how those concepts are connected to each other.

conductivity the degree to which a specific material can conduct electricity.

confidence belt a range of values within a regression, bounded by a specified probability level, that contains the actual value of a given parameter; similar in concept to the confidence interval, but applied to a regression.

confidence interval a range of values within the data distribution that defines the probability that a particular value would fall within that range.

constant unchanging; in mathematics, any value that remains fixed.

continuity equation a fundamental equation that describes the transport or flux of a conserved property (such as mass, momentum, or energy).

continuous numbers that can take any value and are infinitely divisible.

continuous sampling measurements taken without pause or interruption.

control group a group of test subjects in which the variable being tested is either absent or held constant from experiment to experiment.

correlation any relationship between two or more events (or variables) that cannot be explained by chance.

cross-shelf oriented perpendicular to the continental shelf or coastline.

current a fluid parcel moving in a specific direction, distinct from the surrounding fluid.

damping function any function that reduces the magnitude of a value determined by a different function.

data transformation a method by which numerical data may be converted to a new measurement scale by applying a uniform mathematical function to all data within the dataset.

demersal of or relating to the water located just above the benthos.

density the amount of mass of a physical substance, relative to the space (volume) it occupies.

dependent variable a variable whose values are affected by another variable.

descriptive statistics a variety of numerical methods that are used to summarize the available data but without making any conclusions about what those data mean.

desiccate to remove all moisture.

detritus nonliving, particulate organic material.

diagenesis physical and/or chemical changes occurring in rocks and minerals, particularly during the conversion of unconsolidated sediments into sedimentary rock.

diatoms a major taxonomic group of phytoplankton that are characterized by their ability to produce cell walls containing biogenic silica.

dichotomous possessing only two possible values.

diffuse attenuation coefficient an optical parameter that describes the totality of attenuation relative to the direction of the light field.

discrete numbers that cannot be meaningfully divided (such as counts of individuals).

disphotic zone the intermediate stratum of the ocean that receives enough sunlight for shade-adapted organisms to see, but too little sunlight to maintain primary production.

dissolution the chemical process by which a soluble solid substance dissolves in a fluid, forming a solution.

dry weight any measure of biomass that has been completely desiccated.

easting distance measured (or traveled) in the eastward direction.

edge effect changes that occur in community or population structure, specifically at the boundary between two contiguous areas or habitats.

electrolyte any liquid that contains ions.

emigration the movement of organisms out of the area of interest, thereby causing a localized decrease in population.

empiricism the practice or doctrine that all knowledge is ultimately derived from sensory experience.

epifauna animals that live on the surface of the benthos.

equation of state in oceanography, a mathematical model that describes the relationship between the hydrostatic pressure, salinity, and temperature of seawater, particularly as it relates to density.

error, Type I the error that occurs when the analyst accepts the alternative hypothesis when in fact the null hypothesis is true (sometimes called a "false positive" result).

error, Type II the error that occurs when the analyst accepts the null hypothesis when in fact the alternative hypothesis is true (sometimes called a "false negative" result).

estuarine of or relating to areas of transition from fresh water to marine water.

Eulerian a view of fluid motion from a perspective outside the flow, tracking the movement of water parcels past a fixed location.

euphotic zone the surface stratum of the ocean that receives enough sunlight to maintain primary production.

experimental group a group of test subjects in which the variable being tested is present and changed from experiment to experiment.

explicit variable any variable whose values are previously declared or defined.

external boundary any boundary located at the edge of the model domain.

external variable see *explicit variable.*

extrinsic external; defined or coming from the outside.

filtration efficiency a mathematical relationship that defines the speed at which water enters the mouth of a plankton net, relative to the speed of that water passing through the mesh.

finite difference a numerical method commonly used to calculate the difference between two compute nodes, the solution of which is then used to solve more complex differential equations.

fluorescence a photochemical phenomenon whereby energy may be absorbed at shorter wavelengths and then emitted at longer wavelengths.

fluvial of or relating to rivers.

forcing function any function that appears in the equations of a mathematical model and whose solution will impact other functions within the model.

fork length a biometric measure commonly used in fisheries research, defined as the distance from the deepest notch in the tail fin to the most anterior part of the fish's head.

Gaussian in reference either to the distribution of data or the geometric curve that represents that distribution, the Gaussian (normal) curve is bell shaped and symmetrical about the central tendency of the data distribution.

genus a taxonomic category (and naming convention) that is ranked subordinate to family and superior to species.

geomorphology the study of the origins of, and subsequent changes to, the geological features on the Earth's surface.

goodness of fit the extent to which modeled values agree with measured values (in modeling applications), or the extent to which observed data agree with theoretical expectations (in field or lab experiments).

GPS Global Positioning Satellites (or System).

gradient a continuous increase or decrease in the magnitude of a particular property.

gravimetric a method of quantification based on the weight of the compound being analyzed, which usually involves the selective precipitation and subsequent collection and weighing of the precipitate.

grid mesh a geometric pattern that is formed when the spatial dimensions of the model domain have been subdivided, resulting in an array of grid cells, cell boundaries, and compute nodes.

Haversine functions a set of mathematical functions that are used to compensate for the curvature of the Earth and thereby calculate the true distance between two points on the Earth's surface.

heterogeneity the state or quality of being different.

homoscedasticity a synonym for the homogeneity of variances; the special condition whereby two datasets (or two sample populations) possess equal variances.

hydraulic boundary any boundary used to define information exchange between two or more compute nodes that are separated by a barrier to water flow (generally represented as nodes located over water).

hydrography the science of surveying, mapping, and charting bodies of water as well as predicting changes to those water bodies over time, primarily as these features relate to navigation.

hydrology the scientific study of the properties, distribution, and circulation of water on the Earth.

hydrolysis the chemical alteration of a compound due to a reaction with water.

hydrostatic pressure the pressure exerted by a fluid at a given depth, due to the force of gravity.

hypothesis an educated guess made in an attempt to explain some natural phenomenon; in the context of the scientific method, experimental evidence is used to either accept or refute the original explanation.

hypothesis, alternative a statistical hypothesis that represents the alternative to the null hypothesis; specifically, that significant differences do exist between two or more measured populations.

hypothesis, null a statistical hypothesis that represents the default assumption that no significant differences exist between two or more measured populations.

hypothesis, research a specific, testable prediction about the outcome of a research experiment.

hypothesis, statistical a research hypothesis that is testable using inferential statistics.

ichthyoplankton plankton that is exclusively comprised of fishes.

immigration the movement of organisms into the area of interest, thereby causing a localized increase in population.

implicit variable any variable whose values must be estimated or solved within the computational structure of a mathematical model.

independent variable a variable whose values do not rely on any other variable .

inert chemically nonreactive.

infauna animals that burrow and live within the benthos.

inferential statistics a variety of numerical methods that allow conclusions to be drawn from the data being analyzed.

inherent optical properties properties that attenuate light in relation to substances comprising the aquatic medium regardless of the geometric structure of the light field.

interdisciplinary involving or requiring several different academic disciplines, all of which must be integrated in order to solve a complex problem.

interpolate to estimate a missing numerical value within the range of two (or more) known numerical values.

interpolation see *interpolate*.

intrinsic naturally belonging to, or relating to, the natural essence of something .

irradiance the flux (transfer) of radiant energy per unit area, or the density of radiant energy on a given surface.

isobath an imaginary line that connects all points possessing the same water depth.

isobathyal from or belonging to the same depth.

iteration a numerical method involving the repetition of a computational procedure applied to some previous estimate of an unknown value, so as to refine each subsequent estimate and ultimately converge on the true value (or minimize the error between successive attempts).

kurtosis a measure of "steepness" (or "flatness") of a Gaussian (normal) curve.

labile easily altered.

lacustrine of or relating to lakes.

Lagrangian a view of fluid motion from a perspective within the flow itself, using the instantaneous position and velocity vectors of a water parcel.

larva the immature life-stage of an organism (or an individual) that has yet to grow into an adult (plural "larvae").

leptokurtic a condition where the kurtosis is very steep; the distribution of data is tightly constrained about the central tendency.

line transect density estimator a mathematical method by which to determine the number of organisms occupying a defined area or volume, using one or more line transect(s).

linear regression a mathematical model that describes the relationship between the dependent variable (y) and a single independent variable (x).

macroscopic observable with the naked eye; not requiring the use of a microscope.

mathematical model a mathematical description of a dynamic system using equations and arithmetic; synonymous with numerical model.

matter any physical substance that occupies space and possesses mass while at rest.

mean the arithmetic average of a set of numerical values.

mean sea level (MSL) the average level of the ocean surface.

median the midpoint of a range of numeric values arranged in rank order.

meristic data quantitative data that are represented as discrete integers.

mesocosm an experimental space that seeks to replicate a small part of the natural environment under controlled conditions.

metric data quantitative data that are represented on a continuous scale.

Michaelis–Menton one of the best-known mathematical models that accurately describes enzyme (and other cellular uptake) kinetics.

microscopic not observable with the naked eye; related to or requiring the use of a microscope.

mode the most frequently occurring value within a set of numerical values.

model domain the spatial and/or temporal region contained within a mathematical model.

model resolution the degree to which the spatial and/or temporal dimensions of the model domain have been subdivided.

moored system an array of measuring devices deployed on a static platform, usually along an anchor line or on a buoy .

mortality the number of deaths relative to the size of the overall population; death rate.

motile capable of motion; mobile.

multidimensional scaling a process by which certain characteristics or measures can be grouped according to a particular dimension, then visualized in comparison with characteristics or measured grouped according to another dimension.

multidisciplinary involving or requiring several different academic disciplines that may or may not be interrelated.

multiple regression a mathematical model that describes the relationship between the dependent variable (y) and two or more independent variables (x_n).

multivariate of or pertaining to more than one variable.

natality the number of births relative to the size of the overall population; birth rate.

Navier–Stokes equation a set of fundamental equations that define fluid dynamics.

negative correlation a mathematical relationship between two variables such that an increase in the magnitude of one variable induces a decrease in the magnitude of the other.

nekton aquatic organisms that are able to move independently of the currents.

neuston plankton that are constrained to the very surface of the water.

nominal measurements measures of qualitative data that cannot be ranked in any meaningful way.

nonparametric a class of inferential statistical tests that make no assumptions about, and are not dependent on, the distribution of data.

nonpoint source coming from a large, undefined area.

normal see *Gaussian*.

northing distance measured (or traveled) in the northward direction.

objective pertaining to, or characteristic of, the true nature of an object, unaffected by the personal interpretation of the observer.

oligotrophic characterized by the low availability of nutrients.

open boundary any boundary within (or at the edge of) the model domain, through which certain information computed by the model is allowed to pass.

ordinal measurements measures of qualitative data that can be ranked in logical order.

outlier an observation or measurement that falls far outside the expected data distribution.

parameter a mathematical value that is held constant in a specific equation, but may vary in other equations.

parametric a class of inferential statistical tests that make assumptions about, and are wholly dependent on, the distribution of data.

pelagic from or related to the open ocean.

periodic sampling measurements taken at specific intervals.

periodicity the quality or tendency to recur at intervals.

physical boundary any boundary used to define information exchange between two or more compute nodes that are separated by a physical barrier (generally represented as nodes located over land).

phytoplankton plankton that is exclusively comprised of photosynthetic organisms (that is, plankton capable of primary production).

plankton aquatic organisms that are largely incapable of maintaining their position in the water column and typically drift with the current.

platykurtic a condition where the kurtosis is very flat; the distribution of data is loosely constrained about the central tendency.

plot any area or volume with defined limits or boundaries.

plotless without defined limits or boundaries.

plotless density estimator a mathematical method by which to determine the number of organisms occupying a defined area or volume, without using defined boundaries during the survey.

point source coming from a specific location or point.

Poisson in reference either to the distribution of data or to the geometric curve that represents that distribution, the Poisson curve is a special probability distribution that describes the frequency of occurrence of independently rare events when monitored over fixed time scales.

population in the context of statistics, this represents the totality of subjects (or data) that exists.

positive correlation a mathematical relationship between two variables such that an increase in the magnitude of one variable induces an increase in the magnitude of the other.

Practical Salinity Scale the preferred analytical scale of normalized salinity measurements using conductivity, while simultaneously compensating for the effects of temperature and hydrostatic pressure.

precision an assessment of the closeness of agreement among a set of repeated measurements; in statistical applications, represented by the overall range of values from repeated measures of the same variable .

primary production the creation of chemical energy, produced by living organisms (most commonly as a result of photosynthesis), that is subsequently stored in organic molecules and used as food.

prime meridian the Earth's line of zero degrees longitude.

principal component analysis a statistical procedure that converts a set of observations of potentially correlated variables into linearly uncorrelated variables called principal components, which are then grouped (or excluded) according to the calculated distances between those principal components.

principle of constant proportions a principle in marine chemistry that states that the proportion of conservative salt ions does not change, regardless of salinity.

probability level a threshold value sometimes referred to as a "significance level" (α) that is used to define the probability of a specific event or result, or the significance level required for the acceptance (or rejection) of hypotheses being tested using inferential statistics.

probability value see *p value.*

p value a value, calculated by a test statistic, that is compared against the chosen probability level (α) in order to determine the significance of a statistical test result.

quadrat any small plot (usually with four sides) that is used to define the physical boundaries of a survey area.

qualitative relating to, or expressible in terms of, quality (attribute).

quantitative relating to, or expressible in terms of, quantity (amount) .

radians the SI unit of plane angular measure, defined as the angle measured at the center of a circle whose arc is equal in length to the radius; defined mathematically as $180/\pi$ (or about $57.3°$).

range a measure of how far a set of numbers are spread out, regardless of the data distribution.

rank order a process whereby data have been sorted in logical order, either from lowest to highest or highest to lowest.

rarefaction curve a method often used in ecology where cumulative information gained is plotted against sampling effort; also assumes the accumulation of new information will approach some theoretical limit as sampling effort approaches infinity.

recruitment the increase in a population, related to either natality (and subsequent growth) or immigration .

Redfield ratio a well-established ratio of atomic carbon, nitrogen, and phosphorus found in organisms throughout the world ocean to be C:N:P = 106:16:1.

redox the chemical alteration of a compound caused by caused by oxidation/reduction reactions oxidation/reduction reactions.

refractory not easily altered.

regression analysis a statistical process for exploring correlations between two or more variables.

relative uncertainty the uncertainty of a particular measurement, in relation to the uncertainty of any measurement taken with the same device.

relict see *refractory.*

remineralization a biogeochemical process by which organic molecules are transformed to inorganic forms, usually as a result of biological activity.

resolution the smallest increment of change that is reliably detectable in an instrument or measuring device.

rosette an array of several sampling bottles, all arranged in a carousel, whereby individual bottles can be triggered to close separately from the rest.

rugosity the quality or state of having an irregular shape; a measure of the small-scale differences in the height of an irregular surface.

salinity a measure of the saltiness (dissolved salt content) in a body of water.

sample in the context of statistics, a subset of the larger population of data.

scale measurements measures of quantitative data, in continuous order.

scientific notation a method by which numerical values that are too big (or small) to be written in decimal form can instead be written using exponents.

senescence the process by which growth is slowed and/or living biological tissue deteriorates as a result of advanced age.

sessile stationary or attached; immobile.

significant figures the digits that are used to express a numerical value to the indicated degree of accuracy.

significant wave height a value used in physical oceanography to represent the arithmetic mean of the highest third of all measured wave heights.

skewness a measure of asymmetry about the central tendency of a Gaussian (normal) curve.

solution cascade a numerical technique by which the solution of one equation is used to solve subsequent equations.

species a taxonomic category (and naming convention) that is ranked subordinate to genus and represents the most specific category of taxonomic classification.

species diversity a group of mathematical indices commonly used in ecology to quantify (or conceptualize) the proportional abundance of each species within the larger community; generally synonymous with biodiversity.

species dominance the degree to which members of a particular species are more prevalent than their competitors in a community; generally considered to be the conceptual opposite of species evenness.

species evenness the degree to which members of all species are evenly represented within the community; generally considered to be the conceptual opposite of species dominance.

species richness a simple count of the total number of species present in a particular habitat, community, or region.

specific dynamic action the amount of energy required, beyond resting metabolism, for the digestive processing and storage of food.

specific volume the amount of space (volume) occupied by a physical substance, relative to its mass.

spectral the distribution of variable wavelengths of radiation emitted by the Sun.

standard deviation a measure used as a type of descriptive statistic to quantify the amount of variation within a data distribution.

standard error of the mean a measure used as a type of descriptive statistic to quantify how precisely the true central tendency of a distribution is known, based on the standard deviation and the sample size.

state variable any implicit variable that is specifically used to describe the state of a dynamical system.

stencil any geometric pattern used in the grid mesh to interpolate unknown values at select compute nodes, based on the known values in surrounding compute nodes.

stoichiometry a calculation of the relative quantities of atomic elements or chemical compounds needed for a particular reaction to occur.

strata a series of layers.

stratified sampling measurements taken from different strata.

stream order a method of classification of streams based on their relative size and connectedness within a watershed.

subjective pertaining to, or characteristic of, the perceived nature of an object, based on the personal interpretation of the observer.

substrate the material on or from which an organism lives.

synoptic characterized by or relating to conditions as they exist simultaneously, over a comprehensive area.

synthetic analysis a method of study that seeks to define the truth in relation to the natural world.

taxonomy the scientific discipline concerned with the systematic classification of organisms.

terrigenous originating from the land.

test, left-tail a one-tail test that is used to verify whether a particular measure falls below (is less than) a chosen value in the data distribution.

test, one-tail a statistical test in which a particular value is tested to verify whether it falls above or below one particular "tail" of the data distribution.

test, right-tail a one-tail test that is used to verify whether a particular measure falls above (is greater than) a chosen value in the data distribution.

test, two-tail a statistical test in which a particular value is tested to verify that is bounded between the two extreme "tail" ends of the data distribution.

titrimetric a method of quantification that requires a precisely known concentration of a particular reagent (the titrant) necessary to induce a reaction, thereby creating the compound being determined (the analyte).

transect a straight line (or narrow section), situated over a geographic region, along which measurements are taken.

t **test** a class of statistical tests designed to examine the differences between two populations by comparing their descriptive statistics.

underway system an array of measuring devices deployed on a moving platform, such as a vessel.

univariate of or pertaining to a single variable.

universal constant any value that remains fixed at all times and under all conditions.

Van Dorn a water sampling device constructed of a tube open at both ends, each with watertight seals that can be triggered remotely, thereby capturing water collected from a specific depth.

variable a mathematical value that changes.

variance a measure of how far a set of numbers are spread out, as defined by the data distribution (that is, the square of the standard deviation).

variance-to-mean ratio a measure, mathematically defined as the variance divided by the mean, commonly used to compare the dispersion or "spread" of data about the mean.

virioplankton plankton that is exclusively comprised of viruses

viscosity the intrinsic property of a fluid (related to its internal friction) that resists forces that would otherwise cause fluid motion.

vorticity the rate of rotation of a fluid.

water quality the chemical, physical, biological, and aesthetic characteristics of water.

watershed a specific area of land bounded along its periphery, such that all of the water within that area shall ultimately drain to the same location.

wet weight any measure of biomass in its natural state (that is, prior to desiccation).

wind stress the shear stress (a force) exerted by the wind on the surface of the ocean and other large bodies of water, resulting in fluid motion.

zooplankton plankton that is exclusively comprised of animals.

z **score** a standardized statistic that reveals how many standard deviations a particular value deviates from the mean.

Index